SPACE SCIENCE, EXPLORATION AND POLICIES

PRIMORDIAL SPACE

THE METRIC CASE

SPACE SCIENCE, EXPLORATION AND POLICIES

Additional books in this series can be found on Nova's website at:

https://www.novapublishers.com/catalog/index.php?cPath=23_29&seriesp=Space+Science%2C+Exploration+and+Policies

Additional e-books in this series can be found on Nova's website at:

https://www.novapublishers.com/catalog/index.php?cPath=23_29&seriespe=Space+Science%2C+Exploration+and+Policies

SPACE SCIENCE, EXPLORATION AND POLICIES

PRIMORDIAL SPACE

THE METRIC CASE

BERND SCHMEIKAL

Nova Science Publishers, Inc.
New York

For permission to use material from this book please contact us:
Telephone 631-231-7269; Fax 631-231-8175
Web Site: http://www.novapublishers.com

LIBRARY OF CONGRESS CATALOGING-IN-PUBLICATION DATA

Schmeikal, Bernd, 1946-
Primordial space / Bernd Schmeikal.
p. cm.
Includes index.
ISBN 978-1-60876-781-6 (hardcover)
1. Space and time--Mathematics. 2. Space and time--Philosophy. I. Title.
QC173.59.S65S36 2009
530.11--dc22
2009045676

Published by Nova Science Publishers, Inc. New York

CONTENTS

PREFACE

This book is a ricochet against mainstream physics. It sprang out of the idea that outer symmetries of space-time are the same as inner symmetries of matter. In other words, the standard model of physics is a space-time group. The author began to publish about the generative process of space-time in 1996. He presented this idea at the *Fifth International Conference on Clifford Algebras and their Applications* held in Ixtapa, Mexico, summer 1999, organized by Zbigniew Oziewicz where he held a lecture to the memory of Gian-Carlo Rota. Since then he has discussed this issue with many physicists and mathematicians. Some of these discussions were polarizing and were provoking extreme reactions: While Vargas meant "*You are the guy who will go down in history as having shown that SU(3) is spacetime physics. You should have the most important chair in theoretical physics in Austria*", Robert Kiehn simply stated "*You have been brain-washed with prejudices*", as Kiehn prefers topology before geometry and Lie Groups. Consequently, in the first chapters, the author recalls his first experiences in the HEPhy group with Walter Thirring.

Reading the book does not need much pure mathematics, but rather some practice in Clifford algebra and a little knowledge about differential forms. Sometimes when polarized isotropic multivectors are calculated, it is gone back to the roots laid by Élie Cartan. The whole approach has thorough philosophic foundations. The Austrian philosopher Gerhard Frey had respected and transcended the principles of constructivism. He was convinced, only by a debasement of all statements to a pre-given objective world, we obtain by subjective reduction ('Epoche') the pure experiences of our consciousness, the universe of phenomena. In accord with such philosophy, *only in the rarest cases there will be recoverable a complete lawful relation, so that we can state a collation for any point in time.* Like Frey so Schmeikal states that a cosmological principle cannot be proven. It is difficult to count back to the big bang. He shows in mathematical terms why such rigor is too inexact. It is this calculation which touches topological evolution as was introduced by Robert Kiehn.

If we look below the surface we encounter surprising realities. We come upon a thermodynamic equivalence between time and volume correlating with angular momentum algebra. We find out the Pauli principle in strong interaction is an escape from chaos. The state of a particle turns out decomposable into a pure state, primitive idempotent, and a nilpotent representing an isotropic direction field. To the traditional dynamic element of a harmonic oscillator we can add some space-time discontinuities. Finally, the quantumchromodynamics of gluon fields and quarks turns out to be a continuation of the 4-space Maxwell combinatorics into the 16-dimensional Clifford algebra. We really obtain a new cosmology if we enlarge our view to the graded constitutive Lie groups of Clifford algebra.

Chapter 1

ROOT OF FOREWORD

Space-time is not an object. It is not just there, somehow outside. Outside of what? Outside of cognition, outside human perception, outside culture, outside mind. It is not independent of psychological processes. When I studied development psychology and genetic structuralism, as was initiated by Jean Piaget, I comprehended that a child learns about orientation in space and time before it acquires a precise perception of measures of location, volume and length. Concepts of spatial and temporal order are slowly developed in parallel with the operational structures of cognition. The prenatal psychologist R. D. Laing had said that orientation is part of a morphogenetic structure which we acquire in our mothers womb. The foetus begins to coordinate space according to the points of the compass. We feel what is above and what below, differ between left and right, move forward rather than backward. There is in it some direction of evolution. We would not detect a psychological evolution of concepts of orientation and measure if matter and nature themselves would not contain emotional facts that challenge the emergence of those cognitive concepts. Piaget was among the first who have shown that the operational structures of logic are connected with the most elementary discrete space groups. I would say that those algebraic structures are deeper rooted than those concepts which endow space with a metric. To thought they accordingly appear to be less complicated. If we look at the cross section of a water melon we see symmetric whirls of tissue that build up a tri-star coordinate system. If we look into a brain, we also see two symmetric currents of the cortical hemispheres. We can presume the emergence of polar coordinates and creation of discrete dissipative structures in turbulent flow. Thus we may presume the appearance of measurable metastable structures as resulting from topological evolution. When we look at the almost eternal structures of the Euclidean 3-space from the viewpoint of biology and non-equilibrium thermodynamics, we may be tempted to identify geometric structures as metastable arrangements that are formed in dynamic systems far from thermodynamic equilibrium. This may lead to certain preferences as to what was first, topological evolution or invariant forms of geometry. I regard this question as idle. For in those situations where both currents and invariant structures of orientation are formed, in those acts of creation, the arrow of time is indeterminate.

What can be said with considerable certainty is that sets of events can be subjected to algebraic structures of orientation independent of the existence of dimension, and separate from measurable length, metric and volume. This observation seems to me to be of outstanding importance for both anthropology and science. I have dealt with it in two sections. The first has the title '*Priming Orientation*' and is related to A. N. Whitehead's natural philosophy of events which he proposed as an alternative to the then philosophy of science in the Tarner lectures 1919. From this formalization we can learn how finite sets with elementary topological relations lead to finite non-abelian reorientation groups. The investigation of '*Reorientation of Space-time Algebra*' discovers the foundations of reorientation in terms of primitive idempotents in the Minkowski algebra. We work out why and how orientation of space-time is equivalent to the creation of matter. Some chapters – '*Conformal Timespace*', '*Timespace Analysis*' and the '*Dark van der Waals Gas*' actually deal with the appearance of indeterminate time in HEPhy and cosmology.

Briefly said, this book is about structures and phenomena that are lying hidden underneath the surface of space-time. It begins with a few biographic events, Majoranas legacy, the philosophy of Gerhard Frey and some related anthropological topics which have to do with high energy physics. It continues with a reconstruction of the theorem by Banach and Tarski in Minkowski space. We are making acquaintance with the standard model as a property of space-time. So we are challenging quite unusual actions such as penetration of quarks by a probe. We propose to apply a penetrating function D. Thus measure and basis are connected with the axiom of choice. Think about a field primordial to strong interaction. It lies dormant under the surface. Consider such a field as extended, but not measurable. You cannot reach in and take out a part just as you cannot penetrate a quark. Consider that field as f, a domain or some set of such fields as A and the penetrating function as D. We use D for "*diffusion*" in the sense of penetrating or invading a substance. Then, if penetration were possible throughout the domain, we had to have a peculiar mathematical-logic relation:

There is a D such that $$\forall_{f \in A} D(f) \in f$$

This is the axiom of choice (AC). As the invasion and selection of such a strong interacting element is not possible we have to have the negation

$$\neg\forall_{f \in A} D(f) \in f \Leftrightarrow \neg AC$$

But negation of the AC implies that we cannot establish a basis within a vector space A. Underneath the surface there are many inscriptions. One of them reads as follows. The process of nature involves transitions among four types of spaces that allow for four combinations of properties. We may be confronted with primordial fields which have no basis and no measure. We shall observe measurable spaces having no basis. Some may have a basis but are not measurable. Only in a few rather ideal situations nature will provide spaces with basis and measure. Though I know about the inset of chaos and critical phenomena, it has always consoled me that we live in a world where we can walk a mile, rest, and walk another mile, and we can be sure that we shall have walked two miles altogether. We can put a ton of fuel oil into the tank, and we can add another ton, and that will give two tons and not one and a half. Many fundamental events and actions are so to say guided and stabilized by linearity. Even if fundamental processes were topological or beyond equilibrium, it is definitely a big surprise to see how many stable and linear phenomena guide our living in space and time.

Mathematicians know there are formulas which must not be read from the left to the right. But they are understood best if we begin reading them in the middle. $F = g f X f^{-1} g^{-1}$ is such a formula or E = [[[[[A]]]]]. Greens functions and propagators can be approached by orbiting them. Constructing fractals we are confronted with similar objects. Surprisingly, we come upon this phenomenon when we are writing a book or a foreword. I begin to think about a foreword sometime, somewhere while writing, say, chapter ten. Then I update that. I rewrite it several times. The book itself also grows from the inside out, that is, it begins somewhere in the middle and grows towards both sides. But the readers don't always realize that order. The readers bring in their own order. Don't they?

You may read the book orderly, beginning with the foreword. Or else You may start it up somewhere in the middle or even at the end where You locate Your own interest, find something new or give something top priority. That is alright.

The introductory statement you are reading now is made at the end of the writing. After just having finished the final chapter I am finding out it is difficult to write a foreword as if it were a conclusion. Actually the book has sixty six chapters. Somehow I would like to spare the readers to go into all those pages if they do not feel concerned and are not seriously interested to penetrate this matter. I am interested to open up some windows. I want the readers to enjoy finding out for themselves how they can think best about this matter of primordial space. I regard intelligence and creativity as more important than conviction or habitual tradition. Do not like dogmas, universal laws and all the rest of it. But I also belong to those writers who fear false revolution as much as dogmatic science. To put it brief, I respect traditional thought, but I also question its results as much as I question my own.

Some of us, as a precaution, first of all try to wipe out all those attitudes, streams and theoretic positions that endanger or compromise their own research. Definitely I do not belong to them. For example there is a position taken in by defenders of non equilibrium systems and thermodynamics that seems to allow them to declare Lie groups as superfluous as these would force physics phenomena into the garment of reversible processes. Though I find the work of those authors rather enlightening, I do not share their position. One can design group theoretic rules for a game which brings forth irreversible regime. Also I do not follow any new sort of dogma which cancels out geometry in favour of topology. I am rather interested in the relation between those two. Irreversible topological evolution can bring forth geometric equilibrium structures. Therefore topology can be enriched by geometry, though it may be that in physics topological processes are more fundamental than geometric ones. But that relation is a very subtle one. We must not rush in and diminish great parts of physics while we overlook that we are just fixating our minds to some new crutch which is not much better than the old one.

There are some who say that special relativity is false. But I have met only a few who have repeated the very first calculations in that field, e.g. Einstein 1905, who studied the historic roots, and thus found out, there went something wrong. One of those few who really were ready to go into it and came upon such fine mistakes is my friend Zbigniew Oziewicz. His important finding together with the alternative he had posited in front of us is so subtle that I would never dare to associate it with the predication of a false revolution, as I have said above. Zbigniew found out that a theory describing the relative velocities of massive bodies has to turn over from the orthogonal Lorentz group to a groupoid. Some of us show themselves very serious in their doubts against transformations preserving Lorentz metric (Kiehn), and they should be respected. Kiehn (vol 5, sections 6.9.3-6.9.5), one of the great defenders of *topological evolution*, has thoroughly studied Fock's *Theory of Space Time and Gravitation* and revived the meaning of the Eikonal equation and the propagating discontinuities of electromagnetic fields. Kiehn's work is revolutionary. But, at present, he tends to deny too large parts of physics as dogmatic while mystifying his own territorial dominion. Still I am convinced his work will bring forth significant progress in mathematical physics. We must respect each other and enjoy the other's success. To share success and cooperate with another is more important than competition in theory. If we show more compassion and empathy for each other, we shall probably find better theories and applications as we are doing now. We have no unified theory. In my opinion this is caused by the many conflicts humankind is entangled with since at least hundred years and less by an objective difficulty or discrepancy between relativity and quantum. Probably Kiehn is right when he proposes to understand spinors as scale free phenomena. Spinors are both subnuclear and macroscopic. They form nilpotent direction fields in the sense of Cartan. That is true. But we need not disestablish Lie groups therefore or abolish special relativity. We must be patient and purify our clipped speech.

May be my reader has asked herself if it was possible that the standard model of HEPhy represented by the familiar product of $SU(3)\times SU(2)\times U(1)$ is a mere space-time group. May be she is rated a contrarian. Then she should go into this monograph. I was a contrarian in Walter Thirring's HEPhy group. But then it was much too early to cultivate those thoughts without eliminating oneself, at least in Vienna. We had to follow the dogmas initiated by Einstein, Schrödinger and others more stout-hearted than elsewhere. Anyway, we didn't experience those cognitions as dogmatic. Though I must confess, today I do. But I do so with a little more experience.

The standard model is a space-time group. More accurately, it is what I call a constitutive Lie group for the Clifford algebra of the Minkovski space-time in the (opposite) Lorentz metric. Notice that now we already used four different names of four celebrities just to identify one mathematical object. It was even not enough to speak about 'metric'. But we had to introduce 'opposite metric'. But such are the matters. We need not neglect them. Anyway, the constitutive Lie algebra $\mathfrak{l}^{(2)}$ of that graded group has a peculiarity the elder orthogonal groups which are not graded, do not have: it generates a Spin group that has no rotation group to double cover. Historically, it is so to say the first pure spin group of the space-time algebra.

May be the questions posed in this book are more numerous than the answers I give. I am convinced the best answers are hidden in correctly posed questions. About fifteen years ago I began to ask questions about a presumed *generative process of space-time* in terms of Clifford algebra. I refused to accept Pauli- and Dirac- spinors as the only appropriate representations of state functions. Rather I investigated their constitutive base elements, - the primitive idempotents of the Clifford algebras $Cl_{1,3}$ and $Cl_{3,1}$ - and showed that those elements which we used to construct minimal left (right) ideals were qualified to build pure states in a GNS construction. I have shown this in the »*Lie Group Guide to the Universe*«.

But those pure states needed some group to transform them within their normalized manifold, some Lipschitz group that preserved orthogonality, norm and all the rest of it. This – and many other questions – led me to the construction of transpositions. A transposition is equivalent to an exchange, a bijection that carries a base unit e_1 to some base unit e_2 and reverse. It should even do much more: it should carry a directed line element to a directed area. It should for example turn e_1 into the directed space-time area e_{24} and last not least it should transform two prominent thermodynamic quantities into each other, namely volume and time. As soon as we ask that, we are already in the middle of quantum chromodynamics.

Because the group element that carries out such a strange “rotation” is a trigonal colour or flavour rotation of the *SU*(3).

By the time my calculations resulted in two Lie algebras. The first had dimension 15 and rank 3. This was a direct descendent of the 16-dimensional Clifford algebra. The second had dimension 8 and rank 2 like the *SU*(3). It was a real form of *SU*(3), and as we had a representation in the real Majorana algebra Mat(4, $\mathbb{R}$), a 4×4 matrix-algebra over the real number field and with real entries, I denoted it as a real Clifform of the symmetric unitary group. Every generator of the Lie algebra $\mathcal{L}^{(2)}$ is composed by two base units with different grade. It generates a spin group which is excellently qualified to transform pure states into pure states and mixes, and provides traditional quantum numbers such as spin, isospin, hyperspin, colour, flavour and charge. Nevertheless, it governs rotations in higher spin space without double covering a rotation group. It is indeed a pure spin group and its generators are related to Cartan’s pure spinors or polarized isotropic direction fields.

In the begin each of us who confronted himself with my findings in pregeometry came upon the same paradox. The primitive idempotents had led to the standard model. But the Lie group was more accessible to the analysis than those states that should represent fermions. These primitive idempotents were absolutely practical as pure states. But they were not spinors. While one of the four primitive idempotents fixes the Lie group, the other three form a trigonal *SU*(3) structure. But they cannot directly be compared with polarized spinors. Nevertheless, one of our good friends, Jose Vargas, applied a Dirac-Kähler equation for those fermion pure states and showed one could calculate spectra. What effort! Thanks to Jose! The second problem results from their nonlinear form. Soon it becomes apparent that the primitive idempotents somehow carry all the information about the structure of the isospin spaces, Lorentz transformations and the location of states within those higher spin spaces. But the information about the location in ordinary space-time seems to be neglected or stored somewhere else. How could one construct an appropriate frame for dynamic systems?

This problem could be solved as soon as we used the concept of *pure spinor* as was constructed by Cartan. The pure spinor is conceived as a polarized isotropic direction field with eigenvectors whose norm is zero. This concept can be generalized to form more general isotropic direction fields in Clifford algebra. One can construct ideals whose elements satisfy certain conditions. One of those conditions forces the elements to represent isotropic direction fields. Another I denote as the *Conjecture of Configuration*. It determines a quite general, comparatively weak demand. We assert that the geometric product of a state with itself is in the same space as the state. This represents a mere requirement of algebraic closure. We call the pure state of a quark an inner state. Looking at the form of the primitive idempotent in the Minkowski algebra we derive from that inner state a Cartan spinor or exterior state which is associated with that inner state. At first this inner-outer split is legitimated only by a special viewpoint, namely that of Elie Cartan. A particle seems to be two things at one time: 1.) an element within the maximal Cartan subalgebra of the Clifford algebra – or equivalently: pure state in a colour space – 2.) an element in an isotropic direction field associated with the Cartan algebra. It is a sum of both.

Independent of this problem of spinor representation we can ask the question about the location or place of a fermion state. In a number of chapters we find out that a translation or parallel transposition of the pure state evokes a split of the state into an idempotent and a nilpotent element. Those chapters are essentially “*quarks on travel*”, “*translation split of an inner state*", "*translating primitive idempotent*", "*force sustained places*". It turns out that the translation split corresponds exactly with the dichotomy inner-outer or primitive idempotent versus associated isotropic direction field.

I have just made up my mind to introduce to you in that foreword those results which impress me most. The discourse about field quantization, renormalization and the significance of the harmonic oscillator has led me to the conclusion that the elementary dynamic element of motion is a soliton with a real space-time singularity rather than a harmonic oscillator. The topological approach can be introduced most naturally and without any effort. The multivector components of elementary particles form extremal surfaces and topological discontinuities which ratify the viewpoint of V. Fock and respectively R.M. Kiehn.

In a few concluding chapters I have tried to lay the foundations for a theory of observation which is ready to appreciate the material and cognitive significance of awareness. The '*awareness extension field icons*' and the equivalent graph category of the '*observer field graphs*' should help us to develop an entirely new field of mathematical physics that goes far beyond the traditional separation between observer and observed. This semiotic function in awareness field icons is strongly motivated by radical constructivism and second order cybernetics. It should lead us away from the attitude that any of our models can report some objective mathematical truth about real things existing independent of cognition. Rather we have to respect the structure of reality as being reflected in the structure of cognition. Reacting to the so called real world thought has created an internal structure which reflects the outer by the inner. Independent of the observer things are just what the are. They have no independent objective properties. In the same way the events in nature are neither objectively geometric, nor are they objectively topological. But it is our mind and our collective action that forces us to believe they are either this or that. Independent of our discourse they are just what they are: neither this nor that.

The book concludes with a toolbox which should contain all the important impulses you need to proceed with the rigor in your own way. It begins with quaternions, goes further to Clifford algebra and ends with Cartan-Lie differential forms. By the way, it turned out that angular momentum algebra of the Euclidean 3-space was somewhat incomplete. I have eked it in some sections on *Rotation Roots, Spin Manifolds and Spin Eigenforms*. I am completing this foreword exactly 40 years after the lunar landing of Neil Armstrong, Michael Collins and Edwin Aldrin with Apollo 11. I somehow identify my own life with their adventure. I pity that we have not yet found a properly working drive assembly. I feel responsible that we have not yet read out the unitary model of physical systems underneath the space-time surface. As a sociologist I swear that the reasons for this are not objective matters of science or contradictions between various models and theories. But the reasons for our failure have a sociological, a psychological and a political root.

I have some friends who stayed to me despite of my luxurious attitudes. Jose Vargas and Douglas Torr have carried out a rigor to test some minimal potentiality of the quark model. Jose Vargas has given me important impulses concerning the equation of motion which I have not found until today. His last work inspires me greatly: »*Cartan's Differential Forms: Farewell to the Tensor Calculus*«. Thank you for your efforts, Jose!

Zbigniew Oziewicz has guided me with a peerless science discourse and strengthened me in most difficult situations. He was my unattainable companion during many years. He advised me, found mistakes and difficult thoughts. He shared with me the philosophic attitude and often made me aware of the trifling outfit of my models. But he never embarrassed me.

Recently he sent me a paper with the moderate title »*Isometry from Reflections Versus Isometry from Bivector*«. It carries the hallmark of an unmistakable Zbigniew Oziewicz research memorandum and compiles in my view on a mere 17 pages the recent history of mathematics and physics. I am deeply impressed by the clarity of his mathematics which begins as he goes back to Cartan 1937 – I have it on my table – Dieudonné 1948 and Artin 1957. Each algebra reflection can be expressed by a peculiar conjugation. What, indeed! Zbigniew has worked out very clearly the difference between Cartan's pure spinors, Chevalley's spinors and Hestenes spinors. So he helped me to identify those pure fermion states within the multivector calculus. Since before Zbigniew's challenge, the only legitimation of those primitive idempotent states seemed to be in the fact that they were pure. Thank You, dear friend, for your uncountable, valuable comments, suggestions and contributions!

I also had the opportunity to discuss many questions concerning my derivation of the standard model Lie group with Bertfried Fauser and Heinz Dehnen at the University of Konstanz. Fauser is not only a brilliant mathematician. For me he is also a moral support. When I found a mistake in Jose's calculation, Bertfried showed me, he was right. As this rigor was married with my model, Bertfried's positive vote on the outcome meant an improvement for both of us. When I wrote him a letter in April, lamenting "*sometimes people are being hushed up and their results are worked up without ever quoting them*", he answered: "*There will be progress in mathematical physics, and the assignment to concrete persons (sic Hestenes) is rather a historic coincidence, therefore irrelevant.*" When I bemoaned my old-fashioned mathematical outfit, he consoled me: "*As for your mathematics, technical ability or inability play a role only then, if it hinders one to express himself well and effectively or if it blocks up the solution of the hard problems one encounters on the way. Be sure that I experience this myself over and over again. But what is helping on mathematics and physics in the end are the ideas* ..." Thank you, Bertfried, for your faithfulness.

I had some discourse with someone rather competent in the final topics which concerned topological discontinuities of propagation and space-time solitons – R.M. Kiehn. He handed over five volumes of his »*Non-Equilibrium Systems and Irreversible Processes*«. I did my best and even got the answer: "*The idea of an essential (non-geometric) property appeals to me.*" But ... "*You must realize that the idea the Lie differential (NOT Lie derivative, and NOT the covariant derivative) is the natural process propagator of Thermodynamic Systems. This idea is due to Cartan, NOT due to Ślebodziński, and not due to Oziewicz (who learned about it from my friend, B. Page, and the AMS ABQ meeting, and the Ixtapa meeting in Mexico. This concept was due to my development of the idea of continuous Topological (not geometrical) evolution developed during the period 1964 to 2004, and published in many places. The covariant derivative defines an adiabatic evolution only.*" Yes, alright to the last sentence. But I recalled Fauser's quotation of David Hestenes. Names are not the most important.

I answered: *You are referring to the letter by Zbigniew I quoted (note: in this book). Well, I do not say that this idea is due to Oziewicz or Slebodzinski, and Oziewicz does not either. We all know it goes back to Cartan. In my book I always use the denotation Lie differential. You will not find a spot with 'Lie derivative'. It is perhaps unfortunate to quote this slip of the tongue of Oziewicz. I was unaware of it. Yes, we know the development of topological evolution during 1964 to 2004 is yours. I am, however, interested in topological evolution of geometry. Prehistory of orientation!* – I had asked R.M. Kiehn to write me a foreword. But now I saw that my sympathy for his lifework was an insurmountable proximity. Our relation turned into a topological defect. I must quote Fauser once more: "*I am rather sceptical towards differentiation. I hope to getaway of it. Only if one leaves the whole linearization behind one (i.e. the differentiation) one sees the big whole. Every sphere is locally flat [...]* ". Last but not least Fauser cited a poem by Christian Morgenstern.

As Kiehn had said: „*It is not clear to me that you appreciate the fact that the Falaco Solitons are topological objects that are not things that are necessarily subject to the constraints of Lorentz transformations, or metric, or Clifford algebra, (all though such constraints are – perhaps – of some limited usefulness). Experiments indicate the dimpled surfaces are minimal surfaces.*"

Compiling it all I felt it was right to close the line and give a definite answer: *I do not want to construct a theory for open systems and stay closed for myself. Therefore I do not appreciate your autocracy. We may turn against dogmatic science. But we become authoritarian ourselves. I would like to avoid that and favour cooperation. Sir, it may be natural, if you do not try to write a foreword for my book. That would be less arduous.* – So it came that I wrote that foreword myself, fourty years after the lunar landing of Apollo 11, knowing that Kiehn is a good pilot and quite aware of his extremely valuable and radical work.

Bettina Schmeikal was my wife thirty years ago, and I still love her. She made her carrier and I began with a never ending journey through Greece. Now we are friends. Bettina has live links all over the world. She coordinates my contacts and transactions since many years. Almost everything I read or write passes through her hands. Sometimes we are joking together on the phone. We philosophize about the challenges of and sometimes the idle undertaking of science. She was the first who read the encouragement of my publisher as well as the agreement made between me and Vice President Nadya S. Gotsiridze-Columbus. Thank you, Bettina, thank you Nadya.

Monday, 20th of July 2009

Chapter 2

BEGIN

Some day 1993 I revived an old problem. I was first confronted with it in the HEPhy group of Walter Thirring. That was in 1966. The symmetries of matter resembled some algebraic properties of space-time. Then, some of us carried out geometry with the instrument of Grassman algebra. The significance of geometric algebra was on the increase, but not yet developed enough – I think, it is still not developed enough. I was convinced some important geometric properties could be understood by merely comprehending some root structures of finite groups. I began with a simple rigor that led me astray over some hundred pages, until I decided to ask Pertti Lounesto if he knew what I was doing. He answered that I was just reinventing Clifford algebra and sold me his computer program CLICAL. With CLICAL I began to see clearly what I was after. Continuing with my calculations and developing some sort of new theory where space-time and matter came closer together was a great joy. But although I was able to solve many problems on the way there always arose new ones. Historic approaches and old findings gained importance. My whole view changed by the time. In 1996 I was ready with my first piece on generative processes of space time. Today I am back at Cartan's work on spinors.

Some day afternoon in April 1999 Gian Carlo Rota was found in bed in his night clothes after not having arrived at Temple University in Philadelphia where he was expected for a series of lectures. Apparently he died in his sleep. He had been one of a few mathematicians familiar with phenomenology. So I was invited to present a plenary lecture at some special session dedicated to his memory. That was at the 5th International Conference on Clifford algebras and their applications in Mathematical Physics in Ixtapa, Mexico, June to July 1999.

I tried to explain the meaning of phenomenology in Rotas work and his relation to some other great minds with similar attitudes. Then I spoke, for the first time, about the generative process of space-time, as I called it, and the graded Lie group involved. After the talk Professor Suzuki confessed he was taken aback how all of a sudden the priest dropped his robe and began to deliver pure mathematics. I had invented the syntagmatic group as I thought there existed a link to semantics. I saw a grand unity between geometry and logic, one of Rota's visions, and some paleolithic legacy which later turned out to be of some archaeological and mathematical interest. Then I had the opportunity to speak about these ideas with David Hestenes, David Finkelstein and some other Cliffordians, as we call them. And there were some of us with whom I enjoyed Ixtapa's evenings and dances, Jose Parra, David Finkelstein and Zbigniew Oziewicz. They gave me an impulse and somehow confirmed my work until today.

In January 2002 Heinz Dehnen and Bertfried Fauser from the Institute for Theoretical Physics in Konstanz, Germany, invited me to talk about real spinor spaces and graded Lorentz transformations in the Clifford algebra $Cl_{3,1}$. There was an academic councillor, a joker, who repelled my theory with the argument that this subspace where I located the pure states hung skew in the algebra, and he tried hard to demonstrate that skewness in body language, namely to be as sloppy as a certain quadruple built up by the quantities Id, e_1, e_{24} and e_{124}. The poor guy could hardly stop shaking. Fauser then said to him that he was wrong and that this skewness wouldn't matter. After all he did no less than excommunicate a maximal Cartan subalgebra, I found out later, of the geometric space-time algebra. This Cartan algebra together with the right introduction of isotropic vectors would turn out to be the only way to establish naturally (a) a certain type of pure state in physics and (b) the standard model. Clearly, a pure state is not necessarily a pure spinor, and reverse. But Fauser, then, because of some special german law could no longer hold the appointment. He left for Australia and I engulfed in my own problems. But I never stopped being convinced that what I did, - though it was quite elementary, and apparently without the proper mathematical outfit as is fostered today -, was the only way to settle one of the great disputes of physics.

Recently I realized that there were other researchers into mathematical physics who also experienced that their work was not fully appreciated by the science community, often opposed by people in string theory, like my own former shining example, Walter Thirring. I was consoled by Kiehn's saying that we all grew up with a strong faith in tensor analysis. "But instead of using the orthodox tool Dirac proceeded by a way of his own. He produced an *expression of very unsymmetrical appearance*", said Kiehn, "which he showed to be invariant for the transformations of special relativity theory" (Kiehn 2008, p. 14). Surprisingly, even Élie Cartan (1937, 1966, 1981) maintained some semblance of confession to tensor analysis when he constructed the Euclidean spinor (section 56, p. 44). Apparently only a few of us have read Cartan's »*The Theory of Spinors*«. He invented spinors long before Dirac and derived them from "isotropic vectors". Those are vectors the scalar square of which is zero. By the way, if you try to get the solution by MAPLE CLIFFORD you will be rather disappointed at first. You don't get Cartan's solution. But you get something else which is rather trivial. However, we need that classical approach to spinors. The dynamics of physical fields beyond thermodynamic equilibrium can only be understood in terms of isotropic vectors. But it cannot be comprehended by tensor calculus alone.

It is probably new to most of us that an isospin operator is an observable for isospin, is a graded generator of Lorentz transformations on the one side and an isotropic vector in geometric algebra on the other. It is an isotropic vector that connects time with volume and both with space-time volume. It is subject to a Dirac like wave equation and to its own angular momentum algebra just like a fermion. At the same time there has to be a natural colour rotation which transposes time onto the directed space-time volume. It seems, all those relations required some slanting approaches, - when seen from the viewpoint of the academic councillor.

Chapter 3

REALITY OF PRIMORDIAL SPACE

Primordial space as a mere concept can make the mind of a physicist brim-full of new ideas. Regarding it as a living process, however, can make one careful and contemplative. Why is it that it makes such a vital difference whether we think about primordial space as a concept or as a real living process? The answer is that thought itself, the activity in consciousness, is based on the global activity of primordial space.

To construct a relevant theory of primordial physics implies a certain ethic commitment. Why? There are many answers. The most obvious is that it has material consequences. Quantum field theory predicts that empty space contains a large amount of energy brought forth by the ground state of the quantized oscillators. Sixty years ago Hendrik Casimir (1948) predicted the attractive force between two uncharged, parallel plates. This was first realized by Spaarnay (1958) with 100% measurement error. Meanwhile it can be verified with only a few % error (Lamoroux 1997). The American Institute of Aeronautics and Astronautics (Maclay et al. 2001) has even yielded objective evidence for the appearance of repulsive forces in small square cavity arrays. Certain configurations disturb the process of the vacuum fluctuations such that there come up repulsive vacuum forces. The attractive vacuum force goes like the inverse of the forth power $-1/\delta^4$ of the separation distance δ [nm] between the plates' surfaces. The repulsive force dominates at larger separation distances of the cavities and goes approximately as $+1/\delta$. This may remind us of black holes at small and repulsion at large cosmic scales. Excluding variations of the metric below the Planck length of $l_p \sim 10^{-33}$ cm J. A. Wheeler calculated a vacuum energy density of 10^{94} g/cm³. Thus he created a considerable problem since this would contribute to an attractive gravitational force larger than that of the observable matter, at least unless it was not compensated by some repulsive force. Actually the quantum vacuum energy is a possible candidate for dark energy. Wheelers calculation marks the beginning of pregeometry and quantumgeometrodynamics.

The amounts of vacuum forces and their distributions do not only depend on lengths and volumes of cavities, but also on structures and relations between geometric units. The macroscopic space-time hides a dynamic process of rather dense translocal relationships. It covers up connections which cannot be represented by the strongly disaggregated continuum of the Euclidean or Minkowski space. The points of primordial space, if there are any, allow so to say for a higher degree of connectedness than those of Euclidean space. This trespassing, trans- or interlocal nature of the pico-space can also be found at cosmic scales. Remote regions of de Sitter space can communicate and facilitate a superluminal large-scale transaction. Such translocality is perhaps outlined best by Arthur Clarke's novel "*The Songs of Distant Earth*". Here the discovery of the vacuum energy makes it possible to develop a powerful quantum propulsion engine that carries human beings to remote new worlds.

What has been said up till now denotes the more fascinating and less ethical side of primordial space. Now there comes the "real", less technical side. Suppose you asked me if there were any phenomena primordial to *both* matter and thought, any unique primordial appearance relevant for *both* cognitive concepts and material fields, would I give you a relevant answer? Yes, indeed, I would. Primordial to all phenomena, whether cognitive or physical, language-like or energy-like, conscious or emotional-physical, is a »difference«. The emergence of a perceived difference in no-thing seems to me to precede every evolutionary event or process. As Gregory Bateson said "it is the difference that makes a difference." Therefore, I like to ask what could be the first differences that walk in front of the emergence of space-like and time-like events even before a metric or a conformal mapping can occur? I believe, before that there appears a disaggregation which allows for and challenges orientation. Before pathways and trajectories can be decomposed into separated points, space is oriented.

First phenomena deviating from no-thing allow for orientation, that is, for first differences and (anti)involutive automorphisms of small objects. In the spatial domain those may be denoted by small finite groups of transposition or exchange of domain. In the consciousness domain these phenomena signify logic operations as some sort of precognitive or prelinguistic activity. Therefore we have to be alert to a certain isomorphism between geometry and logic. Primordial to real phenomena there are not only noncommutative global topological structures (Connes 1994), but also translocal intelligence, since the process of nature is not without intelligence, as it is not beyond thought. That is, we have to consider both matter and mind. Both are rooted in primordial fields. Respecting both, we also open up some space to answer questions introduced by the anthropic principle (Carter 1983, Feoli 1999).

In a recent writing from Göttingen there has been emphasized particularly the interval between *bottom up* and *top down constructions* (Requardt 2004). The latter are regarded as somehow inspired by the surface of metric spaces, relativity and so forth. The preferred approach, however, begins on bottom where spatial events do not show many of the surface properties. Requardt follows the viewpoint of Weizsäcker and Wheeler who were convinced that *"space-time is a surface aspect of reality"* and *"quantum physics knows more, not less than classical physics"*. He makes sure *"we favour an approach to quantum gravity which considers quantum theory as a coarse grained consequence of the translocal fine structure of microscopic space-time, being the consequence of the peculiar structure of pregeometry underlying our ordinary space-time manifold."* Consequently there are investigated the strongly connected initial states in terms of topological random graphs, scale-free translocal networks and similar instruments. Nevertheless, it is confessed *"our business is it to reconstruct the concepts of modern continuum physics as, so to speak, collective quantities from this primordial substratum [...]."*

As any two remote regions in the universe may be regarded as potentially connected, so the bottom can be connected with the surface. If we select the proper mathematical instruments, we shall appreciate that the surface geometry contains just the information we need to comprehend the deep structure. If we choose the geometric Clifford algebra "on the surface" we shall find out that the deep view is not at all neglected. Translocal relations can be represented very well. Non-local states are understood just as well as fractal trajectories.

Phenomena of time-wrapping and beyond-time uncertainty can be understood. The Pauli principle turns out as a sub-surface rule which provides subnuclear physics with a means to escape chaos. In brief, in this book we do not prefer the bottom up construction to the top down model. Rather we find out, how and why the root of reality is hidden in the surface geometry. Namely, this even concerns the origin of consciousness, linear writing, logic and similar cognitive processes. In a way, we bring together the dual poles found by philosophers like Baruch Spinoza, René Descartes or even Albert Einstein: res cogitans and res extensa, cognition and spatial extensions both, as was said, being properties of nature and god.

Clearly, deep inside the sub-sub-nuclear regions of the strongest forces it is more appropriate to work with some sort of graph- or network category. I have therefore developed the awareness extension field icons and related to those the dynamic observer field graphs. These mathematical instruments are scale invariant and they incorporate the consciousness category in the sense that they allow for some sort of intelligent universe or cognition-like process of space construction. But I made little use of it up till now.

If we agree with second order cybernetics, any observation of a particle trajectory requires an energetic transaction between an observer and an observed. Therefore, if we claim an electron travelled from point A to point B on a straight line, we are legitimated to ask which of the continuously many points on this line have actually been scanned in an energetic coupling, say, by scattering photons to the observer's eye or measurement device. Which of the points are "real" in the sense that they partook in a real energetic process and which have been supplemented by, say, a graphic device or geometric calculator. There are parts of the physical event which involve a strong observer-object coupling and a great amount of awareness. In those regions the "points" appear as energetic realities which are contacted by the observer. But other regions, - portions of the straight line, - stem from a calculation, from our cognitive concept of space. Those supplementations too are based on an investment of energy and awareness. But here the energy is acting in the consciousness domain. It works in cognition and social groups. If we would see the totality of the process which leads us to our drawing of a trajectory, we would be very surprised. Usually we cut off the whole cognitive, social and cultural process. We restrict our perception to the pre-given space concept resulting in some straight line-drawing connecting A with B, graphically. But motion involves much more. Strictly, therefore, this book is a science mix.

After I had written the last paragraph I recalled what Gerhard Frey said in his book »Gesetz und Entwicklung in der Natur« (»*Law and Evolution in Nature*«): "*For pure states of consciousness the question of simultaneity has no meaning. They can exist only consecutively. Phenomenological two spatial events will be denoted as synchronous, if they belong to the same state of consciousness, if we perceive them together for instance. [...] Between two sequences of events, only in the rarest cases there will be recoverable a complete lawful relation, so that we can state a collation for any point in time. For example, if we want to construct a comparison for a sequence of events we can do this by emitting signals at certain points A_1, A_2, [...] of the sequence of events (optical observations for example) which at arrival at the locus of the other sequence it is compared with (the "clock") coincide with corresponding points B_1, B_2, [...] Even if the signals are given most frequently it is in their nature to be discontinuous. It is therefore questionably if it is at all meaningful to speak about simultaneity in those intervals in which no signals or other lawful relations allow for an immediate comparison of time between the event sequences*" (Frey 1958, p. 112, 116).

Suppose we scanned what we may call "*the whole trajectory*" of a particle by lightlike '*Erstsignale*'[1], as Hans Reichenbach has proposed, to establish a convention of simultaneity in a global positioning system, - could we at all obtain a continuous curve? In his mathematics of the relativity groupoid Oziewicz (2007) points out that 'velocity' is a relation between massive bodies. Taking this very serious we may ask if the appearance of velocity is depending on a flow of photons at all. In a Machian sense, velocity in motion may be efficacious even without light.

We must confess some of our theories locate particles where there are none, in mathematical spaces which are not real. Those are so to speak maximally idealistic. Physics in metric spaces is such an idealistic approach. On the other side there are a few models that keep close to reality. Those afford concepts of space and time which are close to observations in the experiments. The most realistic and mathematically advanced Ansatz sure goes back to Zbigniew Oziewicz who proved the necessity to drop the relativity group and replace it by the relativity groupoid. Any thorough analysis of binary velocity in terms of that ternary relation, which in SR it actually is, shows that the reverse velocity $v(B, A)$ generally turns out as $\neq -v(A, B)$. So our isometric spaces are rather ideal constructs. Relative velocity is not a vector of linear algebra.

Physics has achieved a level of uncertainty where the distance between two points in space turns out less important than the distance between a concept and reality. We need to work out a fundamental theory for the observer-observed relation, something like a phenomenology of scientific experience which,- in this sense top down, - allow us to proceed with a relevant translocal primordial positioning system.

Still, for this Ansatz there must be admitted the analogous claim that it takes a long route to arrive at testable consequences if we work at the same time from primordial space bottom upwards. I have stressed in 1996 that the space-time is not pre-given, but created in a generative process of strong forces. This statement can be upheld even today. It seems most of us agree on this issue. I am sure the HEPhy standard model is hidden best by the space-time surface.

Purpose of this book is not only in showing you some of the surprising results coming on when we take the surface serious, but rather also to help you to think in a new and better way. I confess I feel that some of my friends have gained even more revolutionary and deeper insights into these matters; going ahead of most of us: my friend Zbigniew Oziewicz who has accompanied me in these thoughts since the year 2000. What I say is perhaps less important than the big contributions from Cambridge or Göttingen, but I could not have realized even this little without the friendliness of Zbigniew Oziewicz.

1 Erstsignal = prime signal.

Chapter 4

MAJORANAS LEGACY

While reading the first chapters of this book you will perhaps realize that everything we say in mathematical physics can be said in natural language and real numbers. At the latest, when we represent the standard model of space-time-matter in some Clifford algebra $Cl_{4,2}$, thereby embedding the anti-de-Sitter-Space adS_6, it becomes apparent that we work within a stream of thought launched by Ettore Majorana. The matrix algebra which Pertti Lounesto simply denoted as Mat(4, $\mathbb{R}$) is also called the Majorana-algebra. This real associative algebra is isomorphic with the Clifford algebra of the Minkowski space-time $Cl_{3,1}$. The real Clifford algebras used in our models often also require the matrix algebra Mat(2, $Cl_{3,1}$) which is Mat(8, $\mathbb{R}$). Therefore most of what we found out is said by words and square-matrices of real numbers. This is not by fortune. This stream goes back to the thoughts of an outstanding mind. But it is also a matter of the human history, which is not only a history of heroes, empires and conquests, but also one of collective destruction, despair and fragmentation of the human mind. When the collective psyche is suffering a lot and the mind segregates many special fragments of thought, there are not very many who save the capacity to contact the whole.

It is due to the collective mental fragmentation that science has stayed fragmentary and never could approach that beautiful wholeness that was proposed and demanded by spirits like Louis de Broglie, Erwin Schrödinger, Werner Heisenberg, David Bohm, Carl Friedrich von Weizsäcker, Hans-Peter Dürr and some others. There were only a few who carried physics within their hearts as if it was their own nature. One of them was Ettore Majorana. Apparently he had decoded some of the mysteries of nuclear physics even before Werner Heisenberg and Enrico Fermi. His famous work *"Über die Kerntheorie"* contained corrections of Heisenberg's theory, as he wrote in a letter to his father on 14th of February 1933. Still, Heisenberg supported Majoranas early publication by Karl Scheel (1933) in the leading German »ZEITSCHRIFT FÜR PHYSIK«. He was not jealous, but he responded to the challenges of his time like a philosopher, rather philanthropically with the need of people in mind.

Maria, the sister of Ettore, remembers that in those days, about 1937, her brother often had said: "the physics goes a wrong path" or, she does not exactly remember, "physicists are on a wrong path". Definitely, says his late biographer Leonardo Sciascia (1975, 1978, 1994), Ettore did not refer to research, experiments, results or methods, but perhaps to life and death. May be that he wanted to say the same as Otto Hahn who responded to the idea to "release the nuclear power" that "god could not have wanted that". Sure, our hubris had startling consequences for the quality of theoretical physics.

I remember, when I began to study, our chair led by Walter Thirring at the Philosophic Faculty of the University of Vienna was established as an "Institute for Theoretical Physics". Later it was renamed into "Institute for Mathematical Physics" as if it was worse to have a good theory than to design a good mathematical model. Thirring did not like Heisenberg's approach to quantumfield theory very much. When, in one of our seminaries, I mentioned nonlinear interaction of creation- and annihilation-operators, Thirring argued that this should come later. He was right. It came much later. The situation was somewhat symptomatic for our then prevalent attitude towards nonlinear theories involving self-interaction. But for Heisenberg this seemed natural. He had graduated under Arnold Sommerfeld with a doctoral thesis on stability and turbulence of flowing liquids. This topic naturally involved nonlinear problems which prefigured the later developments of chaos theory in the 70s. Maybe, Thirring did not know that.

After having written the last paragraph I found this somewhat ambiguous attitude of Walter Thirring towards nonlinearity and Heisenberg's way of thinking about it, documented in his german biography "*Lust am Forschen*"[1]. 1950 to 1952 Thirring had the opportunity to meet Heisenberg in Göttingen. As the director of the Max-Planck-Institute, Heisenberg was busy with so many things that he could not dedicate enough time to Thirring and the other visitors. But once a week there was a colloquium: "*There we could learn a lot about the way of his thinking, the more so as he used to do this loudest. I experienced in this way for example, how Heisenberg introduced the 'spontaneous symmetry breakage'. This happened at a lecture about superconductivity and was performed as follows*" (Thirring 2008, p. 89):

> Lecturer: And so I have shown that, in this model, in the ground state there flows no current.
> Critics: But in your model there is at first no distinguished direction. How does the current know where it should flow?
> Lecturer: It flows just anywhere.
> Critics: And why not in the opposite direction?
> Heisenberg: That is probably like with ferromagnetism where the domains adjust themselves somehow or other. Thereby small stochastic disturbances must give a start direction.
> Critics: In this model I do not see small disturbances.
> Heisenberg: But in reality those exist, and that's what counts.
> Critics: I thought the lecturer spoke about a mathematical deduction.
> Heisenberg: Here we must do physics, not mathematics.

Today you can find contributions to quantum theory, the Heisenberg Group and Navier-Stokes Equations side by side in one volume on geometric Clifford algebras (Ablamowicz 2004).

I am grateful, Leonardo Sciascia corrected the former dubious image of Werner Heisenberg in his charming biographic booklet telling us what we actually know about the disappearance of Ettore Majorana *"La scomparsa di Majorana"* (Turin, Einaudi 1975). Indeed, those two scientists belonged to the few who acted faithfully although they did not

[1] I would translate it as "A Yen for Research" or "Delight to Reseach "

display a definite political attitude as for instance Walter Thirring or Albert Einstein. But neither Einstein nor Heisenberg would have believed that when it came to draw a decision on the dropping of A-bombs, the American democracy was not one bit better than Hitler's Reich.

Beyond the nuclear powers there was that fine feeling of Ettore Majorana for electrons, positrons and neutrinos. In his latest publication „Teoria simmetrica dell'elettrone e del positrone“ [2] he laid the foundations for the theory of electroweak interaction of nuclear particles, – twenty years before these facts were discovered. For a long time his knowledge lay dormant until it was received by Tsung-Dao Lee, Chen Ning Yang and Chien-Shiung Wu. Lee and Yang were awarded the Nobel Prize in 1957. Madam Wu instead received the first Wolf-Prize for physics in 1978. She proved by experiment the hard fact of parity violation, exactly as was predicted by Lee and Yang. It should be noted that, like Stanislaw Ulam, an expert in theory, she participated, as a pundit in experimentation, within the Manhattan Project *before* 1945, that is, before Enrico Fermi became one of the additional chief advisers of Robert Oppenheimer in Los Alamos. Possibly my readers have read about the ambiguous relation between Enrico Fermi and Ettore Majorana. So they can follow the intuition of Leonardo Sciascia who has captured some of the feelings and emotions that may have guided Ettore Majorana's deep and revolutionary, though rather sporadic work in theoretical physics. Here, in a way, the circle closes. We come back to the vision which seems purely mathematical.

He designed a standard for the smallest real form of Dirac matrices for the spin-group Spin(3, 1) and respectively Clifford algebra $Cl_{3,1}$:

$$e_1 = \sigma_3 \otimes \sigma_3 ;\ e_2 = -Id_{(2)} \otimes \sigma_1 ;\ e_3 = -\sigma_1 \otimes \sigma_3 ;\ e_4 = i\sigma_2 \otimes \sigma_3$$

Those are real 4×4 matrices providing a signature $\{+ + + -\}$ so that we obtain squared base units from $diag[1, 1, 1, -1]\ Id_{(4)}$ as it must be for the generating flat space, the Minkowski space-time $\mathbb{R}^{3,1}$. From the e_j there can be constructed the 16 exterior products, the Graßmann monomials of grades 0, 1, 2, 3 and 4 as well as their linear combinations, the so called multivectors or Clifford numbers. These are the most general elements of a geometric algebra which has a bilinear or a quadratic form. In the above representation Majorana conceived the Dirac spinor, the neutrino and the mass of a particle. He even proposed to describe the neutrino by his version of the wave equation. This is very beautiful. It uses γ-matrices with real entries only.

Beauty and intuition are important properties which guide a physicist when he encounters a big problem. For a long time he cannot solve it, but it touches him deeply and doesn't let him go. As a freshman in physics I acquired such a deviant problem when I worked in the HEPhy group of Walter Thirring. We were scanning hadron events, searching after K-vertices and the presumed Ω^- having strangeness −3. There were all those fascinating papers coming in, written by George Zweig, Murry Gell-Mann and Yuval Ne'eman. Some of us had a good background in Graßmann algebra and as a scholar of Edmund Hlawka I could go into topology. Soon there arose the feeling that the SU(3) multiplets disclosed a hidden geometric affair.

[2] „Symmetrische Theorie von Elektron und Positron“, Nuovo Cimento, 1937, Bd. 14, S. 171

To me it looked as if the space-time acted like a crystal. Suppose the structure of space-time was complex enough, subnuclear resonances could turn out as some sort of discrete spectra of vibrational states determined by the symmetry properties of that geometric object which we giddily denote as space-time. We felt, the Dirac equation had a geometric meaning which perhaps transcended our concept of space-time by far. May be the multiplets were determined by the symmetries of a crystal wave function, but the crystal wave was the space-time.

Then I could not dare to articulate this thought without risking unfair competition among us youngsters. At least I was not advanced enough, then, to go into it with sufficient precision. Mathematical tools were not yet developed. So I shut up. Some day I met Professors Robert Jungk and Robert Reichardt who asked me to work for an Austrian offshoot of the Rand Corporation. So I took a job as an assistant professor for methodology at Reichardt's sociology department and wrote scenarios for the early "Institute für Zukunftsforschung" also called Institute for Long Range Planning. But that question about the origin of the SU(3) multiplets did not unhand me.

Some day twenty years later, in winter 1993, I went to the library of the Technical University, Vienna. Led by some remote impulse I went to a special bookshelf in the third floor. There I saw a series of books with a white hard cover and a red symbol on it. That symbol attracted my awareness. Two of the books were published by Micali (1992) and Brackx (1993) at Kluwer, Dordtrecht. They contained presentations of the first International Conferences on Clifford Algebras and their Application in Mathematical Physics. Reading some of it, I found out that I had reinvented some parts of Clifford algebra and this was actually the instrument which I needed. Obviously I had found some of what Shaw in the next volume would denote as Dirac groups and finite multivector groups. Those were discrete structures which compiled some of the most important algebraic properties of quadratic Clifford algebras.

Looking at Majorana's 4×4 real Dirac matrices, I saw shining through some representation of the octahedral group and connected with it the generators of a real form of the SU(3). We all know, how to inscribe a hexagon into a 3-cube. The trigons and hexagons of the SU(3) colour- and flavour rotations could almost be figured out by vision within the Majorana algebra and, going into the traditional stuff, - just as well within the Dirac algebra of matrices with complex entries. There was some mystery hidden in this structure and it seemed to have to do with the Clifford algebra of the Minkowski space-time, - sometimes also briefly called Minkowski algebra.

Consider a Dreibein built up by three unit vectors which form three orthogonal edges of a cube. Draw a space-diagonal through the origin and rotate the Dreibein. Every trigonal rotation by $2\pi/3$ about the diagonal brings forth a space congruence. All possible symmetries of a cube form the octahedral group of Euclidean 3-space. But the Minkowski algebra contains 16 base units with a more complicated pattern of commutation-relations and orthogonality. Which were the mathematical objects or instruments to carry a base unit multivector of any grade to any other such multivector? How can one carry, say e_1 to, say, bivector e_{23} or to the "pseudoscalar" e_{124} ? How is a time like standard unit e_4 rotated into a space-time volume or director $j = e_{1234}$? All these questions can be answered.

May be there existed some further fundamental geometric object similar to a Dreibein which induced the hexagonal and trigonal rotations of the SU(3). Suppose we could generate a finite automorphism group of such transposition involutions by means of Coxeter reflections. How could these be represented as elements of the Clifford algebra? If the space-time had its own fundamental wave functions how would they relate to the transpositions? How would a reflection change the pure state, and if the resulting function was another pure state, would this give rise to another basic reflection? All these questions can be summarized in one: how can we constitute space-time by constituting matter and reverse, how can we constitute matter such that this gives rise to the known symmetries of space-time? With the same stroke we would also have to revise the concept of motion. What does it mean to move a particle? Could we say the trajectory would be a straight line or curved from the begin, or would it perhaps be better, more adequate to say that a trajectory emerged locally in a more extensive global nonlinear event. Could we say that a trajectory occurred in a self interacting nonlinear field? Not just because we wanted to say that but because the structure of the geometric algebra implied such events!

The following turned out at first: those reflections which cause involutive transpositions in the graded multivector basis are derived from primitive idempotents of the algebra. These primitive idempotents constitute continuous manifolds of a discrete lattice structure. At the same time they provide a material structure of quantum states which comply with the requirements of pure states.

What is a subnuclear particle, we have to ask. How can we represent an electron, a quark? Are they Dirac-spinors? Since the fundamental work of Chevalley (1954) we know a Dirac-spinor is an element of a complex linear spinor space $S = (\mathbb{C}\otimes Cl_{1,3})\,f$ with a *graded* element f which is a primitive idempotent in the Clifford algebra. In terms of gamma-matrices and the unit imaginary i the f takes the form

$$f = \tfrac{1}{4}(1 + \gamma_0 + i\gamma_{12} + i\gamma_{012}) \qquad (1)$$

I advise the reader to look up this form on page 139 in Pertti Lounesto's »Clifford Algebras and Spinors« (2001). He will find out there that each of the four complex components of one column spinor actually involves one primitive idempotent of the above form, and each of those is orthogonal to any other, that is, they mutually annihilate each other. But if we further survey all the alternatives for different types of spinors given in this book ranging from Crumeyrolle's spinor space over flags, poles and dipoles to Hestenes' unimodular spinors, we may even begin to doubt that classic spinors are the right objects at all to represent electrons and fermions in general. Already in the seventies the number of constructions of spinors and fields rose so rapidly that I could not get rid of the feeling that we were missing or overlooking something very important.

In 1975 I met Professor Hiley and David Bohm at a gathering with the philosopher Jiddu Krishnamurti in Berner Oberland. Hiley showed himself utterly convinced that any construction of a universal algebraic spinor had to be compatible with the procedure of the Gelfand-Naimark-Segal construction (GNS). In the 1980s Hiley was supported by Bohm and Frescura and impelled the algebraization. By 2000 he ended up with a GNS construction in what they denoted as 'generalised Clifford algebra' or 'discrete Weyl algebra', anyway, some special types of *-algebras. At that time, doubts as to whether the Dirac spinor can at all represent a general pure state began to rise. But it was only in the bygone year that I could prove why, at first, not the spinors but the primitive idempotents of the Clifford algebra provide the appropriate mathematical images for subnuclear states, and those pure states supported the GNS-construction in a most natural way.

But pure states of the above form (I) were graded bilinear elements which just as naturally brought forth discontinuous iterators and nonlinear equations of motion. Rather suddenly, most of the theoretical physics of the last century appeared to me still as a complicated attempt to linearize and thereby diminish the fundamental nonlinear phenomena.

Correlating the structure of physical fields with the structure of Clifford algebra it is utterly clear, a subnuclear state, whether electron, quark or neutrino, cannot be represented fully by a spinor. If a particle with certain quantum numbers has a sole local representation at all, then it is rather the sum of a nilpotent and a primitive idempotent, where the idempotent represents a global rather than a local structure of a bundle of preserved properties, a force, and the nilpotent is the locus. This may at first sound somewhat surprising or even artificial, but it is a logical consequence of the properties of geometric algebras with an inner, an exterior and the geometric product. The reason why our models do not work lies in the obstinacy with which most of us try to save grade preserving connexions for the derivative, Lorentz rotors as even elements together with some new version of the general relativity principle, as I am just reading (Hestenes 2005), a *"displacement gauge principle"*.

It is believed that a Cambridge group (Lasenby, Doran and Gull) had a brilliant insight (Hestenes 2005, p. 924) when they introduced gauge theory gravity (GTG). They combined their globally homogeneous and locally isotropic space-time with grade-preserving coderivative and a lot of displacement endowing even vectors. But the problem is that this cannot work, since the Cartan algebras which constitute the forces are graded. And they are graded in such a way that they form sets of different but equivalent, what Hestenes formerly called pseudoscalars, quantities in conformal splits. They form a 4-fold real ring. The Cambridge group has, like most of us, followed a wrong concept. For space-time is exactly that process which connects different grades in such a way that physical laws come upon.

Of course, there is reorientation invariance which asserts that space-time is locally isotropic (Schmeikal 2004, Theorem 1, p. 358). But the laws of physics can neither be derived from gauging, nor from a restricted rotation gauge principle. But there is a universal rotation (UR) invariance which stabilizes local isotropy, and it is this universal rotation that carries a graded fermion state through all its possible grade structures. Because such a pure state can involve grades 0, 1, 2 and 3 with signatures $\{+ + + +\}$ in its standard representation, - but as we learned, - it can even involve grade 4 and all possible signatures, since the UR rotates the wave function through the whole 16-dimensional algebra, though on a sphere (Schmeikal 2006, p. 81). The UR is a graded commutator product of a typical $su_{Cl}(2)$ rotation and a factor $sl_{Cl}(2)$ which provides the Lorentz rotations. A general Lorentz "rotor" or "versor" or whatever you like to call it, is certainly not an even multivector. It is save to say that a general rotor bringing upon Lorentz transformations in the Clifford algebra is a graded element of a 16-dimensional vector space. We must understand what it means that the maximal Cartan subalgebra is necessarily graded.

Once we carry out a thorough review of angular momentum algebra in the Pauli algebra $Cl_{3,0}$ we soon find out that there is the space of spinor eigenforms (Schmeikal 2009)

*-algebra: $E(f) = span_R\{Id, e_3, e_{12}, e_{123}\}$

which provides the whole *Lie manifold* $\mathfrak{L} = \{J_3, J_+(X), J_+(X)^*\}$ of angular momentum operators which generate the $su(2)$ as a Clifform $su(2, X) \subset Cl_{3,0}$ in the Pauli algebra. At the same time those preserve the corresponding coordinate-free idempotent – or *pure state* – and its minimal left ideal. Here the X is a symbol for the general element of the 8-dimensional algebra. Recall that even in this example a rotor or angular rotation discloses graded shift operators in the standard representation

$$J_+ = \tfrac{1}{\sqrt{2}}(e_1 - e_{13}) \quad \text{and} \quad J_- = \tfrac{1}{\sqrt{2}}(e_1 + e_{13}).$$

but as a manifold we have

$$\mathrm{J}_+ = x_2(e_1 - e_{13}) + x_3(e_2 - e_{23}), \; \mathrm{J}_- = y_2(e_1 + e_{13}) + y_3(e_2 + e_{23}).$$

So, if we really proceed reasonably, we cannot possibly believe that a rotor is exclusively to be represented by bivectors or even elements. Once we have recognized this, it is a further step to comprehend the same for special relativity and last but not least for GRT in flat space. For the future of GR is certainly not in curved space. Here those brilliant minds in Cambridge seem to be right: *"The problem with Einstein's GRP is that it is not a true symmetry principle. [...] There is no comparable symmetry group for curved space-time, because each mapping produces a new space-time, so there is no geometric object to be left invariant."* This fact has been pointed at fifty years ago in quite different terms "*On the question of the existence of privileged systems of coordinates Einstein, the founder of gravitational theory, maintained a point of view opposite to ours, denying their existence. This is connected with his aforementioned preference for the local method of discussing properties of space (which method is the basis of Riemannian geometry), and his underestimation of the importance of considering space as a whole. [...] Thus, covariance of equations in itself is in no way the expression of any kind of physical law.*" (Fock 1959, p. xvi)

The graded modules of space-time algebra provide fundamental spaces for the unfolding of physical fields. Once we have comprehended this fundamental unity between space-time and matter we will automatically repeat such important questions as for instance what is the meaning of the general Pauli principle? Why are subnuclear fermions coloured? And the answer is that the Pauli principle offers the only escape from chaos within the chaotic iterators brought in by general rotations and Lorentz transformations of fermions.

We often asked what the origin of the Minkowski space-time is like. Did it play a distinguished role? Remember, some years ago we played around with conformal splits to transform vectors into nilpotents in order to obtain homogenous "real points" in a higher dimensional geometry. If we look at the whole problem more easily we see that there is indeed a distinguished element within the geometric algebra, namely the primitive idempotent (I). This annihilates the whole Lie group of the standard model which rotates the pure states and in this sense forms an algebraic fixed point just like the origin of Euclidean 3-space or the midpoint of some 2-sphere. Also we cannot invert it. But this is not just a point, but there are continuous manifolds of such states which cannot be inverted. We represent this distinguished state f in the Majorana algebra Mat(4, $\mathbb{R}$) and more appropriately even in $Cl_{4,2} \cong$ Mat(8, $\mathbb{R}$). We do the same as Majorana. At first we regard a neutrino as an element of the geometric

Clifford algebra $Cl_{3,1}$ isomorphic with the linear algebra of real 4 times 4 matrices. That does not mean that there is no nonlinear algebra within the sets of the linear one. Since we represent the neutrino by a primitive idempotent which in that algebra, represents a bilinear form. This would not be so, say, in the Pauli algebra. But in the Minkowski algebra it has mathematical and physical consequences.

As a late heritage of Majoranas approach, Kiehn (2008a, p. 18) offers us the Falaco solitons as Majorana particles which appear to be observable as universal topological defect structures. This sudden and somewhat delayed emergence of what I call a new dynamic element – a dynamic root structure just as the old harmonic oscillator – is a real surprise.

Chapter 5

PRIMORDIAL THOUGHT

We are able to think about primordial space and primordial matter. This is itself a mystery. Some say that it is not self-evident because in the beginning there were no observers, no yardsticks and scales, no clocks and no scientists who could have supported any primordial understanding of that which was then. So everything we say today about some supposed begin of the universe ultimately had to turn out as speculative. Everything that we say about anything that we believe to be part of our world is resulting from our experience. Seeing apart from a few exceptions there is hardly anyone who believes in encounters beyond experience. Even the remotest encounters with god and mystic experiences of beyond are denoted by us as *experience* and rather not as beyondness. Metaphysics and transcendence have a difficult life in nowadays science.

It does not necessarily need radical constructivism and second order cybernetics in order to be led to the consideration that it is difficult if not impossible to conceive of an object of space-time or some objective primordial matter independent of the observer. But, quite obviously, we were not there in the early times of the big bang. So we have to ask how we can conceive a primordial object if there were no observers? Shouldn't there be at least some sort of bridge which links some primordial observation with our today experience?

The objects of our world are not metaphysical, but they are constituted in experience. Each of us partakes in this experience. A philosopher like Alfred North Whitehead would perhaps have said that the *passage of nature* appears to our mind as *something that nature is giving to thought*. So he would have coupled the events of nature to thought and experience. But Whitehead had gone even further. He said that *nature is giving something to thought which is for thought only*. That is, nature is not thought. But it has a one-sided relation to thought which is essentially different from a relation which thought thinks it has to nature. There is a mystery in that. Radical constructivists like Heinz von Foerster or Ernst von Glasersfeld say that independent of observations no object can have any quality. But it is interaction with observers wherein such qualities are acquired. Von Foerster (2003, 189) put it this way "*The environment does not contain any information. The environment is as it is*." Without observers the observed is just as it is, it resembles nothingness. So it remains a question if not a mystery who or what was the observing being who experienced primordial space and primordial matter such that the experience could be transported into our collective cognitive system of today? There must be some bridge, must there not? Otherwise we would just speculate.

Before we go deeper into this matter I would like to anticipate a provisional answer to the question what could represent the bridge. My answer is that deep inside us there is a mode of experience, a modality of consciousness and a structure of mind which signifies not only the human observer but matter itself. The primordial experience is not only far back in prehistory, but it is also deep inside in the collective mind.

Personally I am not a strong believer in any sort of total begin, some big bang triggered by god, deity, goddess or godhood once and forever. I prefer to follow my feeling that the big bang happened in a world which was already there, and it might even happen once more or twice or repeatedly. But this is my personal viewpoint. It does not represent the generally accepted view. Nor does it concern that which I shall be saying during the next chapters. Independent of our experience that thinking about the begin of time is somehow mystic, it is a fact that men and women from all different cultures and tribes, even people without writing, have to tell us something about primordial space and primordial matters of existence. Some go beyond that which is and which therefore must not have originated sometime, but is somehow eternal. Those involve some state of Nibbhana and transcendental things beyond space and time. Others speak only of those things we are rather familiar with, the originated ones, not metaphysical, not transcendent, not beyond time and experience. In what follows we shall have to consider both, since human beings have spoken about both. Both emerged in human experience.

Chapter 6

THE PHILOSOPHY OF GERHARD FREY

It is easier to become a good physicist than to become a good philosopher. For philosophy involves such a deep understanding that it cannot be learned in a short time by just accumulating the right knowledge. But part of its wisdom is itself beyond time. Therefore philosophy requires to transcend time every now and then and to make accessible to the mind a fine feeling for that which is neither in space nor in time. In this sense one cannot "become" a philosopher. But one carries a philosopher inside one's heart and mind. It's a practise.

But on the other hand one can use experience very well and so, by the time, can improve one's philosophy. As a physicist in Vienna during the 60s, philosophic interest was so to say occupied by the name of Sir Karl Popper. A theory should offer, he said, the potential to be negated by good experiments and better theories, and that was it. Many of us were mere layman positivists. But for me that was not enough. To really understand the philosophic foundations of scientific discovery and knowledge required much more. Only a few of us thought about Paul K. Feyerabend's or Thomas S. Kuhn's works, and sure no one of us had a sufficient conspectus over the natural science philosophy of Gerhard Frey, though he taught in Austria.

There is no doubt, as Frey pointed out, that we can differ very clearly between phenomenology and the operational construction of an outer reality. There is an inner consciousness domain where phenomena mostly seem to appear one after the other as in a sequence of time, and there is an apparent outer process of events which seem to be located in space. Most of us had accepted, in one way or other, the reality of some objective physical process going on out there all the time.

Thus we also believed in such things like the original explosion, the big bang. The universe was evolving, and it was somehow a question of time until we were fit enough to explain the whole process by our theories. We believed we could approach reality like a skier (like Thirring's father Hans who was a good skier) reaching the valley station, - like a football player shoots a goal or a politician pursues his target. Such belief is connected with the convergence problem.

As long as we believed in the independent existence of one objective reality, the development of natural science could be understood as a continuing approximation to some metaphysical reality. This problem of convergence has always been seen together with the question of the ontological content of scientific theories, said Gerhard Frey. He was perhaps the only philosopher who not only had the necessary background in both physics and philosophy to answer questions which had stayed untreated or unsolved during the last century, but who also designed the necessary fundamental structure of investigation and thought, by the aid of which we could decide whether a certain question can at all be answered or if it was not decidable. Meanwhile many statements, models and theories of theoretical physics and cosmology are very far away from this precise and partly phenomenological basis of science so that our results can only be denoted as exaggerating or fantastic.

Most of what physics is implicitly relying on can be found in Frey's[1] book »*Gesetz und Entwicklung in der Natur*« (*Law and Development in Nature*) from 1958. Some parts of which I regard as indispensable for doing mathematical physics.

It is important to differ between phenomena and psychic objects and human action as operations[2]. The construction of objective knowledge demands both phenomena and operations. *Again and again it has been believed that convergence between knowledge and reality is similar to the convergence of an arithmetic series like in mathematics so that it represents an approximation to some "an sich seiende" onthological fact. Psychic objects are given to us by the psychic phenomena and appear as acts in our consciousness. By the various psychic activities we apprehend corresponding objects. Thus as there are various kinds of psychic acts, we realize by those acts different varieties of objects. On the other hand every kind of human is led by intentionality. Intentionality is in psychic phenomena, it is in perceptions, imaginations, dreams, thoughts, judgements, desires and endeavours. Our theories are not free of intention. The world is so to say my imagination.* We can figure out in the fundamental intentionality of the acts of scientific thought a Buddhist quality, namely inherence and zest for life.

Let me put this in terms of some piece of exchange theory which was once created in a dialogue between the sociologist James S. Coleman and me: *If it does not make any difference for us whether we willingly investigate some events in experience or not investigate them, we could say that we are not interested in such investigation. There would be zero interest, provided we had a utility measure. In this sense intentional action as it becomes rational action, involves not only directedness, but also measure. Our actions become measurable and show a certain social value, a utility. Thus Intentional action involves a kind of symmetry breaking in action: interest.*

The world, said Frey, *is first of all my imagination, a presentation in my senses. What appears to us as nature, is for us only, but is no "Welt an sich"* (Here he seems to agree with A. N. Whitehead). *There is no observed without observer! There is no object without a subject. The world as idea decays into subjects and objects which are both inseparable as well as radically different. In the end they are both appearances of intentionality.* Here Frey approached Schopenhauer who also insisted that such was the essence of the world which, while appearing in both subject and object, is looking at itself.

---

1 It is perhaps easiest for the reader to understand my late pleading if I say that Gerhard Frey was my father in law. As I did not have the opportunity to defend his teaching during his lifetime, - partly because I didn't yet understand enough of it, partly because my own work in physics was not developed far enough, - I feel obliged to make scientists aware of the profound meaning of his work. Lately, I found many of my conclusions legitimated or even forecasted by his philosophy. He is in my memory and in my heart a most gentle and modest man with an incomparable originality and depth of consideration. – In accordance with his daughter I translated freely some of those parts of his early book, which I regard as indispensable for doing mathematical physics.

2 Frey 1958, 71-77, sections a and b

Frey appreciated Husserl's '*phenomenological reduction*' as featuring the basis of our '*pure subjective evidence*' or '*pure subjectivity*' where all scientific knowledge has to begin (p. 73). *That we comprehend things and processes as real and valid, is itself only a phenomenon of validity. To a thing-phenomenon, in our consciousness there is correlated a validity-phenomenon, by the aid of which the 'thing' appears to us as 'real thing'. In our reduction to the domain which offers for us a secure ground of judgements we are neither gaining a science which is valid for us nor a 'being' world which is real for us. We must avoid all statements about such a world. Especially all judgements of the kind that something 'be' in this world, that this scientific sentence be 'valid' expires to this reduction. By such universal debasement of all statements to a pre-given objective world, I obtain by 'subjective reduction' (»Epoché«) this »bracketing« of objective reality, the pure experiences of my consciousness, the universe of phenomena.* Here Frey quoted Husserl (1950). What is required then, in order to obtain intersubjective scientific statements are operationalization and constructed domains of objects. But the whole endeavour begins with a phenomenology of perception and pure subjectivity. Clearly, this provides the platform for every scientific undertaking. This also holds for the concept of time.

The stream of our consciousness, the fundamental fact of the appearance of events in consciousness and their disappearance out of consciousness determine the experience of time. Time is, in this way, a phenomenon of consciousness. As such it is given only in terms of phenomenology and can only be described as such. All inner experiences can appear but within the stream of our consciousness and are therefore determined as time-like. Here Husserl (1928) is quoted. Only by accepting this phenomenological basis as our platform for space-time investigations, can we proceed to the operational construction of a concept of time. He made us aware that it is not easy to present a complete phenomenology of time. But that we can ascertain beyond doubt that those phenomena which pass in our consciousness determine a temporal order of the inner experience. This order of time- consciousness is a definite order in terms of *anterior-later*.

Time is therefore a form of our inner experience. Seen from this phenomenological viewpoint, the 'outer' events – after bracketing of their reality – are mere appearances and phenomena in our consciousness. As such, time represents the form necessary for those 'external events' to emerge in consciousness. It is in this context meaningless to say that we would project time as a form of our inner experience onto the 'outer reality'. Such manner of speaking would only make sense, if the phenomenological Epoché, the bracketing of the world would be abrogated again. (Frey 1958, p. 110)

Frey exhorts us to acknowledge that everything as a content of consciousness is subjected to time as a fundamental form of facts and ideas. Still I wish to add here that Frey at other occasions wrote about the annihilation of the subject-object-fragmentation, namely where the researcher becomes one with the object of observation. The object becomes 'intentionally present'. *The cancellation of the subject-object-split finally demonstrates to us the coherence and consistency of this transcendental-subjective assembly* (p. 76). Clearly, the unification of observer and observed includes the possibility of experience beyond time, that is, the experience of pure presence and oneness. But Frey, at most opportunities tries to avoid clearly and separate off questions of religion or contemplation from those of science. I wonder, however, if this can be kept up so to say with a stiff upper lip in states of perception beyond memory where the inner experience of time fades likewise. Such states are not beyond science. But this would not contradict Frey, but it would rather make the phenomenological side of his analysis more radical. It is also clear that as soon as we speak about a transitive order of time or any other relational structure which is subject of mathematical analysis, time becomes an object of cognition and knowledge. Therefore, we can differ between time as an object of science and time as a form of our contents of consciousness.

All our actions which aim at the states of some pretended outer reality are carried out in time. The 'plot' has a time-like structure. But we also have a space-consciousness which parallels our time-consciousness. However, contrasting time-consciousness, the space-consciousness is not a form of all states of consciousness. *All sensations have a spatial nature. For all inner perceptions it is true that they are space-like only inasmuch as they are representations of external sense perceptions; but in all other cases, inner perceptions are not space-like.*

One could perhaps even consider the spatiality of a content of consciousness as a criterion to differ between inner and outer perceptions (Frey, p. 112). *[...] For pure states of consciousness the question of simultaneity has no meaning. They can only follow each other in a sequence. Two events which are considered spatially separate can be called phenomenological simultaneous, if they belong to the same state of consciousness, for example if we perceive them simultaneously. In this case events are always related to the stream of consciousness. [...] The comparison of events which are not immediately given in my consciousness with the temporal sequence of conscious states, can only be derived and constructed.*

After a considerable number of very precise considerations Frey arrived at a number of striking conclusions which as far as I see are still valid today, as most of the deepest philosophic insights are so. Besides the obvious fact that there exist an unlimited number of exterior measures of time, or outer objective times, there is an answer to the question if there is a beginning to the world and what can be said about it. He has given the following answers in 1958:

> Suppose there is a temporal limit, a beginning to the world. If we accept this hypothesis we can at first think about three possible cases:
>
> 1) The expansion indicates a catastrophic process. There might have been an 'Urexplosion', an original explosion.
> 2) The universe was formerly in an equilibrium state and sometime began to expand. Therefore the velocity of expansion may be taken as increasing.
> 3) But one can also assume that the universe, despite the explosion is in a stationary state.
>
> For each of these three cases there exist various models which are so to say attempts to meet the delineated requirements. A model for the first scenario must in some way make statements about the 'Urexplosion' (big bang). It is clear that all such hypotheses which want to explain this Urexplosion, must remain mere speculations. It is in no way evident how we can experience something which tells us how any assumptions about the nature and constitution of such Urexplosion could be confirmed. But in the second case it would be necessary to presume a dispersing or repelling power. The forces of expansion would, in that case, balance out the gravitation, but the equilibrium would be unstable. [...] The expansion itself is no sufficient proof for the correctness of the assumption of an expansion-force. An example may be seen in the model given by Lemaître (Jordan 1947).
>
> The third case turned out most interesting in this philosophy because it correlated with the 'principle of actuality'. Namely, the assumption of a homogeneous universe is known as

> the cosmological principle. As soon as we postulate such homogeneity as a principle, we must also assume that this property of the universe will not change by the time. But then the universe turned out stationary. Completion of such a cosmological principle would also suggest that the universe is not only homogeneous and stationary where its physical state (density for example) is concerned, but the laws we found should be valid everywhere in the cosmos.

Gerhard Frey agreed with Gold and Bondi, who constructed that principle, that it represented the only possibility to practise cosmology as a natural science, and he added that it goes without saying that a proof is not possible. He then discusses various peculiarities of the problem such as a world with constant universal constants and another with variable ones, he points out the necessity to define some universal time and so forth. It is interesting that he even speaks about theories with a fundamental minimal length (Planck's length) and a discrete space-time having a finite number of points. But every attempt to make valid statements in cosmology and cosmogony turns out to fail an important examination. He comes back to Immanuel Kant who was the first to prove that both statements are possible: 1.) The world has a beginning and 2.) The world has no beginning. Kant even proved them. The situation has not changed significantly since then. There is only one concept of time to which cosmic events can be related. This is the measurable operative time. But using such, there is no way to show that there is one definite universal time. Thus he concludes:

> For this operative time, for which alone our concepts of measure can be defined in a definite way, it makes no sense, as we have seen, to speak about "a universal time". Operative time can be indefinite, equivocal, it is not definite, anyway. This is the reason why in the context of operative time it also makes no sense to speak about a sequence of events where each event follows some others. There is no means for us to define a definite concept of a "state of the world".

Here I agree with Frey. The question whether the world has an initial state or not, whether it is created in a big bang or small bang or no bang at all, is no scientific question. Natural science has no way to answer questions of eschatology. Universal time has no place in science. So it has none in physics. We are, however, allowed to write fairytales, fables stories, myths and sagas about it.

While Gerhard Frey kept a sane distance to the ἐὸν-metaphysics, he did not show surprised by the emergence of metamathematical theorems on decidability, computability and incompleteness. Frey did not only understand the equations of wave mechanics and relativity, logic and metamathematics, but he was also familiar with the anthropological and epistemological roots of arts (Frey 1994). In his work »*Erkenntnis und Wirklichkeit*« (*Cognition and Reality*; Frey 1965) he developed a thought on the perceptibility of the world. He realized: "Nur weil es Symmetrie in der Wirklichkeit gibt, ist diese erkennbar": *It is only because there is symmetry, that reality is discernible*; (p. 110). Frey knew the encompassing meaning of the fundamental works of Herman Weyl on algebraic group-theory and quantum mechanics and he saw their connection with structures of the human consciousness as a whole. In 1918 Emmy Noether had found that any symmetry of space-time is necessarily resulting in a conservation law of physics. Therefore symmetry represents an inner quality of reality responsible for its perceptibility. This is a structuralist viewpoint. We have carried this Ansatz further, so far that the morphogenetic root structure of the standard model symmetry can become visible underneath the surface of space-time.

To some moderate extent this writing follows both the theory of consciousness by Gerhard Frey and the genetic structuralism by Jean Piaget. The root structure of logical thinking or Boolean logic appears as symmetry in geometric algebra (Schmeikal 1989). It is a spatial symmetry that is also basic for the structure of the Clifford algebra generated by the Minkowski space-time in the opposite (Lorentz) metric. In this sense the interval between res cogitans and res extensa is bridged once and for all. But it is bridged symbolically, not physically, but in mere cognition. We do not yet know what that bridge means for extension, that is, for the material world.

In our earlier thought, both Frey and me have been influenced by the works of Hans Reichenbach and Peter Mittelstaedt (1963). We thought in quantum mechanics there would be at work some non-classic logic (Kanitscheider 2004, p. 7). But later we found out that the mathematical apparatus of quantum mechanics and the space of square integrable functions respectively is solely based on classic logic. This is also articulated in my contribution to quantum logic (Schmeikal 2000). Walter Thirring, in his later years, has written an article with Misses Narnhofer where this fact is articulated in a most beautiful title: *Why Schrödinger's Cat is Most likely to be Either Alive or Dead.*

On page 8 Kanitscheider (2004) recapitulates: "Davon ausgehend (Anm.: von Noethers theorem) bestimmt Frey Erkenntnis als ein Ordnen der Fülle der Erscheinungen in Hinblick auf deren strukturale Ununterscheidbarkeit": *Starting from this assumption Frey determines cognition and insight as an organizing of the plethora of phenomena with respect to their structural equivalence*[3]. It has been exactly this procedure that led to the result that the physical phenomena of space-time and matter are structurally equivalent. The equivalence is determined by the elementary organising act of transposing and reflecting fields and spatial elements such as Graßmann monomials or base elements of Clifford algebra. The whole proof that *space-time is matter* is based on the »*method of reorientation invariance*«.

There is something entirely new coming in. Namely, considering a subnuclear particle as a field, its field trajectory is not a line, but principally engrosses all possible grades from zero to four. Motion turns out as graded motion. To give an example, this means that the dimension of the trajectory of an electron can turn out fractal and, involving both space and time, equal to three. At other time intervals it may appear as dislocated and splayed over the whole space of the Clifford algebra, while at some point of emerging time it may seem to enter the Minkowski area at a definite Euclidean location. There is a model in which it is calculated how such a transition works (Faustmann, Neufeld and Thirring 2006).

Several thinkers have concluded that the algorithmic compressibility of the world has cosmological reasons. Interestingly, Frey's considerations point into the direction of quantumgravity where it is attempted to identify the »cognizable subset of worlds« (Kanitscheider, 9). Asking about the scientific character of metaphysics Frey propounded his original method of resolution. He located its origin in a relation between linear and circular thought. Circular structures rely on a feedback to the initial positions. Reflection is a prerequisite for circular thought. Formally, this signifies circular structures of transpositions. Such a relation between reflective transpositions and cycles is fundamental for the geometry presented here. It is a model which is in a perfect correspondence with cognitive structure.

[3] Exact verbal translation is: „structural non-discernability"

The model itself turns out to be a geometric Clifford algebra of the standard model of physics. This special mathematical form is in good agreement with the constructivist comprehension of natural philosophy: "*Our acquaintances with the objects of nature imprint a form onto experience which involves an idealizing invasion into the organization of the object world*" (Kanitscheider, p. 9).

Frey was interested to find out about the relation between history and natural history (Geschichte und Naturgeschichte). On the last pages of his book on law and evolution in nature he writes: *With the projection of history onto physical, operatively determined time, - especially the linear time, - that is, mechanic Eigenzeit, there corresponds an Eigenzeit of the human historian coherence of events. Nobody will contest that there is a reality of processes. But as that which for us is history, the events are comprehended noetic, that is, in acts of consciousness. The historic processes are for us only 'Noemata' (cognitive events). Equally there might be historic Eigenzeiten. Sure, such an Eigenzeit cannot be defined in an operative way. Because what should be measured there? Rhythm and periodicity in History can only be experienced as mere forms of order (patterns) within our consciousness. Historic Eigenzeit can only be determined cognitively. The historic events are projected as such onto the subjective experience of time.*

Here we should end this chapter by quoting a famous statement coined by Vilem Flusser in the mayoralty of Vienna. This is in perfect resonance with Gerhard Frey's consideration: „*Es gab keine Naturgeschichte, so lange es nicht Geschichte gab*". That is, first we had to develop linear writing and historic consciousness. Only then could we invent natural history. *There was no natural history as long as we had no history.* In other words it is hard if not impossible to prove that there was a big bang before there was linear writing. This may sound surprising. But it resembles the deeper truth of reality.

Chapter 7

CULTURAL MYSTIC SYSTEMS

CHASSIDIM IN JUDAISM

It is interesting to at least prod to different cultural belief systems. The kabbalistic mystic tells us about the indefinable and undetermined primordial light "*Ain Soph*". Ain Soph by contraction and expansion (*Zimzum*) brings forth creation. The *light of uncreated light* – by all means similar to the Buddhist *Uncreated* and *Unoriginated* – enlightens human beings and squires the evolution of humankind. This spiritual system "totius mundi" can be found represented in *Die kabbalistische Lehrtafel der Prinzessin Antonia in Bad Teinach* (Betz 2000). This picture had come into being under the impression of Antonias' experience of conversion. It was conceptualized by the aid of an erudite brain trust beginning in 1652 and was drawn up and painted by Johann Friedrich Gruber (1620–1681) painter at the principality of Stuttgart in the years 1659 to 1663. In 1673 it was posed in the picture shrine of the trinity church in Bad Teinach-Zavelsstein. In this village the royal family members spent their holydays. Together with his wife Isolde Betz, born Schnabel, Otto Betz published a monography about the *kabbalistic lecture board* carrying the title »*Licht vom unerschaffnen Lichte.*[1] *Die kabbalistische Lehrtafel der Prinzessin Antonia in Bad Teinach*«.

In the kabbalistic tree of life *Ain Sopf* is posed above the *Sephira Kether* (crown). In many representations the primordial light is a trinity of *Ain*, *Ain Soph* and *Ain Soph Aur*. *Ain* resembles the Nibbhana, *Ain Soph* is a "rotating no-thing" and *Ain Soph Aur* the "enlightening no-thing". Some Kabbalists equate Ain Soph with God, the absolute only contingent on itself incomprehensive cause of creation. Clearly, nowadays physicists can see in the *Zimzum* of *Ain Soph* an image of second field quantization or Higgs-mechanism and in the rotating no-thing a black hole. In a phenomenological approach to physics construction such a view is not at all far fetched. But it reflects our general experience. However, there are serious reasons why we do not follow such analogy. *In reality* Ain Soph cannot resemble Nibbhana. But in thought, - in cognition, - it can.

1 »Light of uncreated light«, kabbalistic lecture board = kabbalistische Lehrtafel

THERAVADA BUDDHISM

Nyanatiloka Mahâthera describes to us »*The Immutable*« translating and explaining the words of the Buddha as follows: *Truly, there is a realm, where there is neither the solid, nor the fluid, neither heat, nor motion, neither this world, nor any other world, neither sun nor moon. This I call neither arising, nor passing away, neither standing still, nor being born, nor dying. There is neither foothold, nor development, nor any basis. This is the end of suffering. There is an Unborn, Unoriginated, Uncreated, Unformed. If there were not this Unborn, this Unoriginated, this Uncreated, this Unformed, escape from the world of the born, the originated, the created, the formed, would not be possible. But since there is an Unborn, Unoriginated, Uncreated, Unformed, therefore is escape possible from the world of the born, the originated, the created, the formed.* (Nyanatiloka Mahâthera 1952).

For scientists constructing physics it is important to contact this deep state of meditation in reality and not in theory. It is neither from this world nor from any other world. It is perceived as transcending the real and is beyond imagination or vision. But it is not constructed theory. Only what we say or write after the encounter may be seen as theoretic. Yet it may be written in a state of Mahâmudra too.

MAHAYANA AND TANTRAYANA

Gendün Rinpoche (1993, p. 89) describes the *deep clear light* as a state where the mind is resting in Mahâmudra-meditation during bedtime. Since this state is beyond any duality it is so deep that we can neither experience a determinable mind as some subject nor some experience as an object. There is neither consciousness, nor experience, nor dreams, and when the mind is awakening we are in the end still in meditation. We feel happy and relaxed in body and mind. Gampopa said even our skin looks delicate and lucidly clear. Those are signs which tell us that asleep we have rested in clear light.

Contacting and understanding the Immutable requires an insight into the transcendence of mind. Namely, a tranquil mind which allows for complete cessation of thought is not the end of mind. A mind void of thought is not no-mind. It is not just an annihilation of all cognitive functions of the brain, but rather a contact of our mind to its ground state. That there is such a ground state represents an important input to all phenomenological, constructivist and systemic theories of physics. A proper construction of physics and cosmology requires an insight into the transcendence of mind, that is, clear primordial light. In his lecture to Naropa, Sri Tilopa says:

> "The things we have created are without substance. Therefore, search for the essence of the highest. The pattern[2] of the mind cannot perceive the meaning of the transcendental mind. The way of doing cannot discover the meaning of not-doing. If you want to put into realization the transcendent mind and the non-action, cut through the root of thought and let consciousness stay uncovered. [...] Once you perceive the open space, the steady conceptions of centre and firm margins dissolve. [...] Even though we may say that space is empty, the space cannot be described. Even though we may claim that the mind is permeating radiance

---

[2] Dharma translated as „law", "pattern" and "way"

by such denomination we cannot prove its existence. Space doesn't know any local fixation. Likewise the free mind inheres nowhere. Stay resting loosely in this original state without changing. No doubt that your chains will unfasten. The essence of mind is like space." (Trungpa 1994, p. 152 ff.) In »*The Tibetan Book of the Dead*« Lama Anagarika Givinda in part I on the "*Realization of the Nature of primordial light in the intermediate stage of the deathhour*" confirms: "*The mind as light is not a metaphor, not an image. It is an inner experience of being, if we follow the reports of mystics of various religions. The Buddhist light-meditation is above all a sinking in ones mind. When all activities of the mind, this uninterrupted inner chatter, have ebbed away, when the mind is clear, unclouded by thought, then it appears as clear light.*" (Govinda 1980, p. 84ff.)

The mind nature partly constitutes the quality of consciousness, namely the conscious mind. But in its essence it goes beyond it. Usually the mind is active. It is no more than cognitive action. But when mental activity is unbound from the objects of observation, the force of cognition turns inside, and in its centre the mind meets itself. It encounters the light void which is the nature of the primordial mind. Our insight into the nature of mind is not a separate, dualistic cognition which stands opposite to the perceived object, but it experiences itself as identical with the nature of that which is the content and the object of the insight. The primordial light is the nature of open unlimited and unconditioned space. Here it is where mind and space are meeting. This is the living basis of religious experience (von Weizsäcker 1971, p. 24ff.). But I would add it is not just religious experience, but experience itself.

Most of us, if we are scientists, have grown up with a special duality, namely the division between cognition and extension as space-time or res extensa versus res cogitans. Thought is not matter. But thought has meaning whereas matter is extended. Those are the properties of cognition and matter, and those were identified as important properties of god. In modern times René Descartes made this difference between the res cogitans as a feature of the mind and the physical res extensa. But this duality has already been questioned by Baruch Spinoza. Spinoza claimed that God involves both cognition and extension. In this writing we are always operating at the interface of mind and matter, light and space-time. The construction of physics requires both the extended fields of energy and the intelligent energy of mind. The clear primordial light is the bridge to the big bang. Without understanding that bridge we cannot do physics.

Hinduismus and Bhagavadgita

Interestingly in Sanskrit the word *Guna* originally means *thread*, *cord* and *string*. Later it was translated as *property*, *predicate* and *quality*. *Guna* appears to be a string that keeps things together: it's the qualities that integrate things. In the Indian Samkhya-philosophy Guna represents the forces which compile primordial matter Prakriti. Prakriti is compounded by three *Gunas*, *Tamas* representing inertness, darkness and chaos, *Rajas* representing restlessness, motion and energy and *Sattva* being clarity, goodness and harmony. As physicists we cannot but marvel the similarity of *Tamas* with dark matter, of *Rajas* with Baryonic matter and *Sattva* with photons and bosons. We are aware that George Gamow one of the founders of our postmodern theory of Big Bang used the Aristotelian denotation *hylê* or *Ylem* for the hot primordial neutron matter. Gamows theory describes the development of chemical elements in a hot neutron- and protongas. Even today this assumption is at the basis of our theory of primordial nucleosynthesis. While Alpher, Bethe and Gamow (1948) claimed all elements originated in the big bang, we know today that only the heavy isotopes of hydrogen, that is, deuterium and tritium as well as helium, lithium and beryllium originate in the early universe. All the other heavy elements were born in stars and supernovae. They correctly predicted relative frequencies of hydrogen- and helium-isotopes.

CHRISTIAN MYSTIC AND THE ABBA

In Christianity the deep insight, as Eckehart said: the real poverty, has been suppressed and the other dimension substituted by a personal god. Only a few have attained the deepest perception, the best known master Eckehart (1260 - 1328), earlier perhaps the Abba, the desertfathers, Antonios the Great, "Father of Monks" (about 251 - 356), Hilarion of Gaza (291 - 371), Arsenius the Great (354 - 450), Makarios from Egypt (300 – 390) and Onophrios the Great (320 – 400). The Mystic teaching of master Eckehart is often seen in connexion with the substance ontology of Thomas of Aquin (1225 – 1274): "*deus est esse*" is contrasted with "*deus est intelligere*". In usual interpretations the intelligere is translated as thought or cognition, and brings forth the statement "*god is thinking*". But *intelligere* should be better transposed to "*reading between the lines*".

In »Quaestiones parisienses« Eckehart represents a universal process at the basis of which there appears god as primordial origin of phenomena. God produces the existing by stepping out of himself and relating the emerging phenomena to himself. So creation becomes a dynamic self-relationing of god. As god is creating in the presence he cannot at any time have stopped creating. In his 50th homily Eckehart says "*There is no becoming, but a being, a becoming without becoming, a being new without renewal, and such becoming is the being of god*". God is a universal ground that has to be kept free of all determinability. Eckehart differs between god and godhood. As in Buddhism the Buddha mind is impersonal, so in Eckeharts teaching the godhead is impersonal. It appears as an "abyss of no-thing" and does not partake in our theories about being and becoming. In his *book of divine consolation* he states that god has created the world in such a way that he creates it without intermission. Creation is constantly happening in the here and now. To be constantly aware of this means an ending the search after god. Above our reason that is searching, says the mystic, there is another reason that is no longer searching.

With a mind who has found can we see a primordial light? "*Sometimes I see*", says Hildegard von Bingen (1098-1179), "*in that light* (annotation: the 'cloud of lively light') *another light which I call the 'vital light itself' [...] And when I contemplate it there vanish all sorrow, all pain from my mind, so that I am not an old woman but alike a young girl.*" (Vita, page 16)[3]

There have been some more mystics after Hildegard who experienced the 'unio mystica', best known Franz von Assisi (1181 or 82 – 1226), also Amalrich of Bena († 1206), Heinrich von Friemar (1250 – 1340) who was influenced by Eckehart, further Johannes Tauler († 1361) and his penfriend Heinrich von Nördlingen who was a pastoral worker for nuns. Many

3 Liber vitae meritorum (1148-1163); http://www.medieval.org/emfaq/composers/hildegard.html (English discography); the above quotation is my own translation.

of those who defended Eckehart became suspicious of heresy like Heinrich Seuse (1295 – 1366) and Marguerite Porète (1250/60 – 1310). Porète is known as initiator of the mystic monography »*Spiegel der einfachen Seelen, die nur im Wunsch und in der Sehnsucht nach Liebe verharren*« ("Mirror of the simpleminded souls who poise only in the wish and longing for love") (Louise Gnädinger 1987). She was handed over to mundane court as a relapsed heretic on Mai 30th and burnt at the stake on With Monday 1st of June 1310. Why some mystics were accused or murdered, while others survived opens up a question for sociology. Mechthild von Magdeburg (1207/10 – 1282/94) who wrote »*My flowing light divine*«[4] descended from a noble family. With twenty she went to Magdeburg and became a member of the "Collegia Beguinarum" for thirty years. Her Dominican confessor Heinrich von Halle strengthened her attempts to write poetic verses about her delights and agonies of her mystic experience with god. Both Hildegard and Mechthild were supported by a favourable social background.

These truths of vital divine light are experienced by any one who wants to come into a close living contact with creation. Some say that it is difficult today because time and thought are dominating our living. But it is possible today as it was possible some thousand years ago. In the end, that is, after many years of practise, there are quite unusual states of experience such as motion beyond time or extension of feeling without space and limits. I shall report about some of my own because they are rarely mentioned in the literature. These records deal with deeply hidden phenomena which can be experienced as some sort of bridge between the deep and high subconscious.[5] I regard these perceptions important for emancipation in every concern.

4 My own translation of »Ein vliessende lieht miner gotheit«

5 The Italian psychologist Roberto Assagioli developed Psychosynthesis as a form of transpersonal psychotherapy. He was familiar with the Kabbala, with Christian, hinduist and Buddhist scriptures, knew Graf Herman Kayserling, Victor Frankl, Eric Ericson, Abraham Maslow, Taitaro Suzuki and Anagarika Govinda. Assagioli collected many views similar to those given by me. He made a difference between deep, middle and high subconscious.

Chapter 8

PERCEIVING NO SPATIAL EXTENSION

We are not speaking about cognition or features of res cogitans in contrast to res extensa. Rather, as we speak of experience we must include perceptions in meditation. After forty years of meditation there emerged a state that did not exist earlier. Either it was not perceived or not recorded. It could not be described after the experience was made, in the same way as will be done in the following paragraph. This state is definitely bound to the action of Kundalini or inner bioenergetic motion, the flow of inner light. It is beyond time as it does not involve any experience of psychological time and is beyond space as it does not involve orientation and measure. There is, however, a strong force field. One perceives force in extension as a deep preconscious sensation. The sense involves our inner sense of touch. Thought is silent. Concepts have ceased. Still there is force, a very strong sensation. There can be perceived different regions of body awareness. The awareness can switch from an area to another. But the area is not directed and cannot be measured. If we say it is complex or chaotic, that does not meet the point. We may say, the force sensation is cloudy but beyond measure. It has no above or below, no left and right. It is possible that there occurs a remembrance of arms, legs, head and so on. Those regions of the field which bring forth the concepts melt away. The awareness may switch and then collects it all in one force field. All that we may call structure, system, relations actually is not there. The features of thought, cognitive appearance have faded. The mind is a strong sensitive force field. But it is utterly beyond our habitually experienced sensations of oriented and measurable, comparable feelings. The usual body pattern with all its details is not there. At several occasions there appears weakly in this a deep clear light. The appearance of that strength in the field of awareness can bring about a strong pain and can exhaust the body. Clearly those heavy sensations belong to the deep and middle layers of the subconscious. But as we begin to investigate the interaction between our concepts, our occasionally emerging thoughts, and the movements of the field, we can also be aware of the fine flows of feeling and emotions that are communicated in a collective field which goes beyond the limitations of one individual body and mind. This reminds me of what I can read in Roberto Assagiolis writings. Namely, we are all connected, not only on the social and physical level, but also by the cognitive and emotional flows which penetrate each other. Feeling love, understanding and compassion for each other are the links between our hearts. I am utterly convinced that science and especially physics need such insight in order to bring forth relevant knowledge for mankind. This has also been emphasized by Friedrich von Weizsäcker, Hans-Peter Dürr, David Bohm, Heinz von Foerster and many others. Especially mentioning von Foerster, there is a challenge that becomes apparent, namely we have to understand that metaphysics and constructivism are not two separate approaches or world-views. But one brings forth the other. Just as metaphysics must be seen as constructed we can see that radical constructivism in its essence is metaphysical. It cannot switch off the essential beyond of the root of creation.

Chapter 9

CONSTRUCTION OF METAPHYSICS

Metaphysics refers to an experience above or beyond nature. It can be represented by the Buddha's *Unborn, Unoriginated, Uncreated and Unformed* which was translated by Nyanatiloka Mahâthera as the »*The Immutable*«. In our western society it is perhaps compiled best in a short statement, - namely, the "unasked basic question" by Martin Heidegger (von Herrmann 1996, GA 9, p. 122): "Why are there phenomena at all and not rather nothing?" (*Warum ist überhaupt Seiendes und nicht vielmehr Nichts*?)[1]. Every human mind who searches for enlightenment must encounter that question. Every enlightened mind comes into contact with the unoriginated and naturally also asks Heidegger's question. For some of us metaphysics is the most basic discipline of philosophy. But this is not true. Metaphysics is not just a concept. It is not theoretic. Rather it is rooted in experience. Ancient greek philosophy knows the central concern of metaphysics questioning as an act of experiencing and comprehending the whole world by some primordial principles called arché. They came into contact with the cognitive paradox of an "unoriginated origin". At the origin there is what is neither born nor originating, neither extended nor unextended, not here, not there. This transcendental appearance was meant by Parmenides of Elea. He said: *to be is the same as perception*. Appearance of phenomena is the same as perception. He was perhaps the only thinker who had the power to map the experience of the immutable into the domain of thought without leaving a rest. Parmenides realized the appearance of an unoriginated in thought. This is real transcendence in cognition, real, not theoretic.

It is often said that science can only approach and explore those things which are measurable. I am convinced that this belief cannot be upheld, although it seems so evident. Physics cannot be unfolded without getting lost in the fragmentary chaos of cognitive processes if it does not begin with that ground state of unborn perception beyond time. It has to be ready to encounter the paradox. There is experience of a total beyond to experience. This is perception beyond time as the Buddha describes it. The paradox consists in the mere fact that that which is neither from this world nor from any other, neither heat nor motion, neither in the consciousness domain nor in the non-consciousness domain, can be encountered and perceived as primordial essence of everything experienced. Therefore creation is always in the here and now. It is beyond time and thought just as the presence of phenomena is beyond time and thought. So what is it essentially that is transcendent to the

[1] This English statement is my own preferred translation of the original in the brackets.

existing? It is the awareness. Awareness is the core of material processes. It signifies matter as holy.

It may be that physicists can hardly follow my argument. But as a matter of fact, we have done incredible violence to nature by separating thought as an independent observer of objects. We must not only stop violating living beings, as the Buddhists demand, but we must also stop violating matter as a whole. Because matter is holy.

Chapter 10

METAPHYSICS OF CONSTRUCTION

There is no doubt that the teaching of the Buddha is a philosophy which involves metaphysics. The episode wherein the enlightened mind experiences *a realm, where there is neither the solid, nor the fluid, neither heat, nor motion, neither this world, nor any other world, neither sun nor moon [...] neither arising, nor passing away, neither standing still, nor being born, nor dying* signifies the end of suffering, the third noble truth in the teaching. Non-orthodox writers in our days have also called it the »*noble truth of the cessation of stress*« "The remainderless fading and cessation, renunciation, relinquishment, release, and letting go of that very graving" (Thanissaro Bhikku 1996, p. 299). To the surprise of normal researchers such remainderless fading of stress offers a transcendental view beyond the existing physics. Such observation denotes the interval between science and religion. The force field fades into Nirvana, the flowing light divine into the ending of suffering. This is not a physical event? What, indeed! This event means ending the violence we are doing to nature. It is not just a decision. But it means clarity without search, that is, a total transformation of the human mind.

In a chapter on "The challenge of Understanding Radical Constructivism" (Glanville and Riegler 2007, p. 167) I found a section titled "*A Possible Parallel with Buddhist Thought*". He refers to Nagarjuna who laid the foundation stone of Mādhyamaka, the middle way, whereby he wanted to restore the Buddhas teaching (Weber-Brosamer 2005). The author of that chapter, Dewey I. Dykstra Jr. is currently a Professor of Physics at Boise State University in Boise, ID, USA and was confronted with the writing of Ernst von Glaserfeld in 1989. He states "*the middle way appears to have encountered and continues to encounter challenges very similar to those faced by radical constructivism. [...] The central idea in the Middle Way when first translated into English was labelled as 'emptiness'. This word is still used in the literature. What it refers to is the notion that when we attempt to go beyond the conventional existence of anything, we find no ultimate essence. The consequence is that the conventional existence of something has a beginning, middle and end. [...] This [...] sometimes put as arising, existing, ceasing applies to all things: objects, ideas, etc. Thus conventional existence as impermanence is an expression of emptiness*". A similar statement, says Dykstra, has been made by Ernst von Glasersfeld. But we have to be aware: Nagarjuna does not refer to "*the notion that when we attempt to go beyond the conventional existence*", but he refers to an actual experience. There is a paradox to cognition. To unfasten the knot we have to be very clear. We can perceive that which is utterly beyond the existing, but we cannot experience it. Namely that which is denoted as Nirvana, which has no begin, no development and no end, can enter our perception space while meditating. But it is a perception, not an experience, and therefore beyond space and time. Awareness as presence transcends time. So we become aware that radical constructivism neither transcends nor spaces out metaphysics. The Buddha mind, - conscious mind, void mind, - is there before the stars are born and is there while they are born. When we step out of silence into the domain of thought, our cognitive image has to respect that.

Chapter 11

SELF RELATION FIELD

IDEMPOTENT: BECOMING WITHOUT BECOMING

The first primordial field we can conceive of has a symbol a. This can have a relation to itself $a \circ a$. The a resting in itself, not stepping out of itself brings forth a. It reproduces itself in some autocatalytic self relation. Its equation is rather trivial:

$$a \circ a = a \qquad \text{fulfils an idempotency relation.} \qquad (1)$$

The basic primordial field is an idempotent a in some algebra $(\mathcal{A}, \circ)$. Physically it represents an 'even parity field'.

Clearly, the observing intelligence that writes down $a \circ a = a$ has to have means complex enough to manage the meaning it ascribes to a physics equation. The a can be represented by an extremely – even infinitely – complex term in some universal algebra $\mathcal{A}$, it may even be as complex as the world is. The a can embed it all, but still we had to have $a \circ a = a$, that is, a is resting in itself. Interacting with itself it stays invariant. This is possible. The a as a mathematical symbol is an idempotent.

Chapter 12

ODD PARITY FIELD

As soon as the primordial field allows for an awareness of back and forth or below versus above or any other parity flip, we state that $\bar{a}$ relating to itself brings forth minus $\bar{a}$, that is:

$$\bar{a} \circ \bar{a} = - \bar{a} \quad \text{odd autoparity or paripotency} \qquad (2)$$

The field is its own parity operator. Note, parity means 'numerical equality' or 'equal strength'. The field a and its total reverse $\bar{a}$ may annihilate each other by addition: $a + \bar{a} = 0$. Therefore the algebra $\mathcal{A}$ will be over a ring or field and we shall have some measure over a ring of sets. With autoparity, we shall see, we are pretty far away from the innermost primordial fields which may neither have basis nor measure.

Physics is coordinate free. In the form of a primordial idempotent the field may cover a location. But at present we do not know about location. The idempotents a and paripotents $\bar{a}$ are more general than wave functions $\psi(x)$ with a coordinate location x. Clearly, we have

$$\bar{a}^{\,2n} = -\bar{a} \qquad \text{and} \qquad \bar{a}^{\,2n+1} = \bar{a} \qquad \text{with integer n} \qquad (3)$$

Our field acts on itself like a parity operator having eigenvalues +1 or −1. If we have equation (1) valid, we say that the field is 'inversion invariant' or even. However, in case equation (2) is holding true, we say that the field is odd.

Notice, weak interacting fields are not inversion invariant, but have odd parity. Before the 'Wu-experiment' was carried out by Chien-Shiung Wu the prevailing belief was that all laws of nature were inversion invariant.

In our mathematics we have introduced the odd parity already in a second step forth from Nirvana. It is not further away from the unoriginated origin than two steps.

Chapter 13

PULSATING ZIMZUM WITH MEASURE

The oscillation of primordial clear light is a volume pulsation. As a self recursive field it can again be described by a simple equation which does not involve measurable time. The field f is not an idempotent. But the product $f \circ f$ deviates only a little from the primordial field f. Again we have $f \in (\mathcal{A}, \circ)$ and a self-relating equation of motion

$$f_{j+1} = f \circ f_j \qquad \text{or equivalently} \quad f_j = f^j \qquad j = 1, 2, 3, \qquad (4)$$

The field after j recursions is the jth power of f. Independent of its concrete algebra representation f_j has an oscillating volume component V_j and an oscillating scalar part ϕ_j, - the light part. Yet the volume, though directed and ordered, is not located in an oriented and/or metric space region. The primordial volume pulsation occurs beyond space-orientation and measurable space-time. In the primordial field f the space-time either has not emerged or it is decoupled and separate from it.

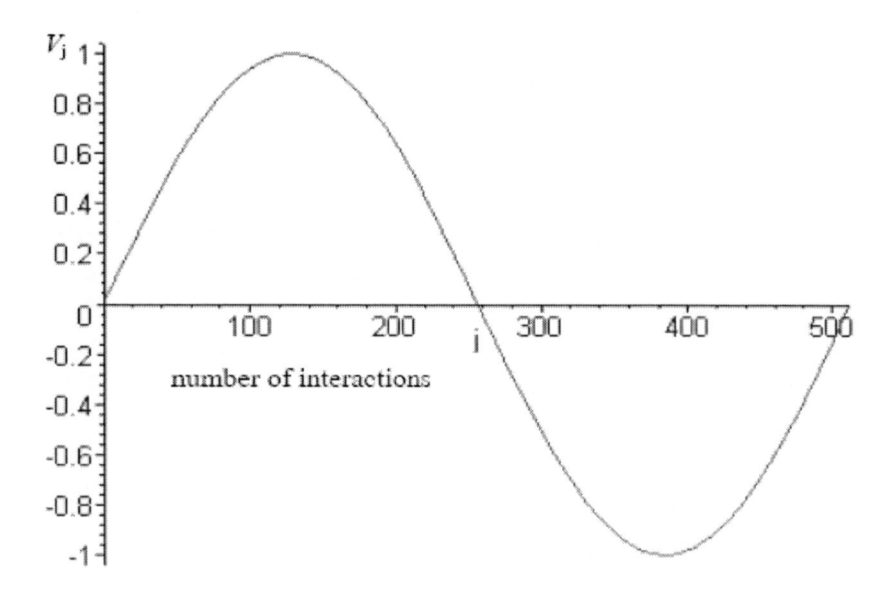

Figure 1. Pulsating volume of primordial field.

REPRESENTATION OF SIMPLE MEASURABLE FIELDS

Let $u \in \mathcal{A}$ a unipotent of the algebra and *Id* the identity. The *u* can also be understood as a "Coxeter reflection" with $u \circ u = Id$. Then the

$$a_{\pm} = \tfrac{1}{2}(Id \pm u) \tag{5}$$

is qualified for representing the equation (1) for idempotents. That is, we have

$$a_{\pm}^{2} = a_{\pm} \qquad \text{with the abbreviation } a \circ a = a^2 \tag{6}$$

On the other hand we define for odd parity fields

$$\bar{a}_{\pm} = -\tfrac{1}{2}(Id \pm u) \qquad \text{with} \qquad \bar{a}_{\pm}^{2} = -\bar{a}_{\pm} \tag{7}$$

$\bar{a}_{\pm}$ is qualified for representing the equation (2) for paripotents.

Finally we consider a field

$$f = \phi + V \qquad f \in PVol \overset{def}{=} span_R\{Id, \omega\} \tag{8}$$

where ϕ represents a real scalar and V a directed volume. The field is an element of a *paravolume space*. Such a space is a vectorspace spanned by a scalar identity *Id* and a unit volume ω with indefinite dimension and obeying the equation

$$\omega \circ \omega = -Id \qquad \text{complex structure} \tag{9}$$

Consider vectorspace $\{Id, \omega\}$ satisfying (9) and take the start value of f equal to $f = 0.99992\ Id + 0.01227\ \omega$. In this case we obtain the figures 1, 2. Note, although the pseudoscalar ω is a directed unit volume it does not belong to a definite vector space with a definite dimension. The ω is so to say an indefinite exterior form on an unknown manifold whose dimension is not known or indecidable. We say that ω and likewise the V represent indefinite orientated extensions on some indefinite manifold. What does the "indefinite" exactly mean? First, it means that we cannot define the Hodge-* operation because we cannot define some orientation-class or any of its representative basis. We have a volume, but no oriented base units. Once the volume can be represented by the exterior product $e_1 \wedge e_2 \wedge e_3$ in Euclidean space (or by any other unit director in a Clifford algebra with a higher dimension) we may claim the field is no longer primordial. But then it has given birth to a variety with definite dimension. That is, in the primordial field we just demand the complex structure (9).

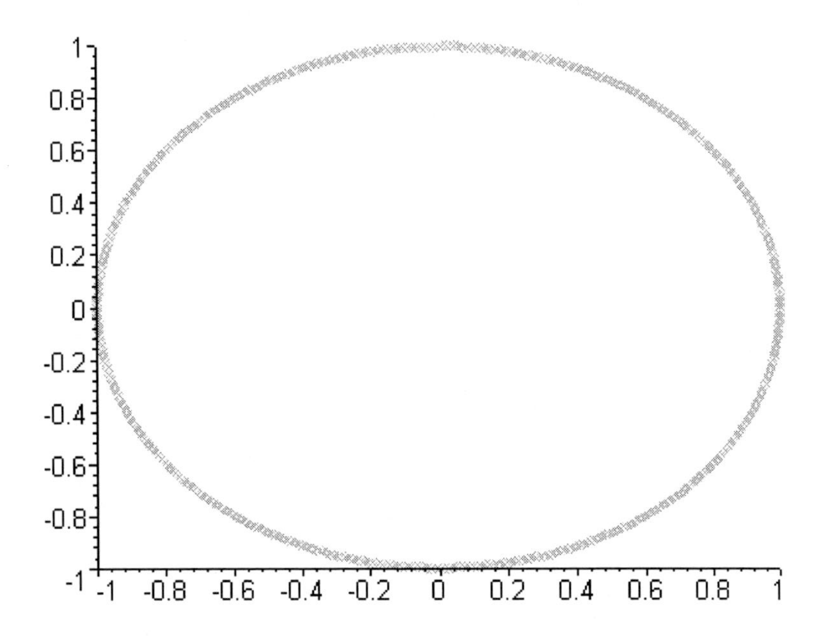

Figure 2. *Ain Soph* a "rotating no-thing" iteration of equation (8): V and ϕ.

CREATION REVERTED

Is not the same as time reversion. But it means the following. In terms of topological evolution we would understand the organization of complexity in terms of a general Pfaff Sequence $\{A, dA, A\hat{}dA, dA\hat{}dA\}$ beginning with a differential 1-form A. The 1-form of action potentials, in nowadays mathematical physics, is built up on a four-dimensional variety of ordered independent variables, say, $\{\xi^1, \xi^2, \xi^3, \xi^4\}$. This variety then supports the exterior differential volume element (Kiehn 2007)

$$\Omega_{(4)} = d\xi^1 \wedge d\xi^2 \wedge d\xi^3 \wedge d\xi^4 \tag{10}$$

We mean by »*reversion of creation*« that cognition, historically, has no other way than to put the process of nature upside down. We base our construction of volumes on the already existing variety of coordinates in a space having a definite dimension. But nature does not do that. Nature is proceeding exactly the other way. Within the primordial fields first there appear action potentials and volumes and only thereafter the "ordered, independent variables" are emerging as stable units of space-time. Therefore we have an A (in the notation of the above section a multiple ϕ of Id) and some $\Omega_{(?)}$ but no variety of $\{\xi^1, \xi^2, \xi^3, \xi^4\}$. Rather we have to respect the phenomenology of primordial fields.

Chapter 14

PHENOMENOLOGY OF PRIMORDIAL FIELDS

Seeking representation for measurable primordial fields we already have put the process upside down. We must begin beyond measure and orientation. We have to correlate physical with mathematical operations in order to see what goes on at the origin, or rather, in the depth of creation. Measure and basis have to do with the axiom of choice. But the axiom of choice applied to a set of physical phenomena, - to a domain of fields, - signifies a certain physical activity. Think about a field primordial to strong interaction. Consider such a field as extended, yet not measurable. You cannot reach in and take out a part just as you cannot penetrate a quark. Consider that field as *f*, a domain or set of such fields as $\mathcal{A}$ and the invading or penetrating function as *D*. We use *D* for "diffusion" in the sense of penetrating a substance. Then, if penetration were possible throughout the field, we had to have, there is a *D* with

$$\forall_{f \in \mathcal{A}} \; D(f) \in f \tag{11}$$

which is nothing else than the axiom of choice (AC). But since the invasion and selection of such a strong interaction element is not possible we have to have the negation

$$\neg \forall_{f \in \mathcal{A}} \; D(f) \in f \tag{12}$$

But negation of the AC implies that we cannot establish a basis within a vector space $\mathcal{A}$. That is to say, the $\mathcal{A}$ cannot, for example, be adequately represented by geometric Graßmann or Clifford algebra. The primordial field rules out the axiom of choice and with it the existence of basis, definite base unipotents, idempotents and all the rest of it.[1] So we have a

- *State I:* fields $f \in \mathcal{A}$ with no axiom of choice and no basis

But we insist with *D* and there occurs a phase transition where we can penetrate the field and take out a bit, say, an *unextended fermion* or neutrino, a wimp. The field reacts chaotic to measurement. Now the choice by *D* is possible, but the field is stardust. The volume cannot be measured because of a theorem by Banach and Tarski (TBT). Therefore

---

[1] It will be a question of the greatest importance, if such primordial fields can exist within the midst of measurable space.

- *State II: fields $f \in A$ with axiom of choice and basis possible, but extension cannot be measured*

The chaos may involve transitions between states without dimension and basis and states with dimension and basis but without measure. The process of nature involves transitions among four combinations of the kind

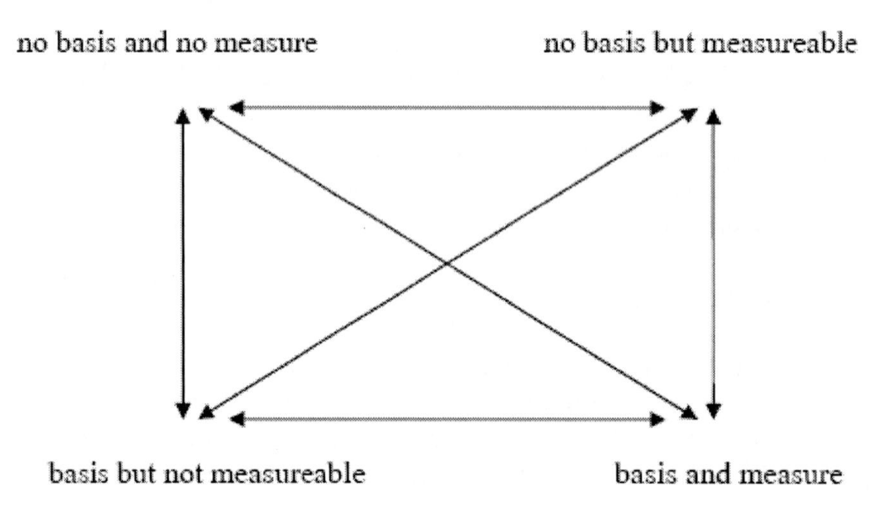

Figure 3. Primordial transitional fields.

Chapter 15

Theorem by Banach and Tarski in Minkowski Space

The theorem by Banach and Tarski demonstrates that our conventional concept of volume V cannot be generalized to arbitrary continuous sets of points. According to TBT a globe can be decomposed und rebuilt into two gapless globes each having the same diameter as the original one. The volume is doubled. But it is not clearly evident how the surplus volume springs up, as the whole construction is based on a congruence transformation. Banach and Tarski gave the proof in 1924 for a ball in Euclidean 3-space showing that decomposition into six parts is sufficient. However those parts are very complicated. They are infinitely porous and without a Lebesgue-measure. They resemble the nowadays fractal discontinua. The existence of such non-measurable continuous subsets of real Euclidean spaces need the AC. Generally, the TBT says that initial and final volumes may differ by an arbitrary real factor. Consider $\mathbb{R}^n$ and two bounded subsets $X, Y \subset \mathbb{R}^n$ with non-empty interior such as finite balls. Then there exists a positive integer m equating the size of an isometric decomposition (congruence classes), and respectively a number of *moves*

$$\beta_1, \beta_2, \ldots, \beta_m \qquad \text{transposing the partition} \tag{13}$$

$$X = X_1 + X_2 + \ldots + X_m \qquad \text{onto the congruent parts} \tag{14}$$

$$\beta_1(X_1), \beta_2(X_2), \ldots, \beta_m(X_m) \qquad \text{such that} \tag{15}$$

$$Y = \beta_1(X_1) + \beta_2 X_2 + \ldots + \beta_m X_m \tag{16}$$

represents an isometric decomposition of Y. We want to show how the TBT can be generalized to the quadratic Clifford algebra generated by the Minkowski space in the Lorentz metric. First we have to be aware of a special requirement in the Banach-Tarski-paradox (1924), namely the existence of a subgroup of GL(3) which is a free group over a set of two generators

$$S = \{\xi, \psi\}. \tag{17}$$

The ξ represents a rotation of Euclidean 3-space about the x-axis by the angle $(\sqrt{2})\pi$ and ψ a rotation about the y-axis by $(\sqrt{3})\pi$. In order to generalize the paradox to the Minkowski space and respectively its Clifford algebra $Cl_{3,1}$ we base ourselves on a rigor which brings together some important orthogonal groups of the $Cl_{3,1}$ which are needed because they constitute the standard model. In essence it is a spin group without a rotation group to double cover.

In a recent work I conjoined the anti-de-Sitter-Space adS_6, having two time coordinates, with the Clifford algebra $Cl_{4,2}$ having a typical Majorana representation by matrices Mat(8, $\mathbb{R}$). I derived the "*bottom up construction*" of constitutive groups beginning with the $SU(2)$ and $SU(3)$ at the bottom and ending on top with $SU(7)$ multiplets in the rank-6 Lie algebra $\ell^{(6)}$ concatenating with the Clifford algebra $Cl_{4,2}$. The bottom is characteristically build up by the rank-3 algebra $\ell_{3,1}$ embedding some rank-2 standard model Lie group $L^{(2)} = SL_{Cl}(3)$ with generating elements from $\ell_{3,1}$. The algebra $\ell^{(2)} = sl_{Cl}(3)$ can be spanned by 8 elements of the space-time algebra $Cl_{3,1}$ (Schmeikal 2006, 2009). Those generate the pure constitutive spin group.

$$\begin{aligned}
&\lambda_1 = \tfrac{1}{4}(e_{34} - e_{134}) && \lambda_2 = \tfrac{1}{4}(-e_{23} + e_{123}) && \lambda_3 = \tfrac{1}{4}(e_{24} - e_{124}) \\
&\lambda_4 = \tfrac{1}{4}(e_3 + e_{234}) && \lambda_5 = -\tfrac{1}{4}(e_{13} + j) && \lambda_6 = \tfrac{1}{4}(e_2 + e_{14}) \\
&\lambda_7 = -\tfrac{1}{4}(e_4 + e_{12}) && \lambda_8 = \tfrac{1}{2\sqrt{3}}(-2e_1 + e_{24} + e_{124}) &&
\end{aligned} \tag{18}$$

The Gell-Mann matrices in $\mathbb{C} \otimes Cl_{3,1}$ are identical with

$$\{-2\lambda_1, 2i\lambda_2, -2\lambda_3, -2\lambda_4, 2i\lambda_5, 2\lambda_6, -2i\lambda_7, 2\lambda_8\},\ i = \sqrt{-1} \tag{19}$$

Verify, the constitutive algebra $\ell^{(2)}$ can be written as a Lie commutator product

$$\ell^{(2)} = sl_{Cl}(2) \times so_{Cl}(3) \tag{20}$$

To construct the free subgroup S we need the factor $so_{Cl}(3)$ having generating elements $\{\lambda_2, \lambda_5, \lambda_7\} \in Cl_{3,1}$. These satisfy the characteristic $so(3)$-commutation relations

$$[\lambda_i, \lambda_j] = \tfrac{1}{2}\varepsilon_{ijk}\lambda_k, \quad \varepsilon_{ijk} \text{ antisymmetric in indices } i, j, k = 2, 5, 7 \tag{21}$$

$$[\lambda_2, \lambda_5] = \tfrac{1}{2}\lambda_7, \qquad [\lambda_2, \lambda_7] = -\tfrac{1}{2}\lambda_5, \qquad [\lambda_5, \lambda_7] = \tfrac{1}{2}\lambda_2$$

Note that these relations correspond one-to-one with the well known bivector relations

$$[L_1, L_2] = \tfrac{1}{2}L_3; \qquad [L_1, L_3] = -\tfrac{1}{2}L_2; \qquad [L_2, L_3] = \tfrac{1}{2}L_1 \tag{22}$$

with bivectors $L_1 = e_{23}$, $L_2 = e_{13}$, $L_3 = e_{12}$. Both $so_{Cl}(3)$ algebras satisfy the inter-commutation relations

$$[\lambda_2, e_{23}] = 0 \qquad [\lambda_5, e_{12}] = 0 \qquad [\lambda_7, e_{13}] = 0 \tag{23}$$

We know that the bivector $L_1 = e_{23}$ causes rotations about the x_1-axis, L_2 about the x_2 and L_3 about x_3 in the Minkowski space. But which ones are the axis left invariant by the λ_i ? The answer is a little surprise. Consider the Clifford manifold

$$\begin{aligned} \chi = \{ & x_1 Id + x_2 e_1 + x_3 e_2 + x_4 e_3 + x_5 e_4 + x_6 e_{12} + x_7 e_{13} + x_8 e_{14} + \\ & + x_9 e_{23} + x_{10} e_{24} + x_{11} e_{34} + x_{12} e_{123} + x_{13} e_{124} + x_{14} e_{134} + x_{15} e_{234} + x_{16} j\} \end{aligned} \tag{24}$$

with directed space-time-volume $j = e_{1234}$.

The element λ_2 leaves the x_2, the x_9 and x_{12} unaltered. That is, it simultaneously rotates the space about base units e_1, e_{23} and directed volume e_{123}. Note those three commute. Thereby it rotates the pairs $\{x_3, x_4\}$ of a plane coordinating base units $\{e_2, e_3\}$ in $\mathbb{R}^{3,1}$. But that is not yet the most general action of λ_2. In Schmeikal (2004, 2005) it has been shown that there are six colour spaces corresponding with isomorphic maximal Cartan subalgebras of the $Cl_{3,1}$, namely

$$\begin{aligned} ch_1 &= \{1, e_1, e_{24}, e_{124}\} & ch_4 &= \{1, e_2, e_{14}, e_{124}\} \\ ch_2 &= \{1, e_1, e_{34}, e_{134}\} & ch_5 &= \{1, e_3, e_{14}, e_{134}\} \\ ch_3 &= \{1, e_2, e_{34}, e_{234}\} & ch_6 &= \{1, e_3, e_{24}, e_{234}\} \end{aligned} \tag{25}$$

Each of those spaces can be generated by two units of grades 1 and 2. Taking out the scalar these are 6 isomorphic Cartan algebras of the rank 3 Lie algebra $\mathfrak{l}_{3,1}$ derived from $Cl_{3,1}$. The pure states within those colour-spaces are the primitive idempotents which represent fermion states.

Consider the well known baryon decuplet. In case of three equal quanta sss = Ω^-, ddd = Δ^- and uuu = Δ^{++} the space-time oscillator shows maximal departure from spherical symmetry and violates the Pauli principle. Therefore it needs a further degree of freedom which is colour (Schmeikal 2009). The state function has the possibility to evade into another space. Which space is it where into the fermion state function from ch_1 quibbles? It's spaces ch_3 and ch_5. That is, the generalized Pauli principle is realized by carrying a space-like unit vector e_1 to a space like directed space-time area, say, e_{24}. This area cannot be interpreted as a (hyper)complex number, but squared it gives +1. It behaves almost like a space unit vector. Almost, because there is a difference, namely, it commutes with the other space unit vectors unequal e_2. Now we can answer what is the most general action of the λ_2. It may rotate (or unfold) any pure fermion state from any space ch_χ out into the whole positive definite subspace

$$Ch = \bigoplus_{\chi=1}^{6} ch_\chi \tag{26}$$

We call it positive definite because of its signature. All its units squared give plus one. Now we can proceed with λ_5 and λ_7. The λ_5 fixes e_{24}, j and L_3 or respectively the x_{10}, x_{16} and x_6-scales. The λ_7 fixes e_{124}, e_4 and L_2 or respectively the x_{13}, x_5 and x_7-axis. So we see, each generator of the $so_{Cl}(3)$ fixes the coordinate of exactly one base unit of a colour space. Further, each fixes one quasi Euclidean bivector with negative square, and last not least, each preserves one general time-space coordinate e_4, e_{123} or j. That is, the $so_{Cl}(3)$ in equation (20) preserves a 3-partition in the Clifford algebra into a positive definite fermion-space, a negative definite time-space and the well known bivector-su(2). From this we conclude that the whole Clifford algebra together with this partition is turned into a Banach Tarski-discontinuous point set by the following two algebraic elements

$$\xi = \frac{\pi}{\sqrt{2}}(-e_{23} + e_{123}) \qquad \psi = \pi\sqrt{\tfrac{3}{4}}(-e_4 - e_{12}) \tag{27}$$

The free group over $\Sigma = \{e^{\xi}, e^{\psi}\}$ represents the necessary piece of evidence to prove the Theorem by Banach and Tarski in the Clifford algebra of the Minkowski space with Lorentz metric. We have to ask about the physical meaning of TBT in $Cl_{3,1}$.

Chapter 16

FRACTAL STRONG FORCE

The free group S consists of words over the 4 -letter alphabet

$$A(\mathrm{S}) = \{\exists, \mathsf{E}, \ni, \in\} = \{e^{\xi}, e^{-\xi}, e^{\psi}, e^{-\psi}\} \subset Cl_{3,1} \quad (28)$$

Every word W has an inverse W^{-1} calculated as usual. We carry out free rotations of elements $x \in Cl_{3,1}$ by forming terms

$$\text{free group rotation:} \quad x \xrightarrow{S} W^{-1}xW \quad (29)$$

We know that the constitutive Lie algebra $\mathcal{L}^{(2)}$ has a manifold of primitive idempotents with one fixed primitive idempotent. This makes the Lie group into a stabilizer group for this state. That is, the Lie algebra $\mathcal{L}^{(2)}$ is annihilated by the primitive idempotent f_ν

$$f_\nu \mathsf{L}^{(2)} = 0 \qquad \text{with} \qquad f_\nu = \tfrac{1}{4}(1+e_1)(1+e_{24}) \quad (30)$$

The pure state f_ν is interpreted as an e^--neutrino field and is thus fixated by its stabilizer group exp $\mathcal{L}^{(2)}$. The other three primitive idempotents

$$f_u = \tfrac{1}{4}(1-e_1)(1+e_{24}) \qquad f_d = \tfrac{1}{4}(1-e_1)(1-e_{24})$$

$$f_s = \tfrac{1}{4}(1+e_1)(1-e_{24}) \qquad \text{represent u, d, and s-fields.} \quad (31)$$

In order to find out what happens to the strong force under a free rotation of the group S it is enough to rotate the strange quark by a progressive iteration just as was done with the pulsating Zimzum with measure (figure 1). It is even enough to iterate the basis unit vector e_1 because that is determining for the path within the manifold. We form the terms

$$e_1,\ \ni e_1 \in,\ \ni\ni e_1 \in\in,\ \ldots,\ \ni^n e_1 \in^n,\ \ldots \quad (32)$$

and obtain an iterated sequence of elements

1 $\quad e_1(1) \quad$ ┌→ $\quad x_2$

2 $e_1(2) = 0{,}444 e_1 + 0{,}497 e_2 + 0{,}497 e_{14} + 0{,}556 e_{24}$

3 $e_1(3) = 0{,}0127 e_1 - 0{,}112 e_2 - 0{,}112 e_{14} + 0{,}987 e_{24}$ a. s. f.

:

This discloses a chaotic and respectively fractal dynamics. Note that the quantities $e_1(n)$ belong to the manifold of base units e_1. Squared they all give the identity of $Cl_{3,1}$.

$$e_1^2(1) = e_1^2(2) = e_1^2(3) = \ldots = e_1^2(n) = Id \tag{33}$$

Carrying out 512 iterations (32) we obtain figure 4. This resembles a fractal x_2. If we calculate an x_2-histogram for the first 1024 elements, we obtain figure 5.

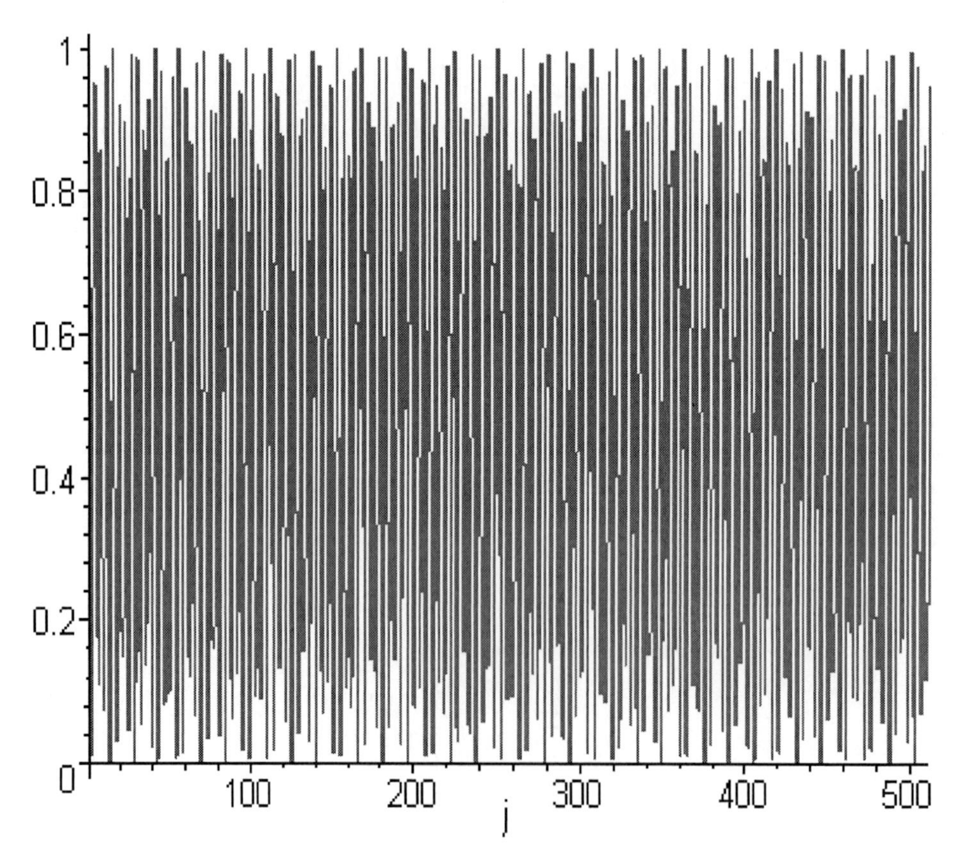

Figure 4. Iteration of discontinuous $e_1(n)$ - manifold, coordinate x_2

Progressive rotation of e_1 by the elements $e^{n\psi}$, $e^{-n\psi}$ gives us a histogram such as in the figure below.

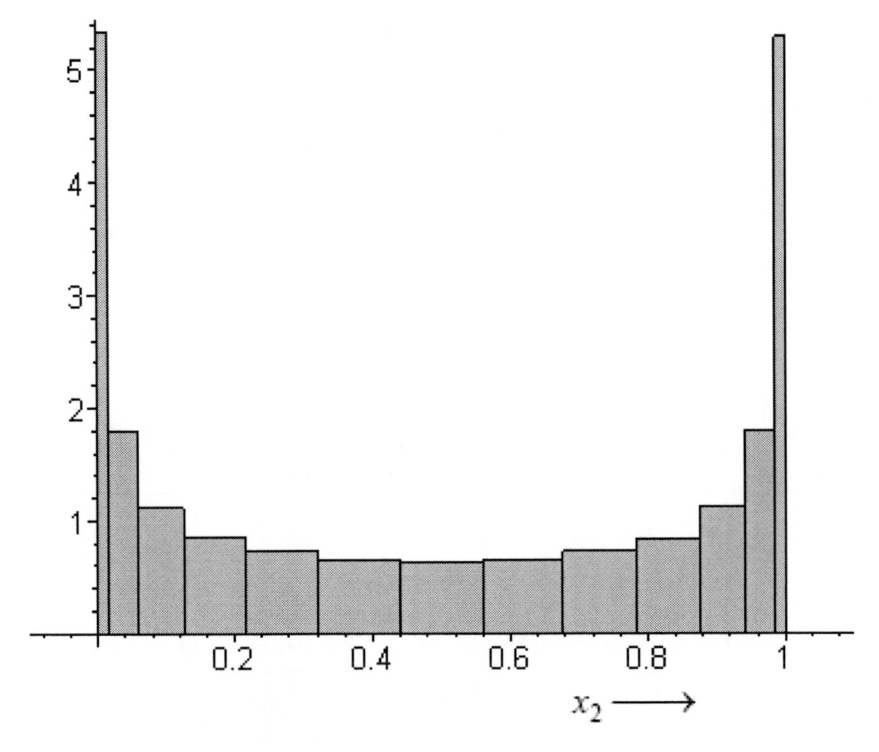

Figure 5. Histogram of discrete x_2 manifold.

Histograms like this one are also brought forth by deterministic chaotic processes such like the "quadratic iterator" and "Feigenbaum-scenario". It has a density function

$$\varphi(x_2) = \frac{1}{\pi\sqrt{x_2(1-x_2)}} \tag{34}$$

In the proximity of the Banach-Tarski discontinuous decay the Fast Fourier Transform (FFT) of x_2 shows a rather strange behaviour. We have chosen the generator in accord with equation (27)

$$\psi = 2\pi\sqrt{3}\,\lambda_7 = -2.720699047(e_4 - e_{12})$$

and watched the close proximity of the constant −2.720699047. At the exact ‚MAPLE-value' of $(\pi/2)\sqrt{3}$ the FFT discloses two tiny spectral peaks at high frequencies as shown in figure 6

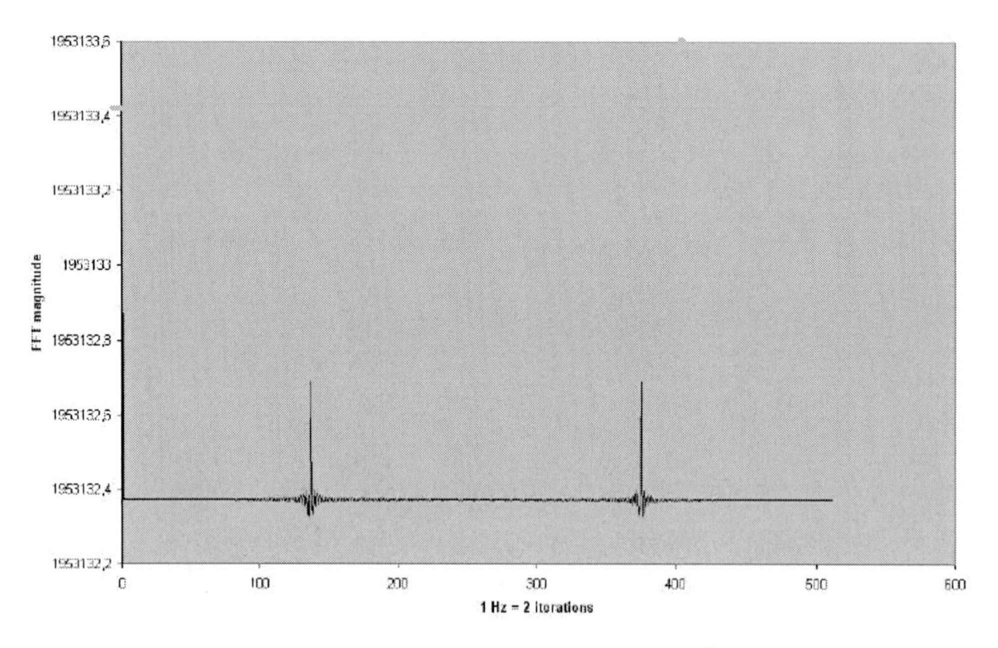

Figure 6. Spectrum of the magnitude x_2 (t) at the critical coeficient $(\pi/2)\sqrt{3}$.

This picture is a mess. It changes significantly if the coefficient in ψ is rounded down to the fifth digit, that is $\psi = -2.72069\ (e_4 + e_{12})$. The rotation brings forth so irregular numbers that the FFT magnitude itself takes the shape of a fractal figure as in Figure 7.

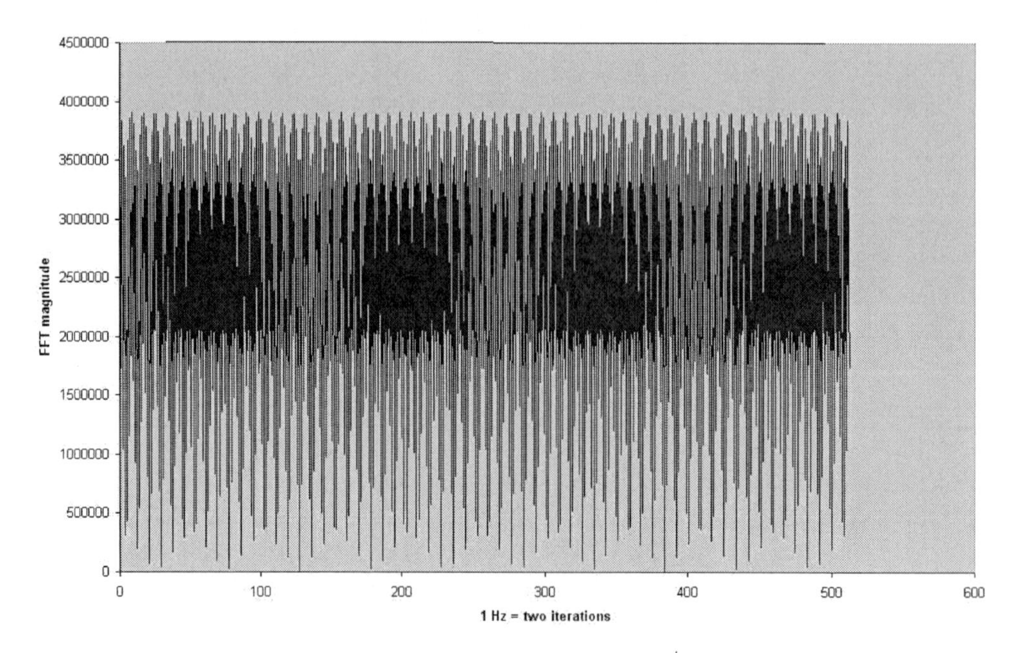

Figure 7. Spectrum of the magnitude x_2 (t) in the proximity of $(\pi/2)\sqrt{3}$.

This tells us that, in general, rotations induced by λ_5 need not be linear operations. But there are circumstances when they become linear in cooperation with some other rotations.

Chapter 17

PRIMORDIAL RELATION BETWEEN QUARK AND NEUTRINO

Rotating the unit vector $e_1(1)$ by a Banach-Tarski rotator $\exp(\psi)$ we obtain $e_1(2)$ as above. With the same iteration we turn a strange quark $f_s(1) = \frac{1}{4}(1+e_1)(1-e_{24})$ into the idempotent

$$f_s(2) = \tfrac{1}{4} - 0{,}028e_1 + 0{,}248e_2 + 0{,}248e_{14} + 0{,}028e_{24} - \tfrac{1}{4}e_{124} \tag{35}$$

It is clear that the primitive idempotent while iterated takes over the chaotic dynamics of the coordinates. But there is one observation of outstanding importance. Namely, the first primitive idempotent f_ν which we have chosen to interpret as a neutrino, does not partake in whichever chaotic movement of the algebra. It is a fix point even in the middle of chaos, because we have

$$e^{-\phi} f_\nu \, e^{\phi} \equiv f_\nu \tag{36}$$

and this holds true also for generators ξ, ψ of the free group. What can we learn from this?

A Banach-Tarski rotation of colour space is a chaotic oscillation in the primitive idempotent manifold which annihilates the Lebesgue measure. Suppose we had a seemingly stable region of space-time wherein we observed the field in self-interaction. The rotation itself would cause annihilation of isometry. Space-time would decay into a fractal manifold. Isometries would perform transitions to conformal regions. Volumes and densities would not be preserved. It is very probable that fields primordial to strong interaction involve such processes that violate special relativity. In such episodes when strong force fermions move rather freely there doesn't seem to be anything like a constant maximal velocity of field-quanta without mass. Investigating into the behaviour of the directed space-time area e_{24} we find exactly the same statistical behaviour of x_{10} as for x_2. The coefficient x_{10} is positive definite for arbitrary iteration numbers. However the irregular motion of the addend $x_8\, e_{14}$ shows a mean value of zero with a typical histogram

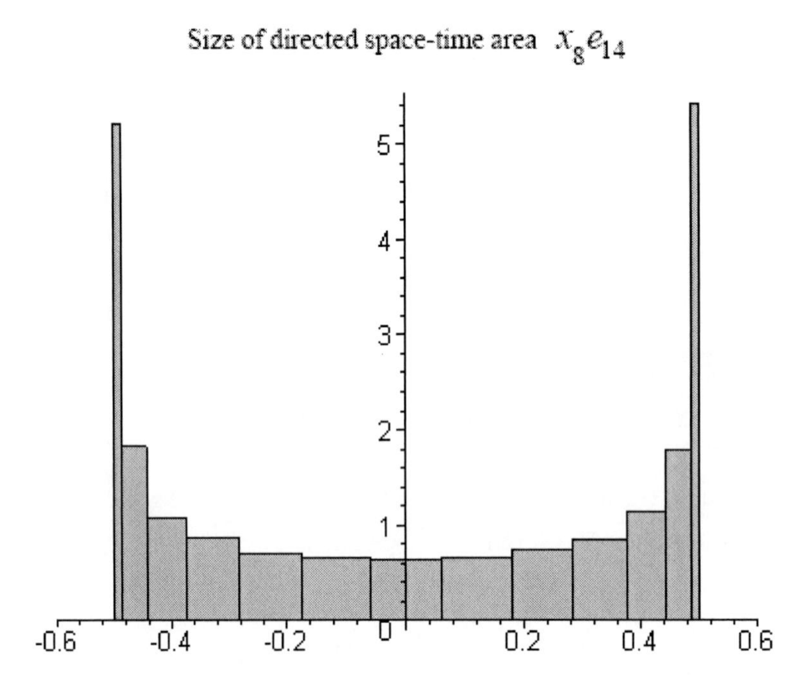

Figure 8. Histogram of x_8 manifold.

While x_2 and x_{10} have a range [0, 1] the size of x_8 e_{14} ranges between $-½$ e_{14} and $+½$ e_{14}. Thus we may observe a chaotic mix between advanced and retarded waves. The space-time oscillator can wrap up forward- and reverted time because reversion of area-direction can cover up time reversion.

Conjecture: Chaotic dynamics and causality violation in primordial strong force fields annihilate the Lebesgue measure. They involve non-isometric transformations. The neutrino field, however, preserves the Lebesgue measure because it does not partake in the fractal dynamics.

It is probable that in deep fields we observe in parallel a measurable neutrino ground field and a relativity violating primordial strong force. After we shall have gone into the question of chaos and the role of the Pauli principle as an antagonist to chaos we shall work out a primer for orientation. Because all these newly formulated questions are based on a phenomenology of orientation.

Chapter 18

ESCAPE FROM CHAOS

THE PAULI PRINCIPLE

Considering 3 equal quarks, sss = Ω^-, ddd = Δ^- and uuu = Δ^{++}, the graded space-time oscillator departs maximally from spherical symmetry. A further degree of freedom is needed in order not to violate the Pauli principle. This quality is colour. The state function has the possibility to evade into other colour-spaces. We said that the fermion state function quibbles from ch_1 to ch_3 and ch_5 and reverse. Next, consider the idempotent f_{54} from colour space ch_5 and a second one f_{34} from ch_3 and, form the product

$$T = (1-2f_{54})(1-2f_{34}) \text{ representing a colour rotation} \tag{37}$$

where $f_{54} = \frac{1}{2}(1-e_3)\frac{1}{2}(1-e_{14})$ and $f_{34} = \frac{1}{2}(1-e_2)\frac{1}{2}(1-e_{34})$

Verify $T f_{12} T^{1} = f_{53}$ is a colour-rotation for the same flavour, that is, of an s-quark. Verify further $T f_{22} T^{-1} = f_{24}$ is a flavour-rotation within the same colour (first index). That means T carries the s-quark pure state from colour space ch_1 to colour spaces ch_5 or to ch_3 in reverse direction. There exist universal trigonal rotations which transport quarks of each family in 3-cycles from colour to colour. Each colour space provides a definite colour and contains three quarks such as u, d, s. Considering u, d, s as one ternary fermion family, spaces ch_1, ch_3, ch_5 provide three different colours for that family. Indeed, colour is defined by structure rather than by any distinguished basis. That is, colour spaces are manifolds just as the pure states are. To decompose the 16-component multivector T according to the $su_{Cl}(2)$ generators, we take the logarithm $\tau = \log(T)$ and verify that

$$\tau = \frac{2}{\sqrt{3}} \arccos(-1/2)(\lambda_2 - \lambda_5 + \hat{\lambda}_7) \tag{38}$$

where ^ denotes main involution. In this case it means reversal of time in λ_7, so that we can also write

$$\tau = \frac{2}{\sqrt{3}} \arccos(-1/2)(\lambda_2 - \lambda_5 + \lambda_7 + \tfrac{1}{2} e_4) \tag{39}$$

This is interesting because the quantity without the ½ e_4 represents another universal trigonal operator of the system.

To understand the special relation between the Pauli principle and the colour-rotations, we have to be aware of the action of components λ_2, λ_5, λ_7. Namely each generator taken alone produces chaotic scenarios. Let us investigate, for example λ_5.

We used a Banach-Tarski rotator ψ (17) to produce irregular motion as shown in figures (4) to (7). But a surprising scenario can also be iterated by the generators themselves. Consider

$$\phi = 2\lambda_5 = -\tfrac{1}{2}(e_{13} + j) \qquad \cong \text{Gell-Mann's } \Lambda_5 = 2i\lambda_5 \tag{40}$$

and iterate the strange $f_s(1)$ quark. We investigate a sequence of n rotations (recall iteration eq. 25) by calculating terms

$$f_s(n) = \left(e^{-\alpha\phi}\right)^n f_s(0)\left(e^{\alpha\phi}\right)^n = e^{-n\alpha\phi} f_s(0)\, e^{n\alpha\phi} \quad \text{for} \quad \alpha = 1 \tag{41}$$

The pure state $f_s(0)$ unfolds into space *Ch* thereby, although staying in a pure state, showing nonzero coefficients of e_1, e_{124}, e_3, e_{234} (figures 9 a - d) depending on the constant factor α=2 of λ_5. These are typical for nonlinear iteration cycles.

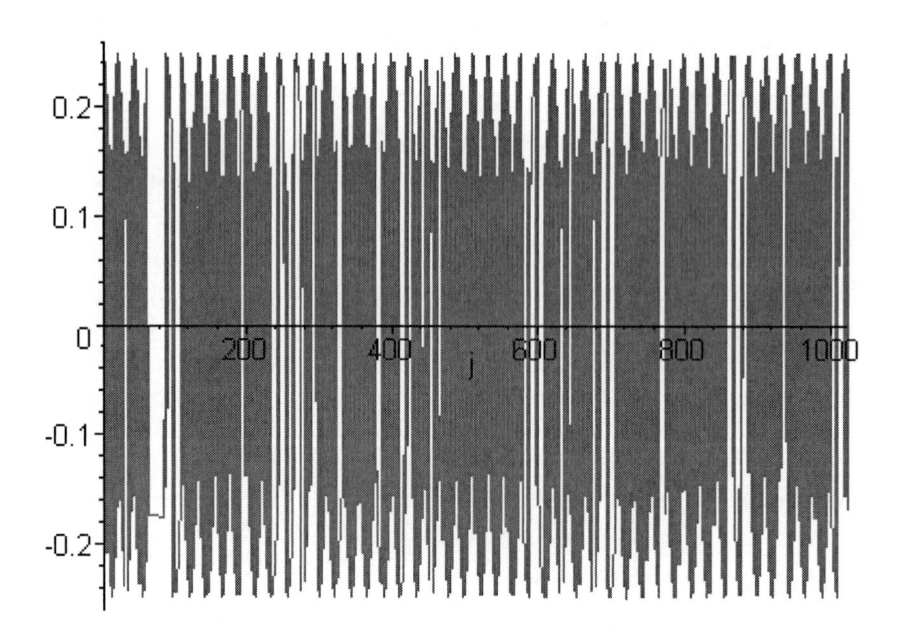

Figure 9a. Iteration of discrete $f_s(n)$ - submanifold coordinate x_2 by ϕ - the component $e_1(n)$ in 1024 iterations.

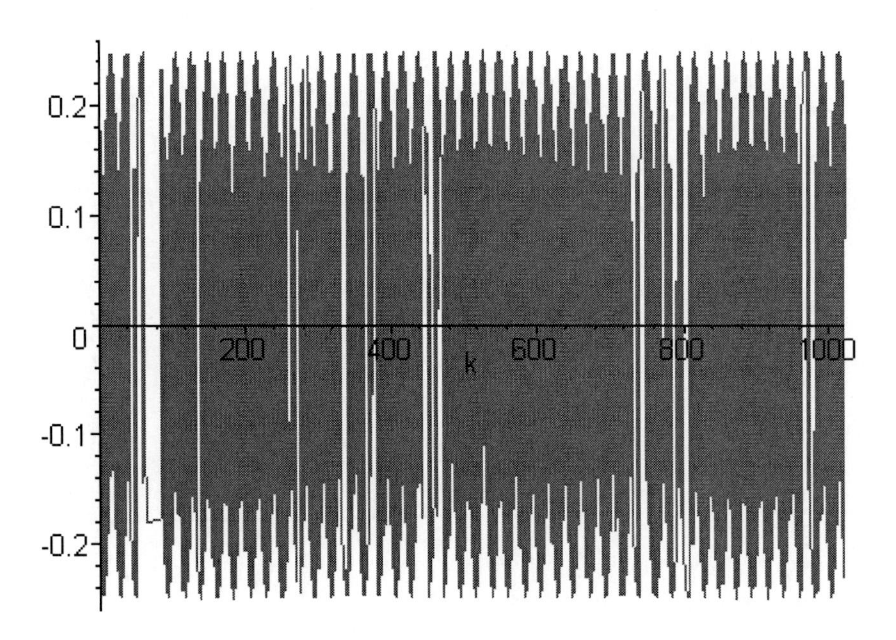

Figure 9b. Iteration of discrete $f_s(n)$ - submanifold coordinate x_4 by ϕ - the component $e_3(n)$.

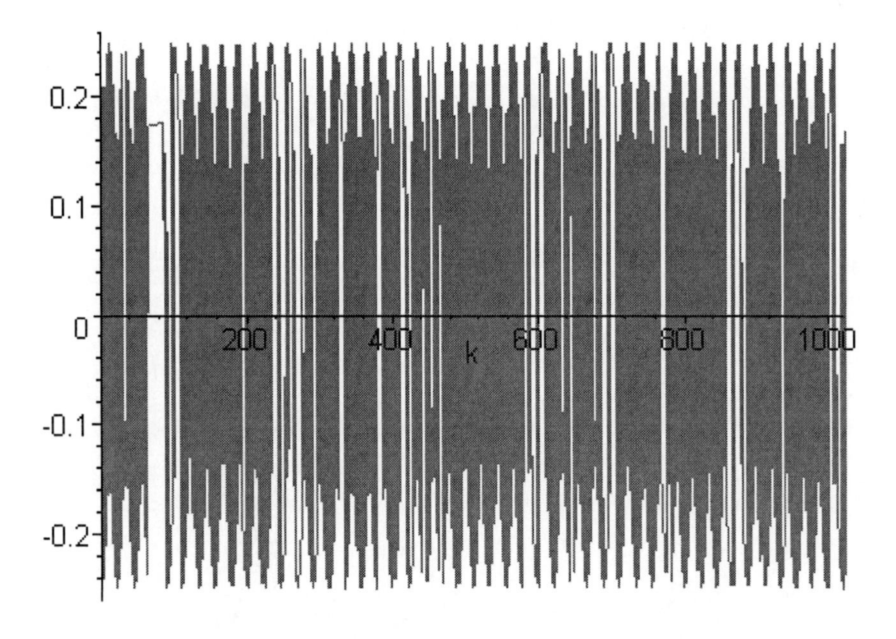

Figure 9c. Iteration of discrete $f_s(n)$ - submanifold coordinate x_{13} by ϕ - the component $e_{124}(n)$.

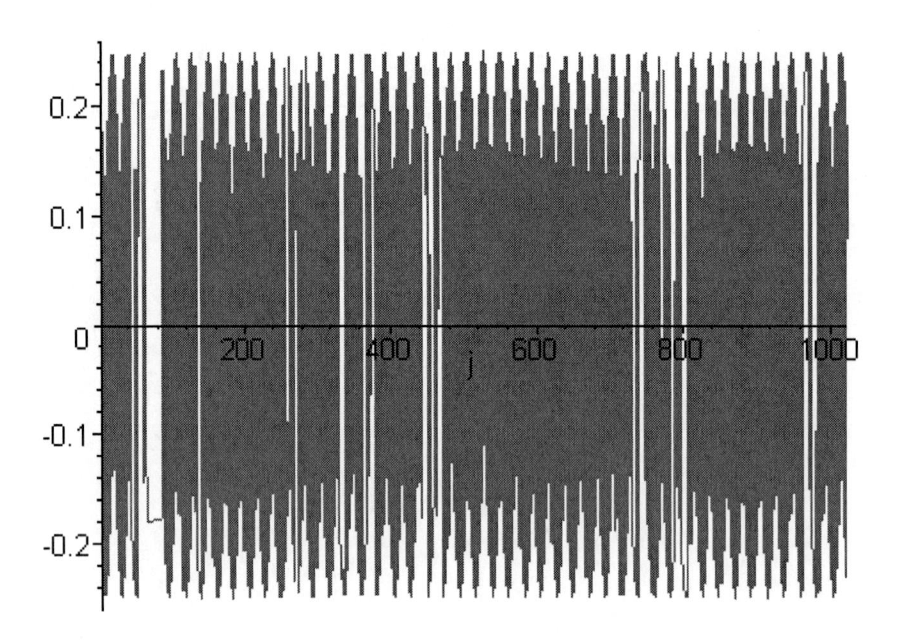

Figure 9d. Iteration of discrete $f_s(n)$ - submanifold coordinate x_{15} by ϕ - the component $e_{234}(n)$.

Histograms of coefficients look as expected (figure 10) and are typical for those irregular processes. We see a superposition of periodic moves.

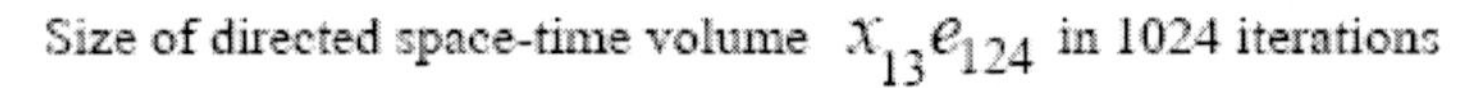

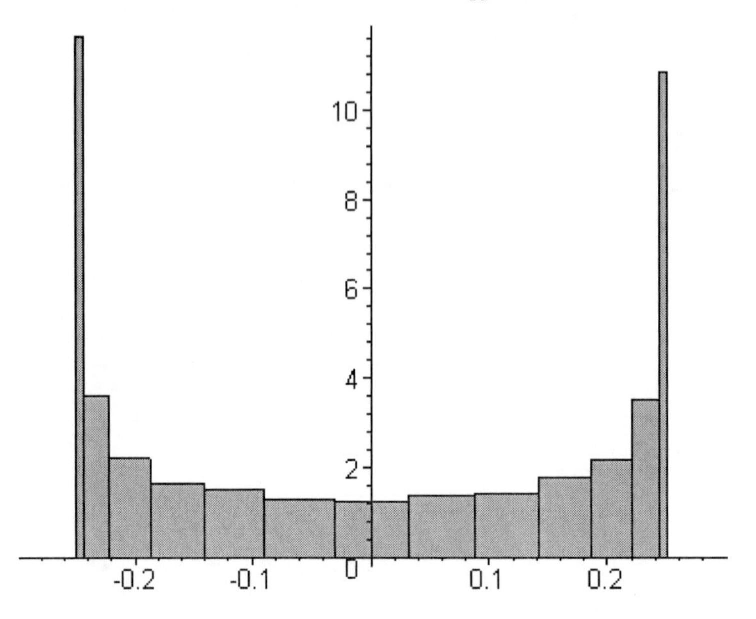

Figure 10. Histogram of discrete x_{124} submanifold.

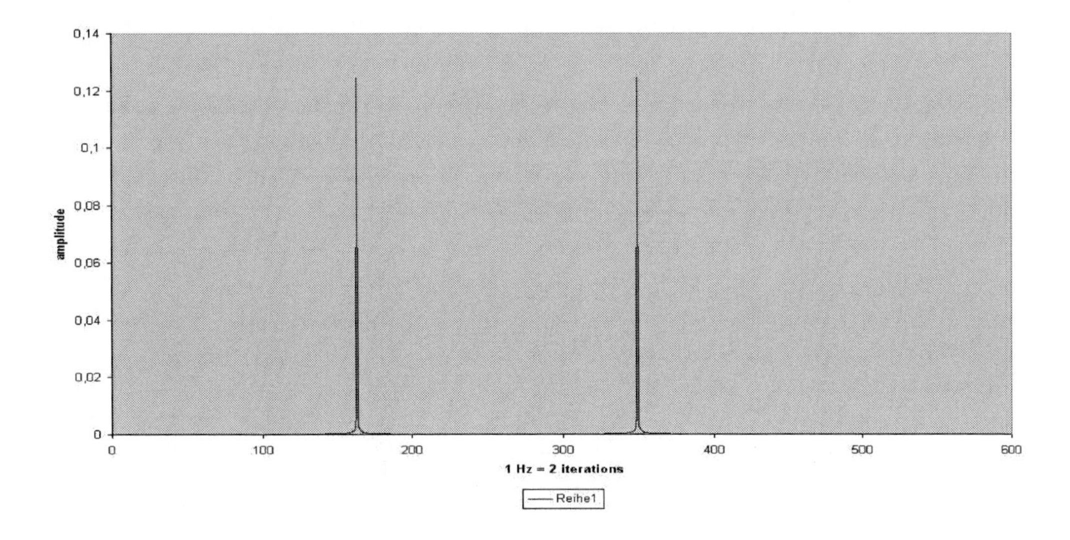

Figure 11. Spectrum of r_n at $2\lambda_5$, lines at 326 and 698 iterations.

The iteration of pure states involves trigonometric polynomials for the coefficients of Clifford monomials. The first iteration turns the $f_s(1) = \frac{1}{4}(1+e_1)(1-e_{24})$ into

$$f_s(2) = e^{\alpha\lambda_5} f_s e^{-\alpha\lambda_5} = \tfrac{1}{2}\eta + \tfrac{1}{4}(\cos\alpha)(e_1 - e_{124}) + \tfrac{1}{4}(\sin\alpha)(e_3 + e_{234}) \quad (42)$$

The quantity $\eta = (\frac{1}{2})(1 - e_{24})$ represents an idempotent in colour space ch_1. This is preserved during the iterations. Therefore we are interested to find out how the remaining part behaves. We consider any multivector having form

$$\xi(n) = r_n(e_1 - e_{124}) + s_n(e_3 + e_{234}) \quad (43)$$

This form is invariant throughout the iterations. The coefficients obey the recursion relations

$$\begin{aligned} r_{n+1} &= (\cos\alpha) r_n - (\sin\alpha) s_n \\ s_{n+1} &= (\sin\alpha) r_n + (\cos\alpha) s_n \end{aligned} \quad (44)$$

leading to trigonometric generative functions with a bipartition of binomial coefficients

$$\begin{aligned} r_{n+1} &= r_1 \sum_{0\le 2m\le n} -1^m \binom{n}{2m} (\cos\alpha)^{n-2m} (\sin\alpha)^{2m} + \\ &+ s_1 \sum_{1\le 2m+1\le n} -1^{m+1} \binom{n}{2m+1} (\cos\alpha)^{n-2m-1} (\sin\alpha)^{2m+1} \end{aligned} \quad (45)$$

The r_1, s_1 are initial values. Elaborating the FFT spectrum for r_n at α= 2 we obtain two cyles with 326 and 698 iterations lenght.

Surprisingly, as soon as we compose a universal trigonal rotator such as T by adding the three generators of the $su_{Cl}(2)$ with equal weight such as in equation (28], the irregular contributions compensate each other and we obtain a quite regular dynamic behaviour. Consider a (1 out of 24 virtual time units) fraction of τ (figure 12):

$$\phi = \tfrac{1}{24}\tau = \tfrac{1}{12\sqrt{3}}\arccos(-\tfrac{1}{2})(\lambda_2 - \lambda_5 + \hat{\lambda}_7) \qquad (46)$$

This iterates regular transitions between different colours of f_s. See the smooth course of number which contrasts the irregular course in figures 9. The colour rotation acts as a linear operator. We first show the real time iteration and in figure 14 the spectrum calculated by a Fast Fourier Transformation. You see the typical low frequency lines showing the natural harmonic frequencies beyond the nonlinar iterations.

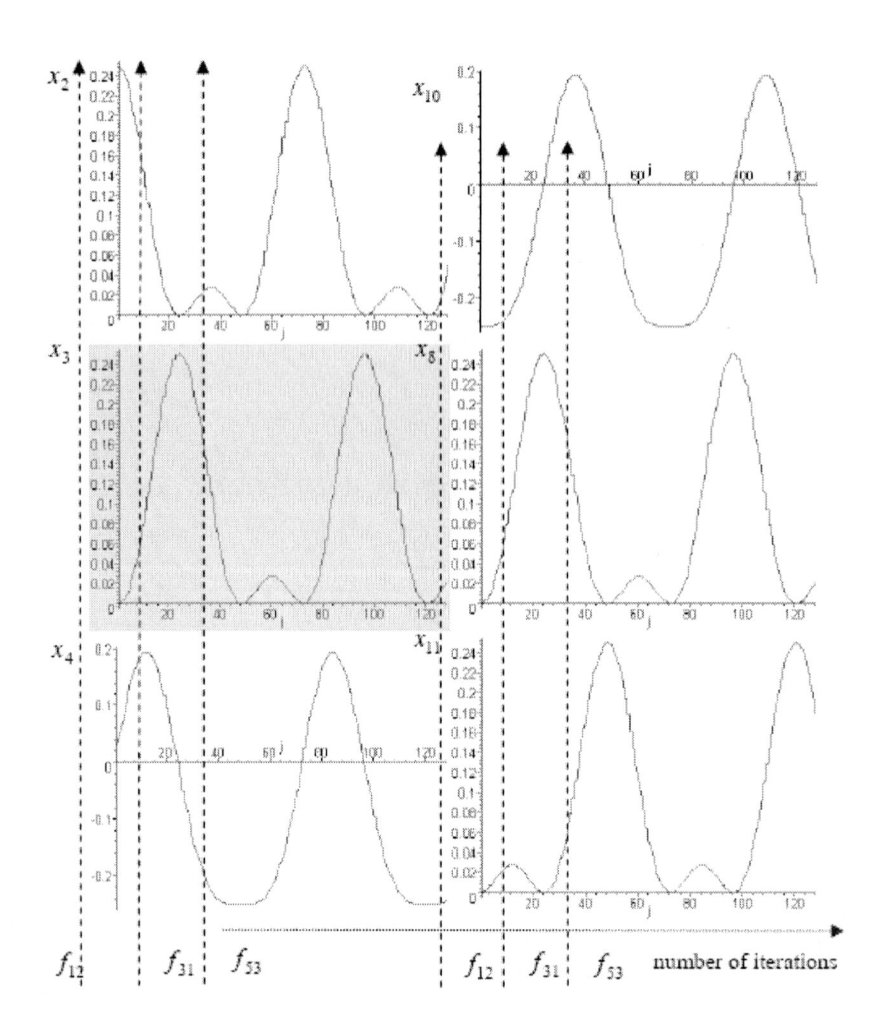

Figure 12. Colour rotation of state f_{22} in 128 iterations by ϕ.

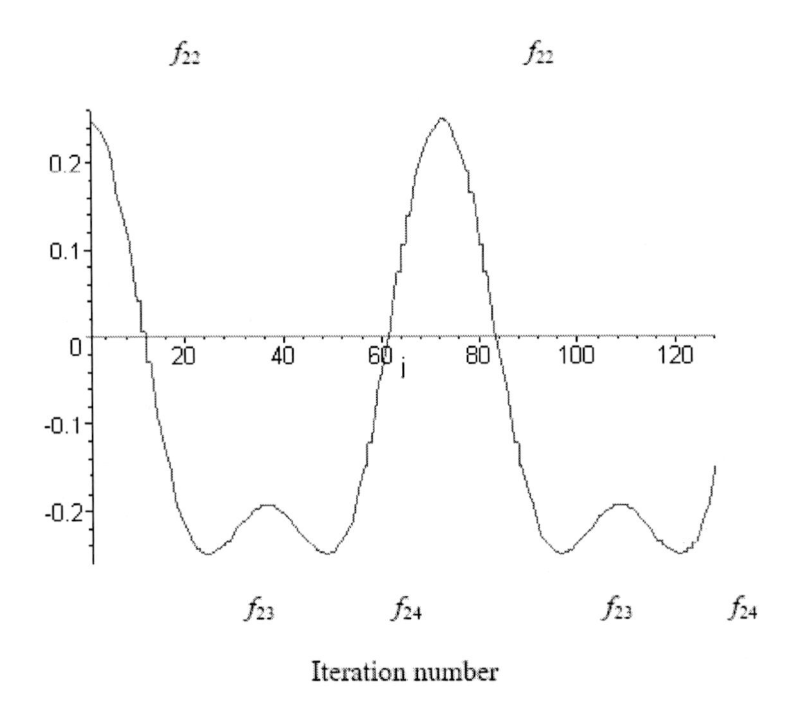

Figure 13. Coordinate x_2 of the flavour-rotated pure state f_{22}.

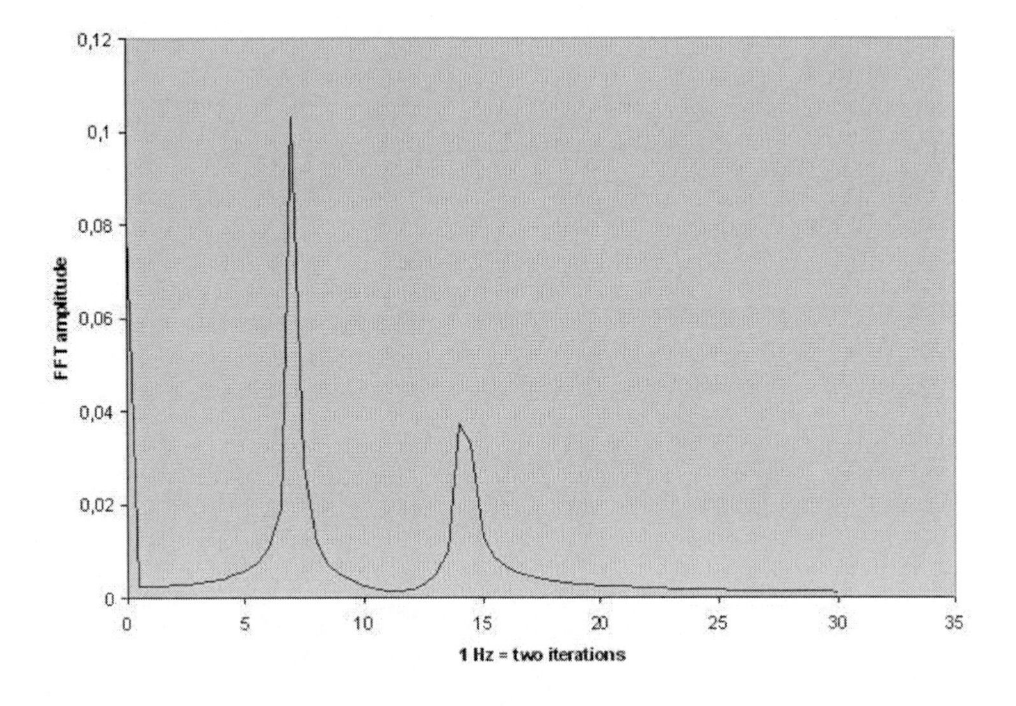

Figure 14. Spectrum of x_2 in flavour rotation of pure state f_{22}, lines at 7 and 14Hz.

It is essentially this special design (38) and respectively of colour rotations in general which eliminates chaos. Thus the Pauli principle can be interpreted as a possibility to escape from noisy into metastable fermion states. Observe, we divided the universal trigonal rotator by 24 (hours). This resulted in the fact that the initial colour, say red, is turned into the second, say green, after exactly 24 iterations. Therefore we can divide the *T* into arbitrarily small parts. It will always stay non-chaotic and its effects can be added linearly. This does not hold true for the single generators λ_2, λ_5, λ_7. The space-time likely reacts to any attempt to apply a single rotator by fluctuations. We can say that the components of the colour operator interact in a harmonic way, namely such as to diminish fluctuations. We called *T* a universal element since it is responsible for both colour- and flavour rotations. Acting on f_{22} the *T* flavour-rotates it within the same colour.

Chapter 19

The Appearance of Orientation

In the evening he went for his last walk on the beach of Adyar. At the end of his walk he turned around and said goodbye to the east, to the south, west and north in this solemn farewell which in old times was called »rotation of the elephant«. [1]

Orientation is a living thing. It is not only a property of space, but a concept in our mind which we have acquired moving around our planet during hundreds of thousands of years.

Let me pose in front of us another attempt to pronounce the unutterable. When thought annihilates on Nibbhana there is no thing existing. There is neither god resting in himself nor godhood. But there is a field neither of this world nor any other. As Mahatero Nyanatiloka (1952) translated: *Truly, there is a realm, where there is neither the solid, nor the fluid, neither heat, nor motion, neither this world, nor any other world, neither sun nor moon. This I call neither arising, nor passing away, neither standing still, nor being born, nor dying.*

This field is not existing, but insisting. Creation happens when, mirrored in the insisting, energy has no other way than to come into existence. Quite obviously, - may this seem true or false to us, - it signifies a duality. This is a duality between existence and insistence, between nothing and something.

If we turn from the surface towards the inner, and if we delve deeper into the inner, the inner discloses more and more inside. Seeing more inside we also see more outside. Definitely this is the most fundamental event of orientation. When you separate in your mind something existing from that which is not, which has never been and is utterly inaccessible, you create orientation. Then you create a feeling and a thought, that is, a basic concept of orientation. Now you can go inside, delve deeper, go forward, look upward and so forth. You see the dimensions of space come in. But the first and most fundamental duality is the inner-outer duality. The difference between inner and outer, between subjective and objective houses the whole drama of human existence. You may smile and say, this is psychology. Yes, it is, but it is also the basis of physics. We construct physics on the basis of this difference.

Most of us affiliate orientation to the points of the compass. Indeed, we may go forth or backward, turn to the left or right, look up or down. Strictly speaking, we connect orientation with a symmetry breaking. That is, we have a fundamental symmetry such as parity, or we have left- and right-handedness. But nature shows us that parity is violated or that natural phenomena involve chirality. Some mirror image cannot be brought to congruence with the

---

[1] Free translation by the author from "*Die Zukunft ist jetzt – Letzte Gespräche*", Jiddu Krishnamurti, engl. „*The Future is Now; Krishnamurtis Last Talks in India*, London 1988.

original by a rotation. A left glove cannot be turned into a right glove. As soon as we decide to clean the floor using a right rubber glove, we have given orientation to our action. We may prefer looking up to the sky rather than to investigate the human society. Then we prefer to study astronomy and not sociology. This gives cognitive orientation to our life. So orientation is a fundamental concept with a profound meaning.

Mathematically orientation is connected with equivalence relations among spaces and algebras. Orientation in Euclidean 3-space is less complex than orientation in Minkowski space. But orientation in the 16-dimensional manifolds of quantum states in the Clifford algebra of the Minkowski space is even much more complicated. Understanding orientation in geometric algebra is actually representing the root of material existence.

When we perform a transition from inner to outer perception we give orientation to cognition. When we turn from left to right, we bring in orientation in outer space. In terms of algebra: there is a difference between this and that, the Galois ground field G_2 eventually defined by numbers $\{0, 1\}$. The algebraic metaphor for a reflection which turns 0 into 1 and reverse is the group $\mathbb{Z}_2$. Preferring a transition $0 \to 1$ before $1 \to 0$ means to give a preferred direction to those transitions. So orientation is resulting from a symmetry breakage. The group which establishes the symmetry may be called the reorientation group.

We shall see that the reorientation group of the Clifford algebra of the Minkowski space with Lorentz metric has a much higher complexity, in terms of grade and structure, than the Euclidean 3-space. A symmetry breakage of the reorientation group of the Minkowski algebra is, we shall find out, an orientation of the standard model of physics. Said in other words:

Orientation of space-time is the creation of matter.

It is important to note the first concepts of orientation. These are very old. The eldest pictograms we found come up from the mesolithic. They have been scarified by our ancestors in rock faces and nummulites perforatus. They roughly look like this

Mesolithic ideogram

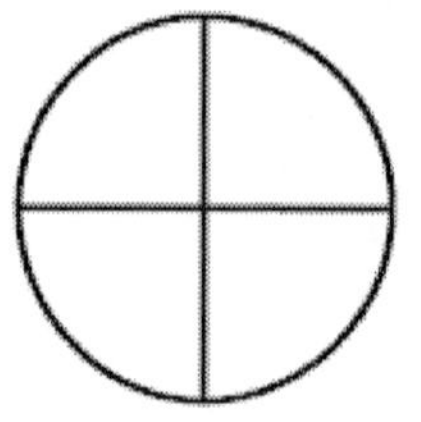

Mathematicians know that this figure has a space congruence group with eight elements. This is the non-commutative dihedral group D_4 or D_{2d} also denoted as spatial congruence group of a square. It involves, among other symmetries, a period 4 -rotation in the plane and flip about an axis. We can represent the period 4 – rotation by the unit imaginary which fits the equations (46)

Rotation of the Elephant

$$i=\sqrt{-1} \quad \Rightarrow \quad i^2=-1$$

$$i^5=i$$

$$\Uparrow \qquad \qquad \Downarrow$$

$$i^4=+1 \quad \Leftarrow \quad i^3=-i$$

(46)

So after 4 times multiplying i with itself we return to i. We can symbolize the clockwise period 4 rotation by a permutation or geometric figure as below in the 7th image. See how 7 can be generated by two reflections. Starting with image 1 we obtain image 8 by a flip or reflection at the diagonal from down left to up right. Taking image 8 and reflecting quadrants on the horizontal axis we reach 7.

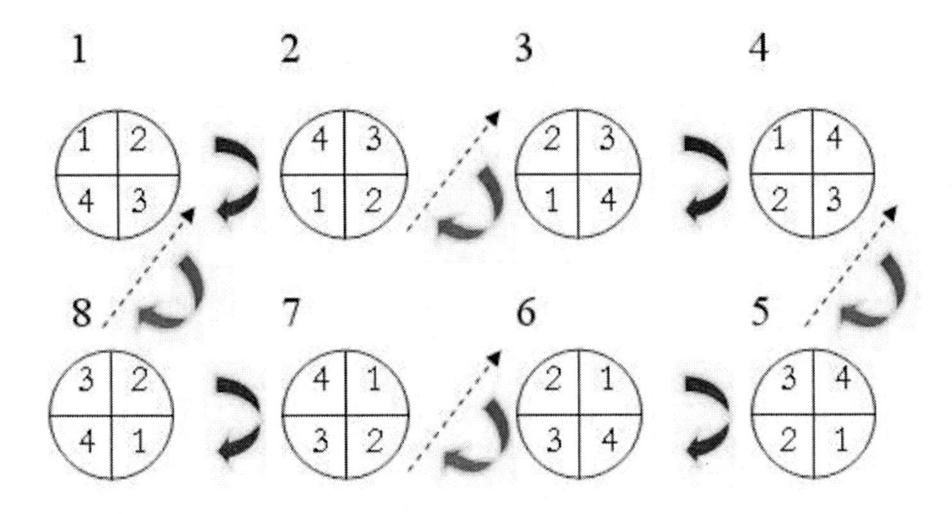

Figur 15. Symmetry operations of the dihedral group.

It is a quite general statement that any rotation can be brought fourth by two reflections. We repeatedly come back to this.

There is a small Clifford algebra which is extremely important for the construction of larger Clifford algebras, for the investigations of the conformal group, for theorems in twistor theory and further models of quantum field theory. This is the Clifford algebra $Cl_{1,1}$ which is generated by the Minkowski plane $\mathbb{R}^{1,1}$. Consider the base unit vectors e_1 and e_2 with their squares $(e_1)^2 = +1$ and $(e_2)^2 = -1$ and their exterior product bivector e_{12} with $(e_{12})^2 = +1$. We obtain the following multiplication table by the Clifford product:

Table 1. Multiplication table for $Cl_{1,1}$ multivector group D_4 by geometric product in the Minkowski plane

1	e_1	e_2	e_{12}	-1	$-e_1$	$-e_2$	$-e_{12}$
e_1	1	e_{12}	e_2	$-e_1$	-1	$-e_{12}$	$-e_2$
e_2	$-e_{12}$	-1	e_1	$-e_2$	e_{12}	1	$-e_1$
e_{12}	$-e_2$	$-e_1$	1	$-e_{12}$	e_2	e_1	-1
-1	$-e_1$	$-e_2$	$-e_{12}$	1	e_1	e_2	e_{12}
$-e_1$	-1	$-e_{12}$	$-e_2$	e_1	1	e_{12}	e_2
$-e_2$	e_{12}	1	$-e_1$	e_2	$-e_{12}$	-1	e_1
$-e_{12}$	e_2	e_1	-1	e_{12}	$-e_2$	$-e_1$	1

This is the multiplication table of the Minkowski plane. But it is also isomorphic with the table for the simple Euclidean plane where both unit vectors e_1 and e_2 squared give +1 but the bivector e_{12} takes in the place of a time-like imaginary unit. This geometric equivalence of a spatial plane with a space-time plane allows for special one-one mappings taking the form of convolution and deconvolution of scale invariant or conformal phenomena of motion. Those I have occasionally called time-wraps because they transform diachronic into synchronic patterns of motion. After the parity involution I regard the dihedral symmetry as the most important pattern of orientation which appears both in so called outer physical world as well as in consciousness as a pattern or "law of thought". Some of the dihedral structure of logic has already been discovered by the development psychologist Jean Piaget.

Shaw (1995) has also used the term "Dirac group" when investigating the discrete groups associated with the base units of geometric algebra. Besides this discrete structure which relies solely on the geometric product we also have discrete structures which are subgroups of spin groups or subgroups which are similar to spin groups, but graded. For such graded groups there does not necessarily exist a rotation group for which the spin group is a double cover. In this case which shall interest us most, we have a Lie group defined by a commutator product and the associated transformation group brought on by a conjugation, that is, a multiplication from both left and right side. This form immediately reminds us of the mathematical structure of self similar patterns or fractal geometric objects.

Chapter 20

ORIENTATION IN THE PAULI ALGEBRA

In universal geometric algebras such as the Clifford algebras we usually define three types of involutive (anti)automorphisms. Those shall be enlarged by one further automorphism, namely the "transposition". Thus we begin with the following abstract table of involutive operations

Table 2. Involutive automorphisms in geometric algebras

	Involutive	Automorphism	Antiautomorphism
Grade involution	$(u^\wedge)^\wedge = u$	$(uv)^\wedge = u^\wedge v^\wedge$	
Reversion	$(u^\sim)^\sim = u$		$(uv)^\sim = v^\sim u^\sim$
Conjugation	$(u^-)^- = u$		$(uv)^- = v^- u^-$
Transposition	$(u^\tau)^\tau = u$	$(uv)^\tau = u^\tau v^\tau$	

The operation has been defined abstractly in a chapter on "*transposition in Clifford algebra*" (Schmeikal 2004, 356). In the sequel we shall concentrate on the concrete implementations. Transposition in the basis of geometric algebra should not be confused with transposition of vectors. Transposition, here, means exchange of standard base units. For transposition of vectors in the sense of linear shift or translocation we could use *translation* or *translocation*. But I prefer to make it clear by the context: whenever we speak of transposition in the context of orientation, we mean a map of some base unit onto some other. Thus it appears to be a generalization of rotation in that the orthogonality relations among idempotents are preserved.

The discrete multivector- or "Dirac group" of the geometric algebra of Euclidean 3-space is the 16-element group generated by the Pauli matrices. It serves as an important example for non abelian groups of law degree with similar properties like the dihedral group.

Transpositions of the base units in the Pauli algebra can be achieved by the reflections of the Weyl group as are defined by the terms

$$u_{ik} \overset{def}{=} \tfrac{1}{\sqrt{2}}(e_i + e_k) \quad \text{and} \quad w_{ik} \overset{def}{=} \tfrac{1}{\sqrt{2}}(e_i - e_k) \quad \text{with } i \neq k. \tag{47}$$

Being reflections they are their own inverses. The u_{ik} transpose by conjugation the e_i to e_k, the w_{ik} to the reverse $-e_k$

$$u_{ik} e_i u_{ik} = e_k \qquad w_{ik} e_i w_{ik} = -e_k \tag{48}$$
$$u_{ik} e_k u_{ik} = e_i \qquad w_{ik} e_k w_{ik} = -e_i$$

The u_{ik} and w_{ik} can also be used to transpose the units in monomials and general elements of the algebra. But the reversion has to be taken into account. For example the u_{12} turns $x = e_1 + e_{123}$ into $\tau_{12}(x) = e_2 + e_{123}$ whereas

$$w_{12}\, x\, w_{12} = -e_2 + e_{123} \tag{49}$$

Both u_{12} and w_{12} preserve the unit director e_{123} in the Pauli algebra $Cl_{3,0}$. These transposition operators even work in the $Cl_{3,1}$ provided that the indices i, k $\neq$ 4. But then we had to have

$$u_{ik}\, j\, u_{ik} = j \qquad \text{but} \qquad w_{ik}\, j\, w_{ik} = -j \tag{50}$$

that is, the sign of the unit space-time volume would be flipped. It is interesting to consider transpositions of unit bivectors onto unit bivectors. The following problem could not be solved by rather advanced Clifford algebra software: transpose the bivector e_{12} onto the bivector e_{13} by a transformation which has the form of a conjugation, that is, we should have

$$T e_{12} T^{-1} = e_{13} \tag{51}$$

but T should not act just on the component e_2 as the Weyl reflections u_{12} and u_{23} are doing, but it should identify the e_{12} as a whole and act on that bivector. Actually this can be achieved by an element λ_2 which we shall identify later as one of the isospin generators of the standard model in geometric space-time algebra. We have

$$\lambda_2 \overset{def}{=} \tfrac{1}{4}(-e_{23} + e_{123}) \quad \text{and} \quad \exp(\pi\lambda_2)\, e_{12} \exp(-\pi\lambda_2) = e_{13} \tag{52}$$

with $\pi = 3{,}14 \ldots$ realize that

$$\exp(\pi\lambda_2)\, e_{23} \exp(-\pi\lambda_2) = e_{23} \quad \text{while} \quad u_{12}\, e_{23}\, u_{12} = -e_{13} \tag{53}$$

We are now in a position to introduce step by step all the important point groups of crystallography which can be considered as most important properties of the Euclidean space. Next we would go further and discuss those features for the space-time algebra. But before we do so, we work out a primordial property of orientation which can be derived even without any Euclidean space congruence and metric measures.

May be many of us have realized in the meantime the importance of space congruence groups of the 'Dreibein' and 'Vierbein' as are given by the universal point groups of the octahedral and hyperoctahedral symmetries. These are connected with a rather simple combinatoric object, namely the symmetric group $\mathbf{S}_n$. In the beginning of the last century, Alfred North Whitehead has proposed a phenomenology for physics which he based on sensual phenomena of events. He proposed a certain event algebra in order to conceive relativity theory anew from a fresh starting point. This phenomenology not only accounted for *outer objective reality* but it was based on typical inner-outer relations as could be perceived only by a living observer capable of sensual perception. Just to give you an idea of Whitehead's construction: His phenomenological physics suggested that statements like "the house is standing in the music" make sense. Any event (such as a train) could incorporate or include other events (such as travellers). So we obtain a reality constituted by events, inclusions, superpositions, disjunctions and all the rest of it. It is possible to construct event algebras and from those primordial symmetries which involve permutation groups and especially the octahedral and dihedral reorientation groups. The following section can therefore be used to become acquainted with the main symmetries of the octahedral group, however in a primordial setting where space and time are not yet.

Chapter 21

PRIMING ORIENTATION

EXTENSION CALCULUS FOR WHITEHEAD » ORIENTATION BY NATURE OF EVENTS «[1]

We perceive time in the passage of nature. Not only events of high energy physics are changing, but also the social, economic and political facts which promote these events. The transience of the creative forces and the permanent decay of awareness caused by the perpetual attempt of thought to reconcile the discerned social fact with the ideals of the discernible physics is at the root of our desire to bring security into a transient lattice of passing events both natural and social. Therefore thought has invented various procedures of abstraction to construct total concepts of order which give us the illusion that the uncontrollable passage, the particularization of transience, can be comprehended and given stability by those concepts. To mention some, before we go deeper, - just a few of those we shall investigate here: serial time, mythological time, mundane time, time of genetic structuralism and developmental psychology, linear metric time, lattice of time, fibres of multidimensional serial time, quantum time, relativistic time, time of scale relative fractional space-time, time derived by the method of extensive abstraction, disconnected oriented temporal order and so forth. All these are constructed by thought.

After having taken notice of such a dozen of concepts let us begin with the last one which is most natural and still very complex. It is perhaps mere fortune to observe that there are four essential topological relations of extension which we list in figure 16.

As I see now, there is some relation between this Ansatz I have worked out in remembrance of Whitehead and the point set topology investigated by Kiehn in section 9.5 in his chapter 9 on “Topology and the Cartan Calculus” (2008, vol 1, 399). May be this should be mentioned. For Kiehn investigates the discrete topology built up by four things.

---

1 This is an old elaboration of mine inserted here to show there are different possibilities to introduce a concept of “orientation”.

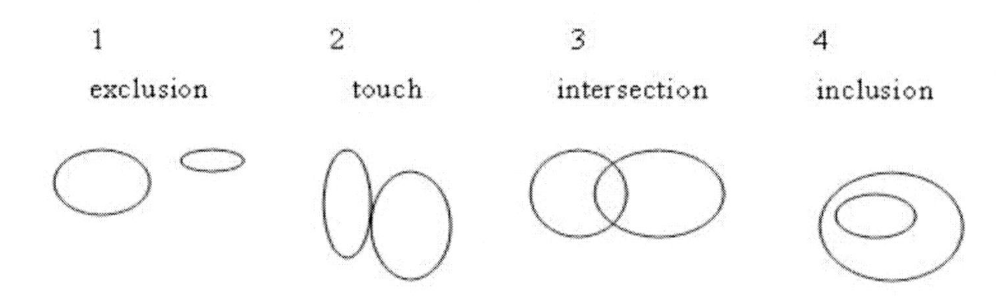

Figure 16. Relations of extension among events.

These four states mainly exhibit the abstract fact of an event extending over other events or being extended over by other events.[2] Those four abstract termini of the relation of extension as are disclosed by awareness are now regarded as four elements of algebra. Namely consider transitions between pairs of the set E = {1, 2, 3, 4}, that is, elements of the abstract relation E^2. Especially we shall allow for symmetric relations between exclusion and touch, between touch and intersection and between intersection and inclusion:

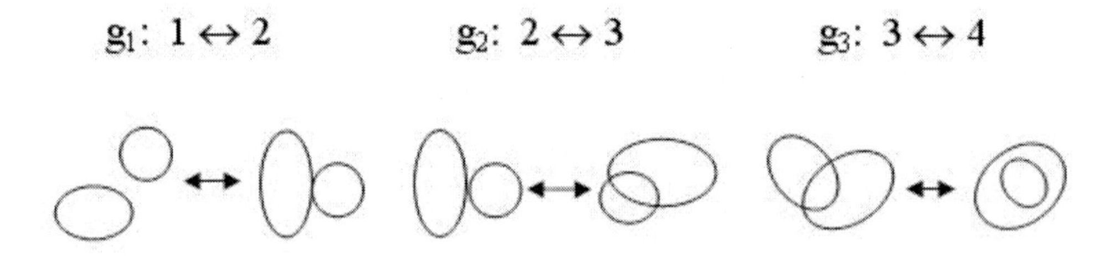

Figure 17. Generators of an event algebra.

The first image (g_1) means that a separation disclosed in the act of awareness in one duration may be disclosed as a touch in another act of awareness in a second duration. (Recall the philosophy of sequential time in thought by Gerhard Frey!) The images represented show some fundamental transitions among the abstract termini of the extension relation. Those three symmetric transitions between the four termini of the relation of extension can indeed be regarded as generators of a small algebra, namely, if we represent them by the symbols of permutations. The symmetric relation g_1 will be represented by the permutation cycle (1 2), g_2 by the cycle (2 3) and g_3 by (3 4). Now all we need to do is to regard g_1, g_2 and g_3 as generators of an algebra of permutations and form all possible products between the generators and the thus obtained new elements. We obtain the following elements which altogether form the set of permutations of the symmetric group $\mathbf{S}_4$:

$\mathbf{S}_4$ = {(1)(2)(3)(4); (1 2); (1 3); (1 4); (2 3); (2 4); (3 4); (1 2 3); (1 3 2); (1 2 4); (1 4 2); (1

2 Here we apply Whitehead's concept of "events extending over events". But we do not use any system theoretic formalism of structuration. The second relation, touch, may indeed be comprehended in terms of the topological concept of open sets. The present formalization, however, does not go beyond the limits of elementary algebra, but instead shows a link with geometry and orientation.

3 4); (1 4 3); (2 3 4); (2 4 3); (1 2)(3 4); (1 3)(2 4); (1 4)(2 3); (1 2 3 4); (1 4 3 2); (1 2 4 3); (1 3 4 2); (1 4 2 3); (1 3 2 4)} (54)

It is striking that these permutations can be interpreted as the reorientation symmetries of the euclidean three-dimensional space. That is, the symmetric group $\mathbf{S}_4$ is isomorphic with the Euclidean octahedral symmetry **O** and thus contains all possible discrete congruences of a »Dreibein« as special coordinate automorphisms of Euclidean $\mathbb{R}^3$ except the main involution C_i (or space inversion of crystallography) which would turn any e_i into $-e_i$. It comprises all possible flips and rotations of periods 4 and 3 (figure 18).

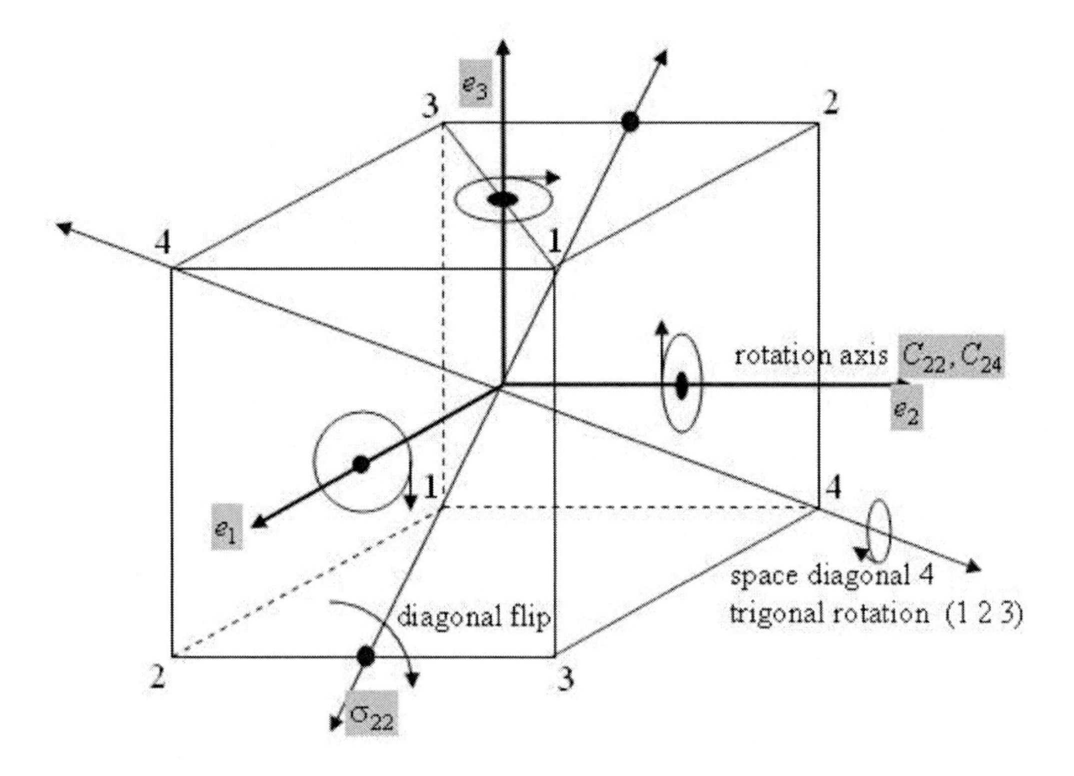

Figure 18. Space congruence of a Dreibein.

Namely, we can write unity = (1)(2)(3)(4); rotations by π (flips) about the main units e_1, e_2, e_3 are respectively

Rotations by π about diagonal units of the form $(1/\sqrt{2})(\pm e_i \pm e_j)$ with $i \neq j$ are

σ_{11}	σ_{12}	σ_{21}	σ_{22}	σ_{31}	σ_{32}
(1 2)	(3 4)	(1 4)	(2 3)	(1 3)	(2 4)

There are 4 space diagonal units labeled 1, 2, 3, 4 of the form $(1/\sqrt{3})(\pm e_1 \pm e_2 \pm e_3)$. Tetrahedral rotations by $2\pi/3$ about them are repesented by cycles of lenght 3

T_{13}	$(T_{13})^{-1}$	T_{23}	$(T_{23})^{-1}$	T_{33}	$(T_{33})^{-1}$	T_{43}	$(T_{43})^{-1}$
(2 3 4)	(2 4 3)	(1 3 4)	(1 4 3)	(1 2 4)	(1 4 2)	(1 2 3)	(1 3 2)

Period-4 rotation by $\pi/2$ about the main units e_1, e_2, e_3 are the cycles of lenght 4

C_{14}	$(C_{14})^{-1}$	C_{24}	$(C_{24})^{-1}$	C_{34}	$(C_{34})^{-1}$
(1 3 2 4)	(1 4 2 3)	(1 2 4 3)	(1 3 4 2)	(1 4 3 2)	(1 2 3 4)

Any of those 24 symmetry operations stands for a rotation of the SO(3) which brings forth a coordinate system congruent with the original one. For instance C_{14} rotates the Dreibein by 90° about e_1 thus turning corner 1 into 3, 3 into 2 and 2 into 4 (see figure 3) which is the cycle (1 3 2 4). The sigmas have to be decoded as follows. The first index indicates the plane to which the diagonal rotation axis is parallel. Further there are two such rotation axis perpendicular to each other. For instance σ_{11} and σ_{12} are both parallel to the plane spanned by $\{e_2, e_3\}$, σ_{21} and σ_{22} are both parallel to the plane spanned by $\{e_1, e_3\}$ and σ_{31} and σ_{32} are both parallel to the plane spanned by $\{e_1, e_2\}$. Further σ_{11} is perpendicular to σ_{12}, σ_{21} to σ_{22} and so on. Thus the diagonal axis σ_{22} is parallel to the plane $\{e_1, e_3\}$ and runs through the midpoint of the edge 2–3. It flips corners 2 and 3, but leaves labels 1, 4 unaltered. Perpendicular to σ_{22} is σ_{21} which flips 1 with 4, but lets 2, 3 unchanged. The σ can also be interpreted as reflections of three different dihedral groups $\mathbf{D}_4$ each of which provides the orientation symmetry of one of the planes $\{e_i, e_j\}$.

A minimal generating basis of the octahedral 'double-group' $_\delta\mathbf{O}$ within the spin group SU(2) of the Clifford algebra $Cl_{3,0}$ which we often use is expressed in terms of the quaternions e_{ij}:

$$s_{11} = (1/\sqrt{2})(-e_{12} + e_{13}) \qquad C_{24} = (1/\sqrt{2})(1 + e_{13}) \tag{55}$$

Consider the representations of termini 1, 2, 3, 4 (figure 4) as

$$1 \ldots x_1 = e_1 + e_2 + e_3 \qquad 2 \ldots x_2 = -e_1 + e_2 + e_3 \tag{56}$$
$$3 \ldots x_3 = e_1 + e_2 - e_3 \qquad 4 \ldots x_4 = -e_1 + e_2 - e_3$$

Using the inverse reflection $(s_{11})^{-1} = s_{12} = (1/\sqrt{2})(e_{12} - e_{13})$ and the inverse period-4 rotation about e_2 given by $(C_{24})^{-1} = (1/\sqrt{2})(1 - e_{13})$ we can calculate the effect of symmetry operations on the four termini. For example we obtain

$$(C_{24})^{-1} x_1 C_{24} = x_2 \qquad (C_{24})^{-1} x_2 C_{24} = x_4$$
$$(C_{24})^{-1} x_3 C_{24} = x_1 \qquad (C_{24})^{-1} x_4 C_{24} = x_3 \tag{57}$$

which represents nothing other than the cycle (1 2 4 3) (see figure 4). Thus we can see that not only has each terminus of the extension relation a geometric interpretation, but also each transformation among the termini can be interpreted as the action of a symmetry operation of orientation in space. We can put this into the form of a mathematical theorem:

The discrete relational calculus of event extension has a complete geometric interpretation and can be fully represented by the orientation symmetries of euclidean 3-space.

The significance of such a proposition can be seen when we understand the following statement by Whitehead: "With this hypothesis (note: that the past partakes in the vividness of the present fact) we can also suppose that the vivid remembrance and the present fact are posited in awareness as in their temporal order." (Whitehead 1964, 67) The theory thus constructed is in accord with Whitehead and is definitely denying the possibility that our remembrance can be organized elsewhere but in the present fact. So there is no other chance for cognition than to base its image of temporal order on some space relation of extension. That is, temporal order can only be based on a calculus of space extension, whether in form of some pre-topology or pre-geometry. What I am using here are basic formulas of relational algebra and the theory of geometric Clifford algebras to demonstrate only the more fundamental aspects of such a theory of time.

It is interesting that in physics we have used a concept of linear motion and system evolution which differs from the concept proposed by Whitehead. The latter is closer to nonlinear physics and biology than to classical mechanics. Namely, in Euclidean 3-space it seems the left side is somehow the same as the right and to go forward can be reverted by turning around and go backward. That does not make a phenomenological difference a priory. But going from generator g_1 over g_2 to g_3 actually means that some event that was separate from some other in the course of evolution is internalized or taken in. Linear motion, in Whitehead's approach, allows for mythological cycles. Some event may give birth to some other. Or it may intersect with some event and absorb it. This gives us a new concept of state space which nevertheless allows for symmetry and orientation just as Euclidean space is providing for vision.

Chapter 22

REORIENTATION OF SPACE-TIME ALGEBRA

The reorientation symmetry of the space-time algebra involves 16 dimensions and has, therefore, a much higher complexity than reorientation in Euclidean space. If we restrict reorientation of 3-space to discrete space congruence elements of the Dreibein or cube, we can sure say that this can be defined best by the octahedral group **O** which eventually could be extended by the total space inversion $\mathbf{C}_i$, so that we obtain a group with a total of only $|\mathbf{O}\times\mathbf{C}_i| = 48$ elements. As soon as we realize the additional dimensions brought in by the exterior product, we also have to take account of transpositions between different grades among monomials with equal signatures such as for example e_1 and e_{24} which squared give $+1$ or time-like e_4 and the unit director j with square -1. Then we have to find a new construction principle which goes beyond the application of the Weyl group and the octahedral group.

In the beginning of this search, almost fifteen years ago, I surveyed the real matrix representation of the Majorana algebra to find some construction principle which would give me the octahedral subgroups as a spill-over. I did so as I knew the standard design of the real 3×3 matrices of the octahedral symmetry elements. I also compared this with the complex matrices of the symmetric unitary group SU(3). Those calculations covered hundreds of pages and in the end led me to a very simple result. The space-time algebra in the Lorentz metric $Cl_{3,1}$ contains a partition of three pairs of lattices of mutually annihilating primitive idempotents from which we can construct corresponding Coxeter reflections as elements of period 2 under Clifford multiplication. Those reflections generate a group which has three times as many elements as the hyperoctahedral subgroup of order four. If we denote the reorientation group as $\mathbf{\Omega}(Cl_{3,1})$ its number of elements is $|\mathbf{\Omega}(Cl_{3,1})| = 3.2^4.4! = 1152$. As soon as we understand the specific actions of these elements, it turns out natural to investigate their relation with the generators of the SU(3) and its real forms and finally also the connection with the standard model. This research was intruding and very serious. It did not have the slightest similarity with a caper as Walter Thirring once was presuming.

Now we know the octahedral group – being reorientation of Euclidean space – is a subgroup of some orthogonal group, namely the SO(3) which is isomorphic with the SU(2). Clearly, as we had found the reorientation not only of Minkowski space, but of the Clifford algebra of Minkowski space, we could legitimately ask the question whether there was any object similar to an orthogonal group or spin group, but obviously with higher complexity than SU(2) and even SU(3) which embedded this hyperoctahedral discrete reorientation group. And the answer is simply that this group is the standard model Lie group of high energy physics as we know it until today and which probably will play an important role even after the successful start of the LHC. But, most surprisingly, it turned out that a trigonal rotation which we usually call colour or flavour acts on the elements of any of the six maximal Cartan subalgebras of the space-time algebra such as $\{e_1, e_{24}, e_{124}\}$. And with such rotation there corresponds most naturally a graded element T of the Clifford algebra which involves all 16 dimensions. This is not by fortune and not an artificial gag of nature, but it is a mathematical necessity which cannot be improved by any other design. The action of this large graded group on pure states is from the left and the right, that is, it is similar to a spin group, but has no rotation group with one-sided action to cover upon. As soon as we have understood the meaning of those six isomorphic maximal Cartan subalgebras of the space-time algebra we have also comprehended the significance of a graded Lie group. It is this graded Lie group which provides all the manifolds of the standard model of HEPhy. Once we comprehend this, it is also clear why there must be a mechanism of time-wrap. Processes with different amounts of energy involve different processes of time-convolution and deconvolution. For time-wrap consumes energy. Now let us sketch how this is working.

Consider the 24 fundamental primitive idempotents of the Clifford algebra $Cl_{3,1}$ which we write as one defining term

$$f_{\chi\alpha} \overset{def}{=} \tfrac{1}{2}(1 \pm e_i)\tfrac{1}{2}(1 \pm e_{k4}) \text{ with ordered pair } i, k = 1, 2, 3 \tag{58}$$

The index χ means "chroma" and is derived from the six ordered pairs of indices out of three. Obviously, the first ordered pair to be considered is given by e_1, e_{24} and provides four mutually annihilating primitive idempotents in the standard notation because of the four sign combinations $\pm$, $\pm$. So the primitive idempotents have two indices the first of which means colour and is derived from 6 ordered pairs of space-like base units while the second stems from the logic sign combinations of the two linear terms. So we obtain 6×4 primitive idempotents. The four (α= 1, 2, 3, 4) of each chroma χ serve as orthogonal pure states because they annihilate each other. They can be added and thereby give us 12 further idempotents which, however are not primitive. Thus for each colour χ we obtain by addition a 16-element lattice of commuting idempotents with *zero* and identity *Id*. Those can be arranged in the form of a 4-dimensional hypercube. So we have six hypercubes or rather three pairs of them, since the pair e_i, e_{k4} is in a way equivalent to the pair e_k, e_{i4} as both give the same wedge product e_{ik4}. This is the reason why in the total reorientation group of the $Cl_{3,1}$ there appears the hyperoctahedral symmetry group $\mathbf{H}_4 \subset \mathbf{\Omega}(Cl_{3,1})$ with index three rather than six. How can the $\mathbf{\Omega}(Cl_{3,1})$ be generated? The answer is that from the primitive idempotents we must turn over to their associated Coxeter reflections which are simply given by

$$s_{\chi\alpha} \overset{def}{=} Id - 2f_{\chi\alpha} \tag{59}$$

Obviously we have

$$s_{\chi\alpha}^2 = (Id - 2f_{\chi\alpha})(Id - 2f_{\chi\alpha}) = Id - 4f_{\chi\alpha} + 4f_{\chi\alpha} = Id \tag{60}$$

which proves that the $s_{\chi\alpha}$ are indeed reflections in terms of group theory. If we denote – just for that moment – the Clifford product by the symbol °, than the algebra $\{\{s_{\chi\alpha}\},°\}$ represents the discrete reorientation group of the Clifford algebra $Cl_{3,1}$. Every element

$$\tau \in \Omega(Cl_{3,1})$$

provides a definite transposition of any base unit monomial via the conjugation

$$\tau e_{(I)} \tau^{-1} = e_{(I')} \tag{61}$$

We can collect the 16 elements of a single chromatic lattice of idempotents in a lattice given by

$$F_{\chi} = \{f_{\chi\alpha}, +\} \tag{62}$$

Then we obtain for example

$$s_{11} F_3 s_{11} = F_5 \tag{63}$$

All states from colour space 3 are carried to colour space 5 by a Coxeter reflection s_{11}. Now observe for completeness the total set of transpositions carried out by the reflection s_{11}:

$$s_{11} e_1 s_{11} = e_1 \qquad \text{we briefly write } s_{11}: \; e_1 \mapsto e_1 \tag{64}$$

We obtain a complete set of transpositions in the standard basis:

$$\begin{array}{llll} \underline{s_{11}}: e_1 \mapsto e_1 & e_2 \mapsto e_{14} & e_3 \mapsto e_{234} & e_4 \mapsto e_{12} \\ e_{12} \mapsto e_4 & e_{13} \mapsto j & e_{14} \mapsto e_2 & e_{23} \mapsto -e_{123} \\ e_{24} \mapsto e_{24} & e_{34} \mapsto -e_{134} & e_{123} \mapsto -e_{23} & e_{124} \mapsto e_{124} \\ e_{134} \mapsto -e_{34} & e_{234} \mapsto e_3 & j \mapsto e_{13} & Id \mapsto Id \end{array} \tag{65}$$

Recall that we have 24 basic reflections. Their products give us further elements of the group such as for example the trigonal rotations of the hyperoctahedral group. There are plenty of them. One of those is

$$T \overset{def}{=} s_{23}\, s_{62} \;\; \text{with} \qquad T^{-1} = s_{62}\, s_{23} \qquad \text{and} \qquad T^3 = Id \tag{66}$$

It transposes the basis

$$e_1 \xrightarrow{\;T\;} e_{24} \xrightarrow{\;T\;} e_{124} \xrightarrow{\;T\;} e_1 \tag{67}$$

in a trigonal rotation among the elements of the first maximal Cartan subalgebra of the Clifford algebra. The word "first" says that we mean index $\chi = 1$, that is, first chroma. But the tetrahedral rotator T not only transposes the base units in such a peculiar way, it also colour rotates the pure states of the fermion system, namely we also have

$$f_{12} \xrightarrow{T} f_{13} \xrightarrow{T} f_{14} \xrightarrow{T} f_{12} \quad \text{and } f_{11} \circlearrowleft f_{11} \text{ preserved} \tag{68}$$

Considering T as the element of the Clifford algebra $Cl_{3,1}$ we have a rather complicated expression

$$\begin{aligned} T &= \tfrac{1}{2}(Id + e_1 - e_{34} + e_{134})\tfrac{1}{2}(Id - e_3 + e_{24} - e_{234}) = \\ &= \tfrac{1}{4}(1 + e_1 + e_2 - e_3 - e_4 - e_{12} - e_{13} + e_{14} + e_{23} + e_{24} - e_{34} \\ &\quad - e_{123} + e_{124} + e_{134} - e_{234} - j) \end{aligned} \tag{69}$$

But if we consider the matrix representation of the $Cl_{3,1}$ which is simply the Majorana algebra of Mat(4, $\mathbb{R}$) as for instance

$$e_1 = \begin{bmatrix} 1 & 0 & 0 & 0 \\ 0 & -1 & 0 & 0 \\ 0 & 0 & -1 & 0 \\ 0 & 0 & 0 & 1 \end{bmatrix} \qquad e_2 = \begin{bmatrix} 0 & 1 & 0 & 0 \\ 1 & 0 & 0 & 0 \\ 0 & 0 & 0 & 1 \\ 0 & 0 & 1 & 0 \end{bmatrix}$$

$$e_3 = \begin{bmatrix} 0 & 0 & 1 & 0 \\ 0 & 0 & 0 & -1 \\ 1 & 0 & 0 & 0 \\ 0 & -1 & 0 & 0 \end{bmatrix} \qquad e_4 = \begin{bmatrix} 0 & -1 & 0 & 0 \\ 1 & 0 & 0 & 0 \\ 0 & 0 & 0 & -1 \\ 0 & 0 & 1 & 0 \end{bmatrix} \tag{70}$$

we obtain for the trigonal rotation

$$T = \begin{bmatrix} 1 & 0 & 0 & 0 \\ 0 & 0 & 1 & 0 \\ 0 & 0 & 0 & 1 \\ 0 & 1 & 0 & 0 \end{bmatrix} \quad \text{and inverse} \quad T^{-1} = \begin{bmatrix} 1 & 0 & 0 & 0 \\ 0 & 0 & 0 & 1 \\ 0 & 1 & 0 & 0 \\ 0 & 0 & 1 & 0 \end{bmatrix} \tag{71}$$

which is essentially equal to the colour rotation matrix of SU(3) if we see apart from the first rows and columns which do not disturb because of the unit entry +1 in T_{11}. It is further a simple but important observation that a multivector element which involves all possible grades nevertheless cannot be more complicated than a real 4×4 matrix. Now that we realize this, we can ask if the other octahedral symmetry elements of the reorientation group do have important relations to the group SU(3) and to its algebra.

Actually, I came upon that question rather subconsciously when I realized that the Gell-Mann matrices were simple linear forms, namely differences of pairs of somehow dual elements of some octahedral subgroup in $\mathbf{\Omega}(Cl_{3,1})$. The observation I made needs some previous knowledge about symmetric spaces and Lie groups. Referring to the compact simple algebra L = su(3, $\mathbb{C}$) = $K \oplus P$ with the 3 imaginary matrices in K, we multiply the generators by the unit imaginary to obtain the eight matrices

$$X_\alpha = iT_\alpha \qquad \alpha = 1, 2, \ldots, 8 \tag{72}$$

and next realize that the $\{X_\alpha\}$ splits into two sets under the involutive automorphism of complex conjugation. It splits the compact algebra L = $K \oplus P$ as P consists of imaginary matrices only. By complex conjugation we obtain a new noncompact algebra L * = $K \oplus i\,P$. Note, we use the eight standard matrices

$$K = \{X_2, X_5, X_7\} = \left\{ \frac{1}{2}\begin{bmatrix} 0 & 1 & 0 \\ -1 & 0 & 0 \\ 0 & 0 & 0 \end{bmatrix}, \frac{1}{2}\begin{bmatrix} 0 & 0 & 1 \\ 0 & 0 & 0 \\ -1 & 0 & 0 \end{bmatrix}, \frac{1}{2}\begin{bmatrix} 0 & 0 & 0 \\ 0 & 0 & 1 \\ 0 & -1 & 0 \end{bmatrix} \right\} \tag{73}$$

$$P = \{X_1, X_3, X_4, X_6, X_8\} =$$
$$= \left\{ \frac{i}{2}\begin{bmatrix} 0 & 1 & 0 \\ 1 & 0 & 0 \\ 0 & 0 & 0 \end{bmatrix}, \frac{i}{2}\begin{bmatrix} 1 & 0 & 0 \\ 0 & -1 & 0 \\ 0 & 0 & 0 \end{bmatrix}, \frac{i}{2}\begin{bmatrix} 0 & 0 & 1 \\ 0 & 0 & 0 \\ 1 & 0 & 0 \end{bmatrix}, \frac{i}{2}\begin{bmatrix} 0 & 0 & 0 \\ 0 & 0 & 1 \\ 0 & 1 & 0 \end{bmatrix}, \frac{i}{2\sqrt{3}}\begin{bmatrix} 1 & 0 & 0 \\ 0 & 1 & 0 \\ 0 & 0 & -2 \end{bmatrix} \right\}$$

It is obvious that K generates the real subalgebra so(3, $\mathbb{R}$) while iP by complex conjugation is turned into the subspace of real, symmetric, traceless matrices. The Lie algebra L * = $K \oplus i\,P$ is now a set of 3×3 real matrices with zero trace. It generates the linear group of transformations by real 3×3 matrices having unit determinant, that is, the group SL(3, $\mathbb{R}$). The split of L * corresponds with the familiar decomposition of a special linear group into symmetric and skew-symmetric matrices.

To come back to the connection between the reorientation group $\mathbf{\Omega}(Cl_{3,1})$ and the role of the su(3, $\mathbb{C}$) $\subset \mathbb{C} \otimes Cl_{3,1}$ we observe the following:

$$c_{14} = \begin{bmatrix} 1 & 0 & 0 & 0 \\ 0 & 0 & 0 & -1 \\ 0 & 0 & -1 & 0 \\ 0 & 1 & 0 & 0 \end{bmatrix} \qquad c_{14}^{-1} = \begin{bmatrix} 1 & 0 & 0 & 0 \\ 0 & 0 & 0 & 1 \\ 0 & 0 & -1 & 0 \\ 0 & -1 & 0 & 0 \end{bmatrix} \tag{74}$$

so that we obtain

$$2T_5 = \frac{i}{2}(c_{14} - c_{14}^{-1}) = \begin{bmatrix} 0 & 0 & 0 & 0 \\ 0 & 0 & 0 & -i \\ 0 & 0 & 0 & 0 \\ 0 & i & 0 & 0 \end{bmatrix} \qquad 5^{\text{th}} \text{ Gell-Mann matrix} \tag{75}$$

In the Clifford algebra the c_{14} is the tetragonal rotation

$$c_{14} = \frac{1}{2}(e_1 - e_{13} + e_{124} - j) \text{ and } c_{14}^{-1} = \frac{1}{2}(e_1 + e_{13} + e_{124} + j) \tag{76}$$

Abstractly, it represents a possible square root of a bivector

$$c_{14} \in \{\sqrt{e_{24}}\} \qquad \text{“in the manifold of square roots”} \tag{77}$$

as we have $c_{14}^2 = e_{24}$. This pattern repeats with λ_2 and λ_7 as should be expected.

$$2T_7 = \frac{i}{2}(c_{24} - c_{24}^{-1}) = \begin{bmatrix} 0 & 0 & 0 & 0 \\ 0 & 0 & 0 & 0 \\ 0 & 0 & 0 & -i \\ 0 & 0 & i & 0 \end{bmatrix} \tag{78}$$

$$2T_2 = \frac{i}{2}(c_{34} - c_{34}^{-1}) = \begin{bmatrix} 0 & 0 & 0 & 0 \\ 0 & 0 & -i & 0 \\ 0 & i & 0 & 0 \\ 0 & 0 & 0 & 0 \end{bmatrix} \qquad \text{similarly with} \tag{79}$$

$$c_{24} = \frac{1}{2}(e_1 + e_4 + e_{12} + e_{24}) \text{ and } c_{24}^{-1} = \frac{1}{2}(e_1 - e_4 - e_{12} + e_{24}) \tag{80}$$

It represents one of the square roots of a trivector

$$c_{24} \in \{\sqrt{e_{124}}\} \tag{81}$$

$$c_{34} = \frac{1}{2}(-e_{23} + e_{24} + e_{123} + j) \text{ and } c_{34} = \frac{1}{2}(e_{23} + e_{24} - e_{123} + j). \tag{82}$$

This represents a possible square root of a unit vector

$$c_{34} \in \{\sqrt{e_1}\} \tag{83}$$

It may seem surprising how the generators that span the algebra are themselves constituted by linear combinations of group elements. An important hint why this can be so is in the fact that the generators, built from the reorientation symmetries in that way, can themselves be represented by the commutator product as in any Lie algebra. For consider

$$A \overset{def}{=} \tfrac{1}{2} s_{61} s_{31} \text{ and } B \overset{def}{=} \tfrac{1}{2} s_{41} \quad \Rightarrow \quad \lambda_7 = [A, B] = -\tfrac{1}{4}(e_4 + e_{12}) \tag{84}$$

The element $i\lambda_7$ represents in the Clifford algebra the generator T_7, that is ½ the 7th Gell-Mann matrix. Analogous commutators do exist for each Gell-Mann matrix. These take the form

$$\begin{aligned} &l_1 = \lambda_1 = \tfrac{1}{4}(s_{24} - s_{23}), \quad l_2 = i\lambda_2 = \tfrac{i}{4}(c_{34} - c_{34}^{-1}), \quad l_3 = \lambda_3 = \tfrac{1}{4}(s_{14} - s_{13}) \\ &l_4 = \lambda_4 = \tfrac{1}{4}(s_{64} - s_{62}), \quad l_5 = i\lambda_5 = \tfrac{i}{4}(c_{14} - c_{14}^{-1}), \quad l_6 = \lambda_6 = \tfrac{1}{4}(s_{44} - s_{41}) \\ &l_7 = i\lambda_7 = \tfrac{i}{4}(c_{24} - c_{24}^{-1}) \qquad l_8 = \lambda_8 = \tfrac{1}{\sqrt{12}}(2c_{32} - c_{22} - c_{12}) \end{aligned} \tag{85}$$

with $c_{j2} = c_{j4}^2$ reflections obtained from period 4 rotations. Thus every Gell-Mann matrix is represented by a complexified difference of elements in the reorientation group:

$$\lambda_i = \tfrac{1}{4}(E_k - F_k) \qquad k = 1, \ldots, 8 \tag{86}$$

where each term E_i, F_i belongs to a 4-dimensional graded linear subspace of the Clifford algebra. For the consideration of a Lie differential it will be interesting to observe the exponential mapping of the components of bilinear terms that constitute an object λ_i . We have a linearity equivalence of algebra- and group level:

$$E_i - F_i = \lambda_i \tag{87}$$

$$e^{E_i} - e^{F_i} = const. \lambda_i \text{ with } \quad const. = 1.0104 \tag{88}$$

Let us recall how every Lie algebra $\mathfrak{l}$ over a field F generates a group through the exponential mapping

$$M = \exp\left(\sum_i t_i \lambda_i\right); \quad t_i \in F, \quad \lambda_i \in \mathfrak{l} \tag{89}$$

We want to investigate the exponential mapping for each of the eight elements λ_i in $Cl_{3,1}$.

Chapter 23

DECOMPOSITION OF EXPONENTIAL MAPS

Recalculate (89) for the field $F \equiv \mathbb{R}$ and verify that

$$e^{2\phi_k \lambda_k} \in \mathsf{S}_k(3) \subset Cl_{3,1} \qquad \text{where by } \mathsf{S}_k(3) \text{ we denote a} \tag{90}$$

4-dimensional subspace of $Cl_{3,1}$ which, despite a scalar, involves 3 different grades. In the four cases of Isospin λ_1, λ_2, λ_3 and λ_8 the $\mathsf{S}_k(3)$ takes the shape of a graded 3 quasi simplex, or what I call a »*grade 3 quasi simplex*«. This is a new geometric object, namely a 3-simplex with relative 1-norm which involves specific Clifford monomials. To show you exactly what that means consider the general element of the algebra $Cl_{3,1}$:

$$\begin{aligned} X = \{ & x_1 Id + x_2 e_1 + x_3 e_2 + x_4 e_3 + x_5 e_4 + x_6 e_{12} + x_7 e_{13} + x_8 e_{14} + \\ & + x_9 e_{23} + x_{10} e_{24} + x_{11} e_{34} + x_{12} e_{123} + x_{13} e_{124} + x_{14} e_{134} + x_{15} e_{234} + x_{16} j \} \end{aligned} \tag{91}$$

keep the indices and consider further the real Clifford element to the first Gell-Mann matrix with its exponential map

$$e^{2\phi\lambda_1} \in \mathsf{S}_I(3) \qquad \text{with space} \tag{92}$$

$$\mathsf{S}_I(3) \overset{def}{=} \{ x_1 Id + x_2 e_1 + x_{11} e_{34} + x_{14} e_{134} \} \text{ and} \tag{93}$$

$$x_1 = \tfrac{1}{2} + a,\; x_2 = \tfrac{1}{2} - a,\; x_{11} = b = \tfrac{1}{2}\sinh(\phi),\; a = \tfrac{1}{2}\cosh(\phi)$$
$$x_{14} = -x_{11}$$

so the sum of coefficients is equal to $x_1 + x_2 + x_{11} + x_{14} = 1$.

This $\mathsf{S}_I(3)$ is an interesting geometric object. It is similar to a 3-simplex. Traditionally a 3-simplex is a tetrahedron. It is generated from a triangle or 2-simplex to which we add a point outside the plane of the triangle. So it has 4 points, 4 faces and 6 edges. It is a tetrahedron. In Euclidean space $\mathbb{R}^n$ a simplex is a convex hull of a set of n+1 points in arbitrary position. Now we proceed to a graded tetrahedral simplex. This is created by a generator of the constitutive Lie group of the algebra $Cl_{3,1}$. Convex hulls of d out of n points are now called envelopes or "d-facets". In our case the 0-facet is essentially a scalar. Next we add a vector to a point being x_2 units of measure away from a paravector $x_1 Id + 0\, e_1$ which gives us the 1-facet, a finite paravector. Then we add a directed space-time area with measure $x_{11}\, e_{34}$ and last we add a directed space-time volume $x_{14}\, e_{134}$ the unit of which fits into those directions which are already there, namely e_1 and e_{34}. In some sense this can remind us of a Euclidean tetrahedron. However, we obtain a 3-quasi simplex which is entirely in a space generated by a Cartan subalgebra of $Cl_{3,1}$. Considering the second Gell-Mann matrix, we obtain indeed a 3-quasi simplex which is more similar to a Euclidean simplex as it involves only spatial base units

$$e^{2\phi\lambda_2} \in \mathsf{S}_2(3) \tag{94}$$

$$\mathsf{S}_2(3) \overset{def}{=} \{\xi / \xi = x_1 Id + x_2 e_1 + x_9 e_{23} + x_{12} e_{123}\} \text{ and} \tag{95}$$

$$x_1 = \tfrac{1}{2} + A,\; x_2 = \tfrac{1}{2} - A,\; x_9 = -B = \tfrac{1}{2}\sin(\phi),\; A = \tfrac{1}{2}\cos(\phi)$$
$$x_{12} = -x_9$$

so the sum of coefficients is again $x_1 + x_2 + x_{11} + x_{14} = 1$.

Obviously, each element of the exponential map in the 3-quasi simplex $\mathsf{S}_2(3)$ is an element of a space of euclidean spinor eigenforms $E(f) = span_R\{Id,\, e_1,\, e_{23},\, e_{123}\}$ (Schmeikal 2009} whereas each element of the exponential map in the 3-quasi simplex $\mathsf{S}_1(3)$ is an element of the colour space Ch_2 with a relative unit 1-norm. Next we have

$$e^{2\phi\lambda_3} \in \mathsf{S}_3(3) \tag{96}$$

$$\mathsf{S}_3(3) \overset{def}{=} \{\xi / \xi = x_1 Id + x_2 e_1 + x_{10} e_{24} + x_{13} e_{124}\} \text{ and thus} \tag{97}$$

$\{\xi \in Ch_1 / |\xi| = 1\}$ where $|\;|$ denotes a relative 1-norm, that is, the sum of coefficients in the polynomial

$$\xi = e^{2\phi\lambda_3} = (\tfrac{1}{2} + a) Id + (\tfrac{1}{2} - a) e_1 + b(e_{24} - e_{124}).$$

We have

$$x_1 = \tfrac{1}{2} + a,\; x_2 = \tfrac{1}{2} - a,\; a = \tfrac{1}{2}\cosh(\phi),\; x_{10} = b = \tfrac{1}{2}\sinh(\phi)$$

$x_{13} = -x_{10}$ and $x_1 + x_2 + x_{10} + x_{13} = 1$. Further

$$e^{2\phi\lambda_4} \in \mathsf{S}_4(3) \tag{98}$$

$$\mathsf{S}_4(3) \overset{def}{=} \{\xi \in Ch_6\} \tag{99}$$

and

$$\xi = e^{2\phi\lambda_4} = (\tfrac{1}{2} + a)Id + (\tfrac{1}{2} - a)e_{24} + b(e_3 + e_{234})$$

$$e^{2\phi\lambda_5} \in \mathsf{S}_5(3) \tag{100}$$

$$\mathsf{S}_5(3) \overset{def}{=} \{\xi / \xi = x_1 Id + x_7 e_{13} + x_{10} e_{24} + x_{16} j\} \tag{101}$$
$$x_1 = \tfrac{1}{2} + A,\; x_7 = -B,\; x_{10} = \tfrac{1}{2} - A,\; x_{16} = -B$$

$$e^{2\phi\lambda_6} \in \mathsf{S}_6(3) \tag{102}$$

$$\mathsf{S}_6(3) \overset{def}{=} \{\xi \in Ch_4\} \tag{103}$$

$$e^{2\phi\lambda_6} = (\tfrac{1}{2} + a)Id + b(e_2 + e_{14}) + (\tfrac{1}{2} - a)e_{124}$$

$$e^{2\phi\lambda_7} \in \mathsf{S}_7(3) \tag{104}$$

$$\mathsf{S}_7(3) \overset{def}{=} \{\xi / \xi = x_1 Id + x_5 e_4 + x_6 e_{12} + x_{13} e_{124}\} \tag{105}$$
$$e^{2\phi\lambda_7} = (\tfrac{1}{2} + A)Id - B(e_4 + e_{12}) + (\tfrac{1}{2} - A)e_{124}$$

$$e^{2\phi\lambda_8} \in \mathsf{S}_8(3) \tag{106}$$

$$\mathsf{S}_8(3) \overset{def}{=} \{\xi \in Ch_1 / |\xi| = 1\} \tag{107}$$

$$e^{2\phi\lambda_8} = \alpha Id + \alpha' e_1 + \beta e_{24} + \beta e_{124}$$

with

$\alpha = 2a^2 - 2ab + a + b$, $\alpha' = 2a^2 - 2ab - a - b$, $\beta = 2b(a - b)$ which gives us a sum of coefficients of $4(a^2 - b^2) = \cosh^2(\phi) - \sinh^2(\phi) = 1$.

This quasi simplicial decomposition of the exponential maps is a specific feature of the graded representation within the constitutive Lie group. It holds for both groups SU(3) in the complexified $Cl_{3,1}$ (the ℓ_k) and SL(3) (with the λ_k) in the real $Cl_{3,1}$. The unit imaginary plays no important role where the tetragonal structure is concerned. Also both groups provide the correct root-spaces. This structure is not a property of the Gell-Mann matrices nor of the su(3) algebra, but it is a quality of the internal structure of their graded representation in the space-time algebra. Every group element obtained from a single generator λ_1, λ_2, λ_3 and λ_8 is situated on a specific graded tetragonal quasi simplex such that the relative unit 1-norm is preserved unity independent of the constant factor ϕ of the generator. This gives a special importance to the one norm in those spaces which accompany the six isomorphic maximal Cartan subalgebras which I have denoted earlier as *"extensions"* because they seem to me to incorporate that special quality which gives spatial extension to fields. I am however aware that "extension" in the Mathematical Classification Index means something else. This does not bother me. Sometimes terms should better be changed. At least it should be open for scientific discourse.

Chapter 24

ISOTROPIC FIELDS AND SPINORS

Elié Cartan had become aware of the importance of spinors in mathematics, about fifteen years before Paul Dirac introduced them as physical objects into quantum mechanics. He defined a pure spinor in terms of polarized isotropic multivectors (1966, p. 93) or, as is sometimes said in more elegant words: as a null isotropic direction field. This concept has found access into the modern theory of minimal surfaces (Sullivan 1989) and Osserman (1986) and into Kiehn's life-work on non-equilibrium thermodynamics and cosmology. Spinors bring in polarity, chirality and enantiomorphism. Physics without spinors has become unthinkable. But although physicists all over the world and day by day use spinors in their calculations, the original rigor and its meaning have been forgotten. Cartan's theory does not begin with spinors, but with an isotropic vector (x_1, x_2, x_3) the components of which satisfy the zero length equation

$$x_1^2 + x_2^2 + x_3^2 = 0 \tag{108}$$

which can be satisfied in a nontrivial way provided the three components of Euclidean space are given, for example, by two complex quantities

$$x_1 = \xi_0^2 - \xi_1^2, \qquad x_2 = i(\xi_0^2 + \xi_1^2), \qquad x_3 = -2\xi_0\xi_1 \tag{109}$$

The pair (ξ_0, ξ_1) of complex numbers constitutes a *spinor*. In order to comprehend the mathematics of a strong force field which naturally incorporates the known symmetries, it is necessary to transpose Cartan's rigor into the language of modern Clifford algebra. It is not appropriate to understand the pure state of $Cl_{3,1}$ as a primitive idempotent under all circumstances. The primitive idempotent may act only as a marker or cursor within the context of algebra, - both Clifford and Lie, - this has already been felt by those of us who tried to carry out calculations with my model in the Clifford algebra and its graded constitutive Lie algebra (Vargas and Torr 2003, Wallace 2008). But we have to find the isotropic multivectors associated with a fermion pure state. Those represent the appropriate fields with appropriate baryon numbers and properties of polarization such as charge, isospin and spin. In other words, we have to find out how the cursor, the pure state, must be used.

We have learned from Chevalley (1954) and more recently from Lounesto (2001, p. 60, 226) that square matrix spinors are elements from a minimal left ideal of the associated

Clifford algebra. A spinor in the Pauli algebra is an element taken from the minimal ideal $Cl_{3,0}f$ or equivalently Mat(2, $\mathbb{C}$) f where f is the primitive idempotent ½ (Id + e_3). A quite analogous statement is valid for the geometric algebra of Minkowski space. Now it is important to realize that not every element $\in Cl_{3,1}$ pairs with f such that it gives a spinor. The space $Cl_{3,0}f$ with an f = ¼ ($Id \pm e_1$)($Id \pm e_{24}$) does not only contain spinors. For take the unit scalar $Id \in Cl_{3,1}$ and consider the geometric product $Id\,f$. Quite obviously, this is not a spinor, though it can be used as a cursor to construct an isotropic Clifford number and thus a graded spinor space. The same is holding for the geometric algebra of Euclidean space, that is, the Pauli algebra. In a Dirac like theory, the attention actually centres on the isotropic elements that can be formed from products $\in Cl_{3,0}f$. Therefore we first transpose Cartan's rigor onto the language of Clifford algebra. Cartan had constructed the isotropic vector

$$y = (\xi_0^2 - \xi_1^2)e_1 + i(\xi_0^2 + \xi_1^2)e_2 - 2\xi_0\xi_1 e_3, \tag{110}$$

which satisfies the null extension condition

$$yy = |y|^2 = 0. \tag{111}$$

We can plot the three quantities x_j by using the complexplot 3d-option of Maple.

$$f_1 : (\xi_0, \xi_1) \longrightarrow \xi_0^2 - \xi_1^2$$

$$f_2 : (\xi_0, \xi_1) \longrightarrow -2\xi_0\xi_1$$

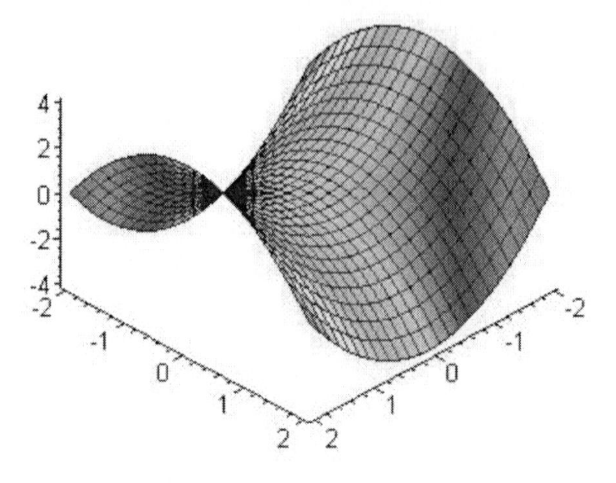

MAPLE » complexplot 3d ([f_1,f_2]) «

Figure 19. Saddle with negative Gauss curvature.

We would expect that we should obtain this result, if we put the defining equation $y = x_1 e_1 + x_2 e_2 + x_3 e_3$ into MAPLE CLIFFORD together with the equation (111]. But instead we get the more trivial result of two solutions $L_{1,2}$ (figure 20)

$$L_{1,2} := \{x_1 = \pm\sqrt{-x_2^2 - x_3^2}, x_2 = x_2, x_3 = x_3\}. \tag{112}$$

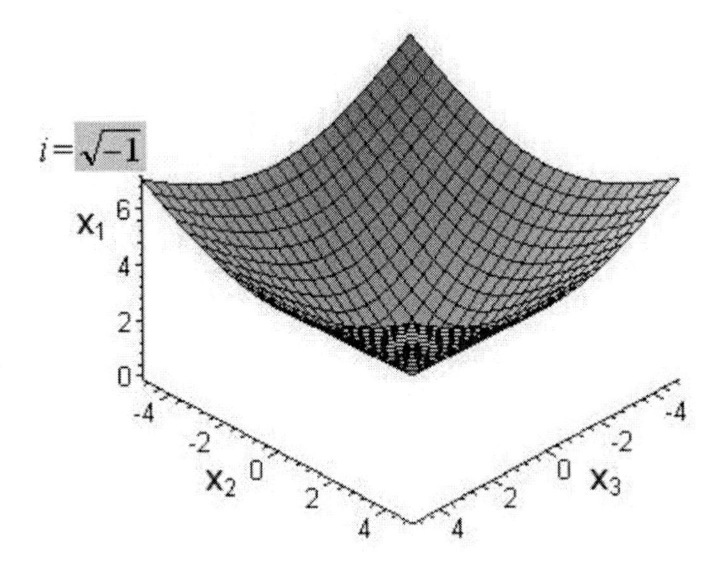

Figure 20. Lightcone solution for zero metric $x_1 = i\sqrt{x_2^2 + x_3^2}$

Clearly, if we substitute equation (110) as a solution, it is recognized as such by the program. We can go a step further. Again the Clifford software does not realize Cartan's approach. We ask for spinorial solutions for an isotropic vector in Euclidean 4-space. That is, we would like to solve the equation (111) for a Euclidean 4-vector with complex components. We have

$$y = x_1 e_1 + x_2 e_2 + x_3 e_3 + x_4 e_4 \in \mathbb{R}^4 \tag{113}$$

Again, Maple just changes the signature indicator by giving back a light-cone solution, but not a spinor

$$L_{1,2} := \{x_2 = x_2, x_3 = x_3, x_4 = x_4, x_1 = \pm\sqrt{-x_2^2 - x_3^2 - x_4^2}\} \tag{114}$$

We go further to introduce Cartan's approach into the Clifford algebra $Cl_{3,1}$. Let us consider the vector

$$y = x_1 e_1 + x_2 e_2 + x_3 e_3 + x_4 e_4 \in \mathbb{R}^{3,1} \tag{115}$$

with the scalar square

$$y\,y = (x_1^2 + x_2^2 + x_3^2 - x_4^2)Id \tag{116}$$

We do not want to get the usual trivial light cone solution for a zero length. To obtain an isotropic vector, we must support the computer by some manual calculations. We transpose the rigor to the geometric Clifford algebra $Cl_{3,1}$. MAPLE is able to calculate the geometric product (116) of y with itself. Thus we try a similar Ansatz as before

$$y = (\xi_0^2 - \xi_1^2)e_1 + i(\xi_0^2 + \xi_1^2)e_2 + (\xi_2^2 + \xi_3^2)e_3 + (\xi_2^2 - \xi_3^2)e_4 \tag{117}$$

which looks quite symmetric, but gives the non-zero result

$$y^2 = (-4\xi_0^2\xi_1^2 + 4\xi_2^2\xi_3^2)Id \tag{118}$$

The computer program can make that zero, and returns 4 possible solutions

$$L := [\{\xi_0 = \xi_0, \xi_1 = \frac{\xi_2\xi_3}{\xi_0}, \xi_2 = \xi_2, \xi_3 = \xi_3\}, \tag{119}$$

$$\{\xi_0 = 0, \xi_1 = \xi_1, \xi_2 = 0, \xi_3 = \xi_3\}, \{\xi_0 = 0, \xi_1 = \xi_1, \xi_2 = \xi_2, \xi_3 = 0\},$$

$$\{\xi_0 = \xi_0, \xi_1 = -\frac{\xi_2\xi_3}{\xi_0}, \xi_2 = \xi_2, \xi_3 = \xi_3\}], \text{ the first of which is} \tag{120}$$

$$y = \left(\xi_0^2 - \frac{\xi_2^2\xi_3^2}{\xi_0^2}\right)e_1 + i\left(\xi_0^2 + \frac{\xi_2^2\xi_3^2}{\xi_0^2}\right)e_2 + (\xi_2^2 + \xi_3^2)e_3 + (\xi_2^2 + \xi_3^2)e_4$$

Now we obtain the geometric product $y\,y = 0$ and for the matrix representation $\det(y) = 0$ and the eigenvalue 0 with algebraic multiplicity 4. Surprisingly, MAPLE CLIFFORD has overlooked another important solution, namely where $\xi_2 = \xi_0$ and $\xi_3 = \xi_1$. We should like to mind this solution

$$y = (\xi_0^2 - \xi_1^2)e_1 + i(\xi_0^2 + \xi_1^2)e_2 + (\xi_0^2 + \xi_1^2)e_3 + (\xi_0^2 - \xi_1^2)e_4 \tag{121}$$

which also gives $y^2 = 0$, $\det(y) = 0$ and 4 eigenvalues (0 0 0 0). Let us for a moment go back to Cartan's very first example (110]. In section 55 he introduced "the matrix associated with a vector":

$$X = \begin{bmatrix} x_3 & x_1 - ix_2 \\ x_1 + ix_2 & -x_3 \end{bmatrix} \tag{122}$$

This matrix has plus minus the scalar length as eigenvalues:

$$\lambda_{1,2} = \pm\sqrt{x_1^2 + x_2^2 + x_3^3} \tag{123}$$

This matrix is essentially what we obtain if we represent the base units e_1, e_2, e_3 by the Pauli spin matrices:

$$x_1\sigma_1 + x_2\sigma_2 + x_3\sigma_3 = \begin{bmatrix} x_3 & x_1 + ix_2 \\ x_1 - ix_2 & -x_3 \end{bmatrix} = \begin{bmatrix} -2\xi_0\xi_1 & -2\xi_1^2 \\ 2\xi_0^2 & 2\xi_0\xi_1 \end{bmatrix} \tag{124}$$

which has $X^2 = 0$, $\det(X) = 0$ and 2 eigenvalues (0 0) as expected for the null vector. We shall see that the angular momentum algebra in Euclidean 3-space has not yet been fully understood, its Clifford algebra not sufficiently exhausted (toolbox 4). This is one of the more implicit reasons why we are not yet entirely satisfied with the solution (124). It is perhaps interesting to observe the switch from the Pauli algebra to the Majorana algebra. If we represent the vector (124) by a matrix in the Majorana algebra, we obtain for example

$$X = \begin{bmatrix} x_1 & x_2 & x_3 & 0 \\ x_2 & -x_1 & 0 & -x_3 \\ x_3 & 0 & -x_1 & x_2 \\ 0 & -x_3 & x_2 & x_1 \end{bmatrix} \tag{125}$$

with 2-fold degenerate eigenvalues $\pm\sqrt{x_1^2 + x_2^2 + x_3^2}$ and squared matrix

$$X^2 = \begin{bmatrix} \sqrt{x_1^2 + x_2^2 + x_3^2} & 0 & 0 & 0 \\ 0 & \sqrt{x_1^2 + x_2^2 + x_3^2} & 0 & 0 \\ 0 & 0 & \sqrt{x_1^2 + x_2^2 + x_3^2} & 0 \\ 0 & 0 & 0 & \sqrt{x_1^2 + x_2^2 + x_3^2} \end{bmatrix} \tag{126}$$

Substituting quantities (110) we get the isotropic vector and zero eigenvalues represented by a 4×4 square matrix. Considering next the general element $X \in Cl_{3,0}$

$$X = x_1 Id + x_2 e_1 + x_3 e_2 + x_4 e_3 + x_5 e_{12} + x_6 e_{13} + x_7 e_{23} + x_8 e_{123} \tag{127}$$

together with six standard primitive idempotents

$$\begin{aligned} f_1 &= \tfrac{1}{2}(Id + e_1) & f_2 &= \tfrac{1}{2}(Id - e_1) \\ f_3 &= \tfrac{1}{2}(Id + e_2) & f_4 &= \tfrac{1}{2}(Id - e_2) \\ f_5 &= \tfrac{1}{2}(Id + e_3) & f_6 &= \tfrac{1}{2}(Id - e_3) \end{aligned} \tag{128}$$

we write down the main heading of the minimal left ideal

$$Y = Xf \qquad \text{with the primitive idempotent } f \equiv f_5 \in Cl_{3,0} \tag{129}$$

that is, we investigate the minimal left ideal and search for the general solution of the equation

$$YY = 0. \tag{130}$$

This equation is satisfied under the most general constraints

$$x_1 = -x_4 \qquad \text{and } x_5 = -x_8 \qquad \text{which gives us} \tag{131}$$

$$Y = \tfrac{1}{2}(x_2 + x_6)e_1 + \tfrac{1}{2}(x_3 + x_7)e_2 + \tfrac{1}{2}(x_2 + x_6)e_{13} + \tfrac{1}{2}(x_3 + x_7)e_{23} \tag{132}$$

Would we have chosen $f = f_6$ instead of f_5, the constraints (131) had to be altered to $x_1 = x_4$ and $x_5 = x_8$. In any case, we are led to conclude that the most general isotropic field in the Pauli algebra associated with the spin vector e_3 has the form

$$Y = \xi_0 e_1 + \xi_1 e_2 + \xi_0 e_{13} + \xi_1 e_{23} \tag{133}$$

which, indeed, has zero square.[1] As a 4×4 matrix

$$Y = \begin{bmatrix} \xi_0 & \xi_1 & \xi_0 & -\xi_1 \\ \xi_1 & -\xi_0 & \xi_1 & \xi_0 \\ -\xi_0 & -\xi_1 & -\xi_0 & \xi_1 \\ \xi_1 & -\xi_0 & \xi_1 & \xi_0 \end{bmatrix} \tag{134}$$

Any general multivector in that subspace which is not isotropic would, however have the form

$$Y = \xi_0 e_1 + \xi_1 e_2 + \xi_2 e_{13} + \xi_3 e_{23} \qquad \text{and as a 4×4 matrix} \tag{135}$$

$$Y = \begin{bmatrix} \xi_0 & \xi_1 & \xi_2 & -\xi_3 \\ \xi_1 & -\xi_0 & \xi_3 & \xi_2 \\ -\xi_2 & -\xi_3 & -\xi_0 & \xi_1 \\ \xi_3 & -\xi_2 & \xi_1 & \xi_0 \end{bmatrix} \tag{136}$$

This does not reproduce ± the scalar norm, but instead we have eigenvalues for the multivector Y

---

[1] Verify this by hand or using a Clifford algebra calculator.

$$\lambda_{1,2} = \pm\sqrt{\xi_0^2 + \xi_1^2 - \xi_2^2 - \xi_3^2 + 2i\xi_0\xi_3 - 2i\xi_1\xi_2}$$
$$\lambda_{3,4} = \pm\sqrt{\xi_0^2 + \xi_1^2 - \xi_2^2 - \xi_3^2 - 2i\xi_0\xi_3 + 2i\xi_1\xi_2} \quad (137)$$

This can be seen as a form more general than (124) and (125]. Thus we abstract from the quadratic form and provide a general definition for isotropic multivectors

Definition: the element Y is an isotropic multivector iff $Y\ Y = 0$ and for the matrix representation $\det(Y) = 0$. It has 4 fold degenerate eigenvalue zero. (138)

Now realize two things, namely 1) the isotropic biparavector Y is a more general null element than the isotropic vector (110) and 2) Y has a most peculiar form: It does not contain the bivector component e_{12}, nor does it contain the classic 'spin vector' e_3. »*In the Lie group guide to the universe*« it was shown that the space of Pauli spinor eigenforms

$$E(f) = span_R\{Id, e_3, e_{12}, e_{123}\} \quad (139)$$

is brought about by the geometric product of two general angular momentum operators J_+, J_- which satisfy the angular momentum algebra in $Cl_{3,0}$, namely

$$J_+ = x_2(e_1 - e_{13}) + x_3(e_2 - e_{23}),$$
$$J_- = y_2(e_1 + e_{13}) + y_3(e_2 + e_{23}) \qquad \text{with geometric product} \quad (140)$$

$$\tfrac{1}{2}J_+J_- = (x_2y_2 + x_3y_3)(Id + e_3) + (x_2y_3 - x_3y_2)(e_{12} + e_{123}) \quad (141)$$

So we realize that the product ½ J_+ J_- gives a space $E(f)$ of spinor eigenforms in the Pauli algebra. This is algebraically generated by the *spin-vector* e_3 and the associated *spin bivector* e_{12}. The space of *isotropic elements* $\Psi = \{Y\}$ is the most general subspace of the Pauli algebra which is 1) correlated with the minimal left ideal $Cl_{3,0}\ f$ and 2) is complementary to $E(f)$.

The space of isotropic polarized multivector fields and the space of Pauli spinor eigenforms complement each other. So we have to be aware of a quite general duality principle in the mathematics of isotropic spaces and spinor eigenforms. This fact we have to keep in mind together with the announcement of an incompleteness of the investigation of Clifford algebra $Cl_{3,0}\ f$. The richness of geometric algebra has not been, - as Hestenes also often pointed out, - fully exhausted. In the next step we drive the rigor a little further in order to obtain at first the isotropic elements of the space-time algebra $Cl_{3,1}$.

Chapter 25

ISOTROPIC FIELDS IN $CL_{3,1}$

Our next exercise consists in searching the general solution for the form of isotropic elements in the Clifford algebra $Cl_{3,1}$ associated with a colour space and respectively the isomorphic maximal Cartan subalgebras of the Clifford algebra. We shall carry out the rigor for the colour space ch_1. Most precisely we begin by giving the start data

$$\begin{aligned} X &= x_1 Id + x_2 e_1 + x_3 e_2 + x_4 e_3 + x_5 e_4 + \\ &+ x_9 e_{23} + + x_{10} e_{24} + x_{11} e_{34} + x_6 e_{12} + x_7 e_{13} + x_8 e_{14} + \\ &+ x_{12} e_{123} + x_{13} e_{124} + x_{14} e_{134} + x_{15} e_{234} + x_{16} j \end{aligned} \tag{142}$$

of the general element $X \in Cl_{3,1}$

The four mutually annihilating primitive idempotents

$$\begin{aligned} f_1 &= \tfrac{1}{2}(1+e_1)\tfrac{1}{2}(1+e_{24}) & f_2 &= \tfrac{1}{2}(1+e_1)\tfrac{1}{2}(1-e_{24}) \\ f_3 &= \tfrac{1}{2}(1-e_1)\tfrac{1}{2}(1+e_{24}) & f_4 &= \tfrac{1}{2}(1-e_1)\tfrac{1}{2}(1-e_{24}) \end{aligned} \tag{143}$$

span a linear subspace $ch_1 \in Cl_{3,1}$, the so called colour space associated with the maximal Cartan subalgebra $\{e_1, e_{24}, e_{124}\}$ of the space-time algebra. This rank 3 subalgebra can be correlated with the largest symmetric unitary group which is embedded by the geometric algebra, the real form of $SU(4)$.

The general element of the colour space ch_1 has form

$$\xi = x\, Id + y e_1 + z e_{24} + u e_{124} \tag{144}$$

As we have done before for the Pauli algebra, we write down the main heading of the minimal ideal

$$Y(f) = f X \tag{145}$$

with primitive idempotent $f \equiv f_i \in ch_1 \subset Cl_{3,1}$ and X as in (142).

That is, we investigate the minimal left ideal and search for the general solution of the equation

$$YY = 0\,. \tag{146}$$

We have written $Y = Y(f)$ to indicate that the space of isotropic numbers ✋ = { Y } is associated with and definite for a given primitive idempotent f_1, … , f_4, a given lepton or fermion state. Consider the first primitive idempotent. Solving equation (145) leads to the general isotropic null multivector $Y(f_1) = Y_1$:

$$\begin{aligned}
\text{(i)}\quad Y_1 &= \tfrac{1}{4}(x_3 - x_5 + x_6 - x_8)(e_2 - e_4 + e_{12} - e_{14}) + \\
&+ \tfrac{1}{4}(x_4 + x_7 - x_{15} - x_{16})(e_3 + e_{13} - e_{234} - j) + \\
&+ \tfrac{1}{4}(x_9 + x_{11} + x_{12} + x_{14})(e_{23} + e_{34} + e_{123} + e_{134})
\end{aligned}$$

Solving further for given pure states f_2, f_3, f_4 we obtain

$$\begin{aligned}
\text{(ii)}\quad Y_2 &= \tfrac{1}{4}(x_3 + x_5 + x_6 + x_8)(e_2 + e_4 + e_{12} + e_{14}) + \\
&+ \tfrac{1}{4}(x_4 + x_7 + x_{15} + x_{16})(e_3 + e_{13} + e_{234} + j) + \\
&+ \tfrac{1}{4}(x_9 - x_{11} + x_{12} - x_{14})(e_{23} - e_{34} + e_{123} - e_{134})
\end{aligned}$$

$$\begin{aligned}
\text{(iii)}\quad Y_3 &= \tfrac{1}{4}(x_3 - x_5 - x_6 + x_8)(e_2 - e_4 - e_{12} + e_{14}) + \\
&+ \tfrac{1}{4}(x_4 - x_7 - x_{15} + x_{16})(e_3 - e_{13} - e_{234} + j) + \\
&+ \tfrac{1}{4}(x_9 + x_{11} - x_{12} - x_{14})(e_{23} + e_{34} - e_{123} - e_{134})
\end{aligned}$$

$$\begin{aligned}
\text{(iv)}\quad Y_4 &= \tfrac{1}{4}(x_3 + x_5 - x_6 - x_8)(e_2 + e_4 - e_{12} - e_{14}) + \\
&+ \tfrac{1}{4}(x_4 - x_7 + x_{15} - x_{16})(e_3 - e_{13} + e_{234} - j) + \\
&+ \tfrac{1}{4}(x_9 - x_{11} - x_{12} + x_{14})(e_{23} - e_{34} - e_{123} + e_{134})
\end{aligned} \tag{147}$$

From that we can learn five things: First, the most general expression for an isotropic nilpotent element Y in the Clifford algebra $Cl_{3,1}$ associated with space ch_1 can be written as

$$Y = xn_1 + yn_2 + zn_3 \in ✋_1 \tag{148}$$

where the three quantities n_i represent themselves nilpotent elements in particular subspaces

$$\begin{aligned}
n_1 &\in N_1 \subset span_R\{e_2, e_4, e_{12}, e_{14}\} \\
n_2 &\in N_2 \subset span_R\{e_3, e_{13}, e_{234}, j\}
\end{aligned}$$

$$n_3 \in N_3 \subset span_R\{e_{23}, e_{34}, e_{123}, e_{134}\} \tag{149}$$

We have for every single number y $\in N_j$ in a subspace N_j of nilpotent Clifford numbers the geometric product $y\,y = y^2 = 0$. The manifolds of isotropic nilpotents in those linear spaces spanned by the right hand sides are only a small part of all the elements. Secondly, we learn the following: Most elements are not nilpotent. First of all, the pattern of signs effectuates only one half of the possible sign combinations, so to say, the 'better half' provides isotropic multivectors. Thirdly, there is a simple recipe to construct them. It depends on the signs of the real components. Two of them must be positive definite, two must be negative. Only then we obtain nilpotent elements. If we have the coefficient of, say, e_3 positive and the other numbers for e_4, e_{12}, e_{14} negative, the element in N_1 cannot be isotropic. Further the magnitude of the numbers is relevant. They must obey the specific logic symmetry that can be seen in the design of equations (147). And there is a forth thing to be seen, namely the components of the colour space ch_1 cannot be found in the nilpotent manifolds. But the latter reside in the complementary space, that is, the basis of that space is complementary to the basis of the colour space. Fifth, we can easily verify, the sum of four quantities Y_1 derived from the general element (142) of $Cl_{3,1}$ gives

$$\begin{aligned} &Y_1 + Y_2 + Y_3 + Y_4 = x_3\, e_2 + x_4\, e_3 + x_5\, e_4 + \\ &+ x_6\, e_{12} + x_7\, e_{13} + x_8\, e_{14} + x_9\, e_{23} + x_{11}\, e_{34} + \\ &+ x_{12}\, e_{123} + x_{14}\, e_{134} + x_{15}\, e_{234} + x_{16}\, j \end{aligned} \tag{150}$$

from which we conclude that

$$X - \sum_{k=1}^{4} Y_k = x_1\, Id + x_2\, e_1 + x_{10}\, e_{24} + x_{13}\, e_{124} \tag{151}$$

which is indeed the general element in the colour space ch_1.

Chapter 26

INNER AND OUTER

ISOTROPIC NUMBERS IN SMALL SUBSPACES

We explain: the denotation “inner” and “outer” is used here for complementary subspaces of inner colour spaces and outer spaces 🏳$_\chi$ that include spaces ✋$_\chi$ of isotropic polarized multivectors. That is, the basis of 🏳$_\chi$ is a complementary basis of ch_1 in $Cl_{3,1}$ so that we have dim 🏳$_\chi$ = dim $Cl_{3,1}$ – dim $ch_1 = 12$.

$$Cl_{3,1} = ch_\chi \oplus 🏳_\chi \text{ with } \chi = 1, \ldots, 6 \tag{152}$$

It is not used in the sense of inner and outer product. The inner space brings in the pure states of flavour as a cursor towards the outer isotropic spinor. The »inner« of ch_1 gives rise to the small first order differential operator

$$\nabla \overset{def}{=} e_1 \frac{\partial}{\partial x} + e_{24} \frac{\partial}{\partial y} + e_{124} \frac{\partial}{\partial z}$$

and a decomposition of the geometric product $\nabla\phi$ into the divergence

$$\nabla.\phi = \frac{\partial u}{\partial x} + \frac{\partial v}{\partial y} + \frac{\partial w}{\partial z}$$

and the inner *colour field curl*

$$\nabla \wedge \phi = \left\{ \left(\frac{\partial v}{\partial z} + \frac{\partial w}{\partial y} \right), \left(\frac{\partial u}{\partial z} + \frac{\partial w}{\partial x} \right), \left(\frac{\partial u}{\partial y} + \frac{\partial v}{\partial x} \right) \right\}$$

that supports the phenomenology of a velocity field for a »rigid body displacement« as we shall work out in a later section. The »inner« provides those geometric elements which can be subjected to the standard model symmetries. The inner substantiates the derivation of the constitutive Lie group $\iota^{(2)}$. But the outer, - being comparatively larger than the inner, -

provides those polarized isotropic multivectors which can be subjected to an equation of motion much more capacious than the classic Dirac equation. Through equation (152) the »outer« is utterly adjusted to the »inner« via the traditional minimal ideal procedure. With equations (149), (150) we have proven that the most general polarized isotropic element $\in$ ✋$_{\chi}$ in the space-time algebra is constituted by three components from subspaces of isotropic fields. Their exterior squares equal colour space ch_1.

$$N_1 \subset span_R\{e_2, e_4, e_{12}, e_{14}\} \text{ with } N_1^{\wedge 2} = ch_1$$
$$N_2 \subset span_R\{e_3, e_{13}, e_{234}, j\} \qquad N_2^{\wedge 2} = ch_1$$
$$N_3 \subset span_R\{e_{23}, e_{34}, e_{123}, e_{134}\}. \qquad N_3^{\wedge 2} = ch_1$$

However, the precise forms of N_1, N_2, N_3 have not been given. We have to complete this now. We have to solve the nilpotent equation (146) especially for those three small subspaces. They have dimension 4 and are embedded into the total space . Every product of any pair of unequal base units gives back a base unit of ⚐$_{\chi}$. However solving for (146) gives four possibilities for the resulting null-vector Y. Therefore we use a first index for space and a second index i = 1, …, 4 for the logic types of isotropic solutions

$$\begin{aligned} Y_{1,1} &= xe_2 + xe_4 + ye_{12} + ye_{14} \\ Y_{1,2} &= xe_2 - xe_4 + ye_{12} - ye_{14} \\ Y_{1,3} &= xe_2 - ye_4 - xe_{12} + ye_{14} \\ Y_{1,4} &= xe_2 + ye_4 - xe_{12} - ye_{14} \end{aligned} \qquad N_1 \overset{def}{=} \bigcup_{i=1}^{4} \{Y_{1,i}\} \tag{153}$$

You see, although space N_1 has dimension 4, at most two components x, y are needed to obtain an isotropic magnitude.

We obtain analogous solutions for the space N_2 of nilpotents

$$\begin{aligned} Y_{2,1} &= xe_3 + xe_{13} + ye_{234} + yj \\ Y_{2,2} &= xe_3 - xe_{13} + ye_{234} - yj \\ Y_{2,3} &= xe_3 - ye_{13} - ye_{234} + xj \\ Y_{2,4} &= xe_3 + ye_{13} - ye_{234} - xj \end{aligned} \tag{154}$$

We know the trigonal multivectors that carry out flavour- or colour-rotations as inter- and intra-relations in the domain of colour spaces. In part 2 we gave the explicit expression for the flavour rotating element. In part 3 we derived the colour rotation that operates within a single colour space, that is, for example ch_1. We said, the pure states in ch_1 would only be the »cursors« for the spinors, inner labels for the polarized isotropic direction fields. If that would have any meaning, we should be able to find a relation between inner and outer in terms of the trigonal »$SU(3)$« rotations.

With the explicit flavour rotation from equation (69)

$$T = \tfrac{1}{2}(Id + e_1 - e_{34} + e_{134})\tfrac{1}{2}(Id - e_3 + e_{24} - e_{234})$$

we calculate

$$TY_{1,1}T^{-1} = -ye_{23} - xe_{34} + xe_{123} + ye_{134} \tag{155}$$

which is indeed an element of N_3. For if we solve the equation (146) for $span_R\{e_{23}, e_{34}, e_{123}, e_{134}\}$ we obtain solutions

$$\begin{aligned} Y_{3,1} &= xe_{23} + xe_{34} + ye_{123} + ye_{134} \\ Y_{3,2} &= xe_{23} - xe_{34} + ye_{123} - ye_{134} \\ Y_{3,3} &= xe_{23} - ye_{34} - ye_{123} + xe_{134} \\ Y_{3,4} &= xe_{23} + ye_{34} - ye_{123} - xe_{134} \end{aligned} \qquad N_3 \overset{def}{=} \bigcup_{i=1}^{4} \{Y_{3,i}\} \tag{156}$$

and obviously the element in (155) belongs to the third class of spaces $\{Y_{3,4}\}$. Now, let us be aware that the isotropic vector Y_2 from equation (147)(ii) is associated with $f_2 \in ch_1$. This contains 3 components one of which we have subjected to a colour rotation. The outcome is an element from space N_3, that is, from the fourth “logic subspace” $\{Y_{3,4}\}$.

It is therefore signalled to understand the element $Y_{1,1}$ as a representative of a fermion spinor with flavour and colour, and it is as reasonable to take $Y_{3,4}$ as a spinor for another flavour with the same colour. As we took the f_2 to represent a strange quark, say, with colour red, the $Y_{3,4}$ should represent the isotropic spinor of a u-quark. Thus it is clear that we proceed from the pure states which are primitive idempotents in colour space to the polarized isotropic multivectors in the complementary spaces which represent the associated real pure spinors of fermions. It is clear that the above rigor for the solution of the null equation (146) can likewise be carried out for the third class of small spaces with isotropic Clifford numbers. The reader is advised to carry out this calculation as a simple exercise.

Example: beyond the thermodynamic branch it is often useful to consider the real or complex matrix representation of a null multivector. To give an example, we would be interested to look at the matrix of a (147)(iii) representation of an $|s\rangle$ to calculate a minimal hypersurface

$$\begin{aligned} Y_3 = {} & \tfrac{1}{4}(x_3 - x_5 - x_6 + x_8)(e_2 - e_4 - e_{12} + e_{14}) + \\ & + \tfrac{1}{4}(x_4 - x_7 - x_{15} + x_{16})(e_3 - e_{13} - e_{234} + j) + \\ & + \tfrac{1}{4}(x_9 + x_{11} - x_{12} - x_{14})(e_{23} + e_{34} - e_{123} - e_{134}) \end{aligned}$$

We take $x_3 = \frac{1}{16}, x_4 = \frac{1}{8}, x_5 = \frac{3i}{16} + \frac{1}{4}, x_6 = \frac{1}{4}, x_7 = \frac{5}{16}, x_8 = \frac{3}{8}$,

$x_9 = \frac{7}{16}, x_{11} = \frac{1}{2}, x_{12} = \frac{9}{16}, x_{14} = \frac{11}{16}, x_{15} = \frac{3}{4}, x_{16} = \frac{13}{16}$ and calculate the matrix for Y_3 according to the last representation (Part 3, equ. 25) as

$$Y_3 = \begin{bmatrix} 0 & 0 & 0 & 0 \\ 0 & 0 & 0 & 0 \\ -\frac{1}{8} & \frac{5}{16} & 0 & \frac{-1-3i}{16} \\ 0 & 0 & 0 & 0 \end{bmatrix}$$

This matrix has eigenvectors

$$\begin{bmatrix} \frac{5}{2} & 1 & 0 & 0 \end{bmatrix}, \begin{bmatrix} 0 & 0 & 1 & 0 \end{bmatrix}, \text{ and } \begin{bmatrix} \frac{-1-3i}{2} & 0 & 0 & 1 \end{bmatrix}$$

with all eigenvalues equal to zero. The 2nd order nilpotent is represented by the null matrix

$$Y_3^2 = \begin{bmatrix} 0 & 0 & 0 & 0 \\ 0 & 0 & 0 & 0 \\ 0 & 0 & 0 & 0 \\ 0 & 0 & 0 & 0 \end{bmatrix}$$

Example 2: look at the matrix representation of a "small Clifford number" from the second space

$$Y_{2,1} = \tfrac{1+i}{4} e_3 + \tfrac{1+i}{4} e_{13} + \tfrac{1-i}{4} e_{234} + \tfrac{1-i}{4} j \qquad \text{which is}$$

$$Y_{2,1} = \begin{bmatrix} 0 & 0 & 0 & i \\ 0 & 0 & 0 & 0 \\ 0 & 0 & 0 & 0 \\ 0 & -1 & 0 & 0 \end{bmatrix}$$

It has two real eigenvectors [1 0 0 0] and [0 0 0 1] with single and fourfold degenerate null eigenvalues. The geometric product of $Y_{2,1}$ with itself is represented by the resulting null matrix as before.[1] – All this can make us confident that we can find valid representations of standard model fermions by isotropic numbers.

Energy Partition and Direction Field

In this section we shall clarify the relationship between the colour spaces and the isotropic direction fields of fermions. To do so we have to say good bye to a number of habits. At the beginning of the previous chapter I have said that the primitive idempotent acts only as a marker or cursor within the context of algebra. What does that mean? Throughout a large number of sections we used the primitive idempotent of the geometric algebra as pure

1 Here I took the Clifford algebra over the complex field in order to demonstrate that the equations stay valid.

state. Why should it be a cursor to some other subspaces which also seem to represent those states?

Think about a wave and its wave equation. We can differ between a vector- or spinor space and the field amplitude. In either case, relativistic or not, we have a definite element that represents the energy of our wave. In the Schrödinger equation energy is given by $i\hbar\partial\psi/\partial t$ which refers to a scalar Laplacian $(-\hbar^2/2m)\ \Delta$. This corresponds with the non-relativistic constraint $-2mE + p^2 = 0$ which, in the relativistic case, must be replaced by the condition $-E^2 + m^2 + p^2 = 0$. This equation holds for the Dirac equation as well as for the Klein-Gordon equation. In any case, we expect the Laplacian that is derived from the Dirac operator to be scalar-valued, real and invertible. This is even expected from the most advanced *Dirac type operators* (Marmolejo-Olea and Mitrea 2004, 94ff.) that are used in nowadays harmonic analysis. Let us differ between the energy shell and the direction field. The Laplace type operator need not necessarily project onto the scalar field. It may just as well turn out to be a 4-fold scalar. Let us slowly work out how that goes.

It is important to maintain in some way the spinor approach for the representation of pure states. Even in thermodynamics, Kiehn had realized that in his 2005 work he had not yet fully appreciated the topological significance of a large class of direction fields that had been called spinors by Cartan, namely the isotropic complex vectors (Kiehn 2008, p. 101f.). As we know, spinors are similar to vectors, but do not behave as vectors with respect to rotations. These elements of the Clifford algebra also generate harmonic functions and are related to conjugate pairs of minimal surfaces.

Those minimal surfaces are given by special symmetric elements of peculiar subspaces of $Cl_{3,1}$. We shall denote those special subspaces as isotropic direction fields or isotropic multivectors.[2] Direction fields are not mere phenotypes of quantum phenomenology, but they are involved at all levels, micro- and macro-, and independent of scale. They play an important role in thermodynamic processes apart from equilibrium. This is valid especially in strong force interaction which by its very nature takes place far from equilibrium in metastable states of material stability in its outer appearance.

As soon as we make a difference between the inner and outer space of (sub)nuclear matter, as we have done, and at the same time appreciate the idea that spinor states reside in minimal (left) ideals of the Clifford algebra, as we have done on the last pages, we are confronted with the fourfold ring of the inner space and therefore with a multiple Laplace type of a second order differential operator. There is no escape from this insight.

Consider the space N with neutral signature given by elements

$$\mathsf{N}_1 = \{xe_2 + ye_4 + ze_{12} + ue_{14}\} \in Cl_{3,1} \tag{157}$$

N is in the complementary basis of ch_1 and covers the minimal left ideal of the pure state f_2. The N contains a symmetric subspace given by the symmetries

$$\mathrm{I}_1 = \{xe_2 + xe_4 + ze_{12} + ze_{14}\}, \text{ that is, } y = x \text{ and } u = z \tag{158}$$

2 Kiehn prefers the denotation of a directionfield over the names complex isotropic vector or null vector that appear in the literature. He states: "The familiar formats of Hamiltonian mechanical systems exclude the concept of spinor process directionfields, for the processes permitted are restricted to be represented by direction fields of the extremal class, which have no zero eigenvalues."

The I represents an isotropic direction field connected with the pure state f_2. The N, however, discloses the special feature

$$\overset{2}{\Lambda} \mathsf{N}_1 = ch_1 \tag{159}$$

This identity has significant consequences for the Laplace type of differential operator. To see this let us take a look at the procedure as usual (e.g. Lounesto 2001, p. 258). Consider a Dirac operator of the familiar type

$$\nabla = e_1 \frac{\partial}{\partial x_1} + e_2 \frac{\partial}{\partial x_2} + \ldots + e_n \frac{\partial}{\partial x_n} \tag{160}$$

The Dirac operator applied twice is equal to the Laplace operator. As the base units are subject to the anti-commutation relations

$$e_k e_l = -e_l e_k \tag{161}$$

the Laplacian turns out to be scalar. This cannot be demanded for a Laplacian derived from a Dirac type operator on N_1. For consider

$$\nabla(\mathsf{N}_1) = \frac{\partial}{\partial x} e_2 + \frac{\partial}{\partial y} e_4 + \frac{\partial}{\partial z} e_{12} + \frac{\partial}{\partial u} e_{14} \tag{162}$$

We obtain

$$\Delta(\mathsf{N}_1) = \nabla\nabla(\mathsf{N}_1) = \left(\frac{\partial^2}{\partial x^2} - \frac{\partial^2}{\partial y^2} - \frac{\partial^2}{\partial z^2} + \frac{\partial^2}{\partial u^2} \right) Id - 2\left(\frac{\partial^2}{\partial x \partial u} - \frac{\partial^2}{\partial y \partial z} \right) e_{124} \tag{163}$$

This may be surprising, but is in no way a problem. It is rather a quite natural and desired solution. For let us bear in mind that a quantity of the form $\Delta(\mathsf{N}_1)\psi$ is (i) an element of the colour space ch_1 and (ii) a deformation of an idempotent which is the sum of the two primitive idempotents, or pure states, $f_2 + f_3$. All we have to do, is understand that the field amplitude is not a scalar in the ring of real numbers, but a number in the fourfold ring ${}^4\mathbb{R}$ of reals. The second order differential operator of the Laplace type is indeed the bridge that leads us from the isotropic direction fields to the colour spaces of fermion pure states. It is remarkable that each of the three constituents N_k of the isotropic subspaces contributes to exactly one of three types of subspaces of ch_1 and respectively ch_χ, namely those spanned by the idempotents

$$f_{12} + f_{13}, \quad f_{12} + f_{14} \quad \text{and} \quad f_{13} + f_{14}; \tag{164}$$

we get

$$\Delta(\mathsf{N}_2) = \nabla\nabla(\mathsf{N}_2) = \left(\frac{\partial^2}{\partial x^2} - \frac{\partial^2}{\partial y^2} + \frac{\partial^2}{\partial z^2} - \frac{\partial^2}{\partial u^2}\right) Id - 2\left(\frac{\partial^2}{\partial x \partial z} - \frac{\partial^2}{\partial y \partial u}\right) e_{24} \tag{165}$$

$$\Delta(\mathsf{N}_3) = \nabla\nabla(\mathsf{N}_3) = \left(-\frac{\partial^2}{\partial x^2} + \frac{\partial^2}{\partial y^2} - \frac{\partial^2}{\partial z^2} + \frac{\partial^2}{\partial u^2}\right) Id - 2\left(\frac{\partial^2}{\partial x \partial z} - \frac{\partial^2}{\partial y \partial u}\right) e_1 \tag{166}$$

In the previous chapter we have derivated that the outer spaces ⚐$_\chi$ that include spaces ✋$_\chi$ of isotropic polarized multivectors have dimension 12. We are led to pose the legitimate question how the general element of the Laplace type would look like that would lead to a wave amplitude located entirely in one of the colour spaces. Without restriction of the general result we can answer this for the colour space ch_1. Any solution is either derived from a Dirac type operator connected with subspace N_1 or from N_2 or from N_3. We obtain conjugate pairs of solutions. So it does not require the 12 dimensions, but 4 are sufficient, and it is precisely four dimensions that are needed, not more and not less. That is, the complementary outer space decomposes into a 3-partition with respect to the inner space. The three outer constituents N_1, N_2, N_3 are algebraically not entirely equivalent. The exterior products of

$$\begin{aligned} &\nabla(\mathsf{N}_2)\wedge\nabla(\mathsf{N}_2) = 0 \\ &\nabla(\mathsf{N}_3)\wedge\nabla(\mathsf{N}_3) = 0 \end{aligned} \tag{167}$$

vanish for all values x, y, z, u where

$$\nabla(\mathsf{N}_2) = \partial_x e_3 + \partial_y e_{13} + \partial_z e_{234} + \partial_u j$$

and

$$\nabla(\mathsf{N}_3) = \partial_x e_{23} + \partial_y e_{34} + \partial_z e_{123} + \partial_u e_{134} \tag{168}$$

But for N_1 we obtain a non vanishing exterior

$$\nabla(\mathsf{N}_1)\wedge\nabla(\mathsf{N}_1) = -2(\partial_x \partial_u - \partial_y \partial_z) e_{124} \tag{169}$$

so that together with (162) there is an inner product having form

$$\begin{aligned} &\nabla(\mathsf{N}_1).\nabla(\mathsf{N}_1) = \nabla(\mathsf{N}_1)\nabla(\mathsf{N}_1) - \nabla(\mathsf{N}_1)\wedge\nabla(\mathsf{N}_1) = \\ &= (\partial_x^2 - \partial_y^2 - \partial_z^2 + \partial_u^2)\, Id \end{aligned} \tag{170}$$

For the isotropic direction fields where $y = x$, $u = z$, we have indeed

$$\Delta(\text{✋}_1) = \Delta(\text{✋}_2) = \Delta(\text{✋}_3) \equiv 0 \tag{171}$$

the Laplace type of operators vanish in all three spaces ✋_k.

Chapter 27

LORENTZ TRANSFORMATION BY ISOSPIN

From the decomposition at hand there follows a special property of the isospin generators, namely, they act as generators of proper orthochronous Lorentz transformations in the ungraded Minkowski space. For instance isospin λ_1 does the following

$$e_3' = e^{2\phi\lambda_1} e_3 e^{-2\phi\lambda_1} = (\cosh\phi) e_3 - (\sinh\phi) e_4$$
$$e_4' = (-\sinh\phi) e_3 + (\cosh\phi) e_4 \quad (172)$$

which represents a boost.

We know, the elements λ_1, λ_2, λ_3 generate the group $SL(2, \mathbb{R})$. They carry out Lorentz transformations among $\{e_2, e_3, e_4\}$ while preserving the scale of direction e_1. (Schmeikal 2004, 368) It might perhaps be preferred to preserve the spin direction e_3 and model a Lorentz boost in direction e_1. This is also possible if we reconsider that the constitutive group of rank 2, the $\iota^{(2)}$, has six isomorphic images that can be obtained from the Weyl group by permuting the spatial base units. Namely, considering

$$\lambda_1^{\tau} = \tfrac{1}{4}(e_{14} - e_{134}) \text{ and transforms } \xi' = e^{2\phi\lambda_1^{\tau}} \xi e^{-2\phi\lambda_1^{\tau}} \quad (173)$$

we obtain the classic Lorentz transformation alongside e_1:

$$e_1' = e^{2\phi\lambda_1^{\tau}} e_1 e^{-2\phi\lambda_1^{\tau}} = (\cosh\phi) e_1 - (\sinh\phi) e_4$$
$$e_4' = (-\sinh\phi) e_1 + (\cosh\phi) e_4 \quad (174)$$

That reflects our usual procedure. We measure the distance to the object as the product of the speed of light and half the time taken by the signal. This is a fundamental statement, and we could repeat here what Fock had said in 1959: "Knowing the speed of propagation of a radio signal – it is in fact equal to the speed of light – one obtains the distance to the object as the product of the speed and half the time taken by the signal. As a matter of principle this method is important because it reduces the measurement of length to a measurement of time intervals and does not make use of the properties of absolutely rigid bodies. The essential assumption is that the speed of light is constant. Here this speed plays the part of a conversion factor from time to length. Its numerical value must be determined by other experiments and these do have to use a standard of length." (p. 2)

One of the enrichments in the present geometric algebra approach is in the fact that light plays the part of a conversion factor not only from time to length, but also from space-time areas to space-areas. This means a convolution of time or what I called time-wrap. The time wrap allows nature to deconvolute a spatial wave pattern distributed over a space area or volume into a space-time-area or -volume pattern and vice verse. In my *Lie group guide to the universe* I have explained the time wrap without showing where precisely the wrap stems from. How is it made? How does time convolution and deconvolution come upon? And the answer is comparatively simple: first of all by a »grade 2 Lorentz transformation«:

$$\begin{aligned} e'_{34} &= (\sinh\phi)\, e_{13} + (\cosh\phi)\, e_{34} \qquad \text{by } \lambda_1^{\tau} \\ e'_{13} &= (\cosh\phi)\, e_{13} + (\sinh\phi)\, e_{34} \end{aligned} \tag{175}$$

In this way the space area on e_{13} is converted onto a space time area e_{34} in the Cartan subalgebra of colour space ch_2. Applying λ_1 instead of $\lambda^{\tau}{}_1$ we would obtain

$$\begin{aligned} e'_{14} &= -(\sinh\phi)\, e_{13} + (\cosh\phi)\, e_{14} \qquad \text{by } \lambda_1 \\ e'_{13} &= (\cosh\phi)\, e_{13} - (\sinh\phi)\, e_{14} \end{aligned} \tag{176}$$

with e_1 preserved and e_3 boosted. I would like to connect this observation with our first lectures on special relativity. Consider an isometry of the Minkowski space. It is called Lorentz transformation and is essentially built up by boosts, parity and time-reversal. Consider only two dimensions such as in equation (174). Let us write it in simple terms

$$\begin{bmatrix} ct' \\ x' \end{bmatrix} = \begin{bmatrix} A & B \\ C & D \end{bmatrix} \begin{bmatrix} ct \\ x \end{bmatrix} = \begin{bmatrix} Act + Bx \\ Cct + Dx \end{bmatrix} \qquad \text{real } A, B, C, D \tag{177}$$

c velocity of light

where x can indeed represent the length x_2 on direction e_1 in the general element $\in Cl_{3,1}$ and ct belongs to e_4. Demanding the invariance of the line-element means that

$$\begin{aligned} (ds')^2 &= (c\,dt')^2 - (dx')^2 = (A\,c\,dt + B\,dx)^2 - (C\,c\,dt + D\,dx)^2 = \\ &= (c\,dt)^2 - (dx)^2 = (ds)^2 \end{aligned} \tag{178}$$

which requires

$$A^2 - C^2 = 1, \qquad D^2 - B^2 = 1 \text{ and } AB - CD = 0. \tag{179}$$

These conditions can best be complied by putting

$$A = D = \pm\cosh\phi,\ B = C = \pm\sinh\phi \text{ so that } \tanh\phi = \tfrac{v}{c} \tag{180}$$

The four sign combinations of $\{A, B\}$ correspond with the four components of the Lorentz group which preserve space- and/or time orientation or revert them. The quantity $\cosh \phi = \gamma$ is the familiar Lorentz factor $\gamma = \left(1 - \frac{v^2}{c^2}\right)^{-1/2}$ and $\sinh \phi = \frac{v}{c}\gamma$.

A classically oriented boost has $A = +\cosh \phi$ with positive cosh and negative $B = -\sinh \phi$. If B appears reverted, (as it does in equations (175) we would have parity -1, that is, spatial reflection of the directed unit vector with measure x. But in equation (175) the x measures a directed area and not a vector, namely the coefficient x_7 of area e_{13} Therefore by this bivector Lorentz transformation the area of the bivector is reverted. Let us work out what that means in connection with the dynamics of the field. Suppose that, in principle, the field can demand all 16 degrees of freedom, - which indeed does not only seem possible, but rather signifies the general case of the dynamics, - and suppose further that motion in direction e_1 as in equation (174) is correlated with the directed area e_{13}, then the classical boost on e_1 is accompanied by a reverted boost on the area e_{13} which implies a reversal on the spin vector e_3 as e_1 is preserved. But, if instead we would, in accordance with equations (172) and (176), let the field travel in the direction e_3 instead of e_1, the classical boost on e_3 would be accompanied by a corresponding boost on the area e_{13} and a space-time area dilatation on e_{14}. This suggests that the field propagates in the spin plane, a fact which has variously been proposed by some who developed models for electron movement. In our case this would concern fermions with baryon number 1/3, isospin given by $\frac{1}{4}(e_{24} - e_{124})$ and spin by $\frac{1}{2}\, e_{13}$. Note, spin commutes with isospin as it should. So while the field travels in direction e_3 it is spinning in the plane e_{13} and dilatates the area e_{24}. This is indeed only a fragment of the whole interwoven dynamics of heavy fermions.

Interestingly, the generators of one definite constitutive Lie algebra $\mathfrak{l}^{(2)}$ induce many kinds of Lorentz transformations in the graded space of the Clifford algebra, but it projects onto the fundamental vector space, - namely the 4-dimensional Minkowksi-Lorentz space, - only Lorentz transformations as are known within a 3-dimensional Minkowski subspace. This may at first seem surprising, but is utterly clear, because the constitutive Lie algebra is a closure on the Lie product $\mathfrak{l}^{(2)} = sl_{\mathrm{Cl}}(2) \times so_{\mathrm{Cl}}(3)$. The commutation relations of λ_1, λ_2, λ_3 bring in the structure constants for the algebra of the Lie group $Sl(2, \mathbb{R})$ isomorphic with the positive component of the spingroup $Spin^{+}_{2,1}$. This restriction of Lorentz transformations to dimensions $\{+ + -\}$ is typical for the connection between the quantum theory and relativity in the space-time algebra. But this distinguished place of a 3-dimensional Minkowski subspace reappears often in nowadays models and theories, as we shall see.

SPACE FILLING FIELDS AND STRUCTURE OF RELATIVISTIC PURE STATES

To understand the holistic view of physics requires several important attitude changes. First we must see that the trajectory of the Schrödinger- or Dirac theory represent only small shadows of the whole event of particle propagation. Formerly it was as if we described the whole mighty ocean by a small fragment of one of its waves. The holomovement, as David Bohm had called it, involves a graded structure of the field, and this cannot be avoided by any means. The phenomenology cannot be linearized, and it cannot be grade-equalized. A Dirac spinor is not at all the proper vehicle to represent a fermion. But such a field is better represented by a manifold of primitive idempotents in geometric Clifford algebra. Next we must understand that the Lorentz-group is also only a shadow of a larger graded Lie group which describes more general Lorentz transformations in the whole domain of the space-time algebra and which essentially constitutes the standard model Lie group of the particles. Then we must comprehend the dynamics is highly nonlinear. Only part of this nonlinearity is met by the Pauli principle as we saw, another part is in the fact that quantum numbers can be definite, integer, with almost no standard-deviation, although the fundamental process which constitutes particles is nonlinear as Kiehn (2008) has tried to demonstrate in terms of his Falaco solitons.

A fermion, as is understood here, has a binary structure. One of its components is Lorentz-invariant while the other is a Lorentz-variant isospin component. So we have arrived at the next demand: we must understand that space variables, field, Lorentz transformations, isospin generators, quantum numbers, even locations and rotations and so forth are all on one level. Even a fermion at rest is not free of structure, but it contains a Lorentz invariant Euclidean-like idempotent factor and an isospin component which is Lorentz transformed in such a way that all relations of that fermion with its environment such as orthogonality, charge and isospin are preserved, and such that that field tends to reproduce itself by self interaction. In traditional quantum mechanics we had a field on the one side, and an operator such as a Lorentz transformation or the isospin on the other: Here there was the field. On the other side there was the observer changing the isospin or measuring charge. Here there was a particle located in some region of space-time and showing a trajectory having dimension 1, or at most fractal dimension ≤ 2. There we had an accelerating electromagnetic field imposing a Lorentz factor on the protons. But we always thought and worked in a small and comparatively illusory tube of the whole geometry. This situation changes as soon as we realize that the accelerating field induces a Lorentz transformation which carries in it an isospin phenomenology which is also contained in the relativistic fermion. As Goethe said: "Wär nicht das Auge sonnenhaft, Die Sonne könnt es nie erblicken." The isospin components of the interacting fields touch each other in all dimensions involved, - in the concrete case of a standard quark in six, - so that there is preserved an invariant idempotent of the Euclidean space which traditionally gave rise to the minimal left ideal of the Pauli spinors. But now the phenomenology involves the *space of the isospinoreigenforms*.

To slowly go into this I would like us to list up the various Lorentz transformations of the base elements of the Clifford algebra and finally investigate the Lorentz transform of a pure state by the corresponding exponential maps. We denote by L_i the Lorentz transformation effectuating a change of Clifford monomial $e_{(I)}$. By $L_i(\lambda_j)$ we mean a Lorentz transformation of the generator $\lambda_j \in \mathcal{L}^{(2)}$, the isospin, *u*- or *v*-spin. In the lines to come, we only investigate the action of one type of graded Lorentz transformations, namely those induced by the λ_1. Thus we briefly write L for L_1 or $L(\phi)$ without index, but respecting the parameter, that is, the angle ϕ involved in the transformation. We abstract from the index of the isospin. Only in a next section where we involve the components λ_2 , λ_3 as Lorentz transformations , we shall give L the corresponding index. All transformations are calculated by exponential map $e_{(I)}' = \exp(2\phi\,\lambda_1)\, e_{(I)} \exp(-2\phi\,\lambda_1)$ in the form of conjugation. There comes in some redundancy because some equations are repeated. We use only one ϕ just to indicate the type of form.

(i) $e_1' = e_1$

(ii) $e_2' = (\cosh\phi)e_2 - (\sinh\phi)j$

(iii) $e_3' = (\cosh\phi)e_3 - (\sinh\phi)e_4$

(iv) $e_4' = (-\sinh\phi)e_3 + (\cosh\phi)e_4$

(v) $e_{12}' = (\cosh\phi)e_{12} - (\sinh\phi)e_{234}$

(vi) $e_{13}' = (\cosh\phi)e_{13} - (\sinh\phi)e_{14}$

(vii) $e_{14}' = -(\sinh\phi)e_{13} + (\cosh\phi)e_{14}$ (181)

To abbreviate terms, let us put A = sinh ϕ, B = cosh ϕ. And we proceed with bivectors

(viii) $e_{23}' = B^2 e_{23} - ABe_{24} - A^2 e_{123} + ABe_{124}$

(ix) $e_{24}' = -ABe_{23} + B^2 e_{24} + ABe_{123} - A^2 e_{124}$

(x) $e_{34}' = e_{34}$ and go on with trivectors

(xi) $e_{123}' = -A^2 e_{23} + ABe_{24} + B^2 e_{123} - ABe_{124}$

(xii) $e_{124}' = ABe_{23} - A^2 e_{24} - ABe_{123} + B^2 e_{124}$

(xiii) $e_{134}' = e_{134}$

(xiv) $e_{234}' = -Ae_{12} + Be_{234}$ until to the director

(xv) $j' = -Ae_2 + Bj$

This sketches the total action of the isospin generator λ_1 on the whole 16-dimensions of the Clifford algebra. Note that it leaves the monomials e_1, e_{34}, e_{134} unchanged, that is, it preserves colour space ch_2. Be aware that there exist tetrahedral flavour- and colour rotators with period 3 which turn states as represented in the following figure

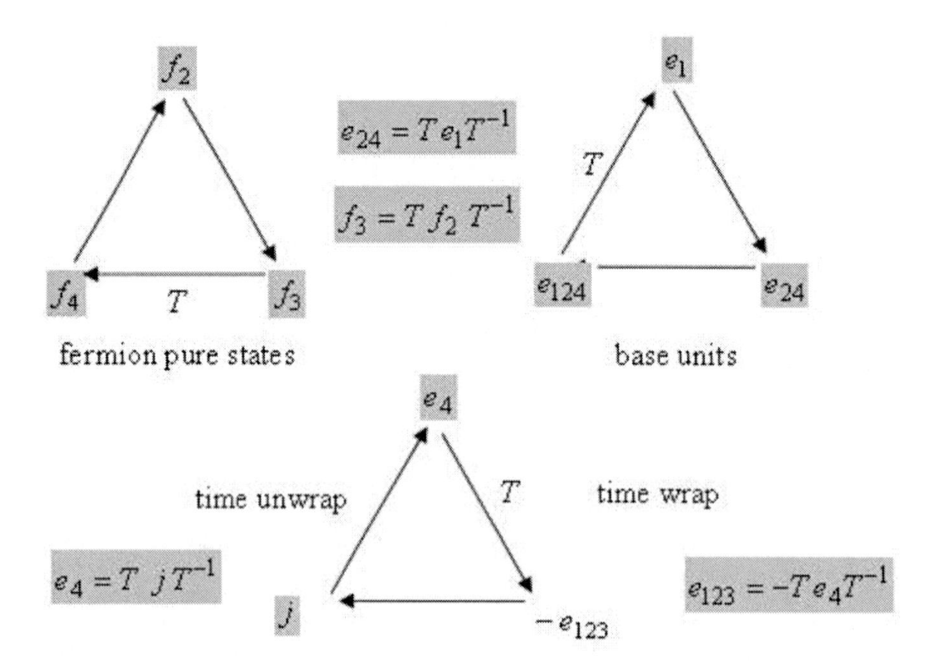

Figure 21. Time wrap by universal flavour rotation T in $Cl_{3,1}$.

This means that in $Cl_{3,1}$ there exists a "time-space" with three time-like monomials that can be colour rotated and fulfil commutation relations similar to those of the Euclidean space algebra (Schmeikal 2009).

Now consider the u-quark with fully expanded polynomial as an element of the colour space ch_1. The simplest representation of the standard element in the manifold is

$$f_3 = \tfrac{1}{4}(1 - e_1 + e_{24} - e_{124}) \tag{182}$$

It transforms by $f_3' = L(\phi) f_3$ like

$$f_3' = \tfrac{1}{4}\Big((1-e_1) - 2AB(e_{23} - e_{123}) - (A^2 + B^2)(e_{24} - e_{124})\Big) \tag{183}$$

We realize that the Lorentz map in a colour space is a linear operation. This is due to the fact that every vector in such a space can be written as a linear combination of its four primitive idempotents. We have

$$L f_3 = \tfrac{1}{4}(1 - Le_1 + Le_{24} - Le_{124}) = \tfrac{1}{4}(1 - e_1' + e_{24}' - e_{124}') \tag{184}$$

The reader can easily test equations (183), (184) by using equations (181) or carrying out his own rigor by a Clifford algebra calculator.

In his fundamental work on associative algebra Charles Sanders Peirce defined a primitive idempotent as one which cannot be represented by a sum of mutually annihilating idempotents. According to this a pure state such as (182) is a primitive idempotent. It is usually represented as a product of two idempotents

$$f_3 = \tfrac{1}{2}(1-e_1)\tfrac{1}{2}(1+e_{24})$$

both of which are not primitive. Note, however, that the first factor η = (½)(1−e_1) is a primitive idempotent in the Pauli algebra, and the second factor (½)(1 + e_{24}) is also a part worth considering. It is a primitive idempotent in colour space ch_1.

Usually mathematicians allow for nilpotents of orders two, three, … , n, …, but they hardly ever use idempotents of, say, order n, even and odd orders greater than some integer. But that is exactly what we can use now. Namely, we can write

$$f_3 = \tfrac{1}{2}\eta + \lambda_3 \tag{185}$$

and bear in mind that $2\lambda_3$ is an idempotent to order 3, and of odd orders greater than 3. In this way the pure state incorporates the third isospin component, and we have

$$(2\lambda_3)^{2m+1} = 2\lambda_3 \text{ and } (2\lambda_3)^{2m} = \eta,\ \text{m nonnegative integer} \tag{186}$$

We now have to define the algebraic structure of a quark, such as f_3 and show how the Lorentz transformation L preserves that design.

Definition 1: The structure of a fermion is now determined by four properties (i) it is a pure state displayed by a primitive idempotent (ii) it can be decomposed into the primitive idempotent of the Pauli algebra halved plus the standard Clifford element λ_3 for the isospin in accord with equation (185). Thus it is (iii) a sum of half an idempotent of order 2 and half an idempotent of odd orders ≥ 3. Further it is one of the generating elements of a 16-element lattice of idempotents 4 of which are orthogonal. That is, (iv) it is in a relation of mutual annihilation with four other primitive idempotents in the same colour space.

Theorem 1: A graded Lorentz transformation $L(\phi)$ preserves the structure of a quark.

We have

$$L(\phi)f_3 = \tfrac{1}{2}\eta + L\lambda_3 \text{ with } L\lambda_3 = \lambda_3' = -\tfrac{1}{2}AB(e_{23}-e_{123}) - \tfrac{1}{4}(A^2+B^2)(e_{24}-e_{124})$$

(eq. (183))

We need for the proof an absorption property which follows from equation (186), namely

$$\eta L = L \tag{187}$$

and the fact that η commutes with L. Calculating the product

$$\lambda_3'\lambda_3' = (\tfrac{1}{2}\eta + L\lambda_3)(\tfrac{1}{2}\eta + L\lambda_3) = \tfrac{1}{4}\eta + L\lambda_3 + (L\lambda_3)^2 =$$

$$\tfrac{1}{2}\eta + L\lambda_3 = \lambda_3' \text{ since } (L\lambda_3)^2 = \tfrac{1}{4}\eta \tag{188}$$

we can verify by a little algebra that the term $2L\lambda_3$ actually obeys the equation (186), that is, it remains an idempotent of odd orders ≥ 3.

$$(2L\lambda_3)^{2m+1} = 2L\lambda_3 \text{ and } \qquad (2L\lambda_3)^{2m} = \eta \tag{189}$$

We also verify the orthogonality relations

$$Lf_3\, Lf_1 = 0 \qquad\qquad Lf_3\, Lf_2 = 0 \qquad\qquad Lf_3\, Lf_4 = 0 \tag{190}$$

Definition 1 can indeed be further expanded. The logic completeness and interdependencies of properties has not yet been sufficiently investigated. But one thing is sure worth to be mentioned. Since long we know that in the Clifford algebra $Cl_{3,1}$ the sum of the two primitive idempotents $f_3 + f_4$ gives the above "Euclidean idempotent " η from which the minimal left ideals of the Pauli algebra and respectively the Pauli spinors are derived, that is, we have $f_3 + f_4 = \eta$. This relation is also preserved

$$Lf_3 + Lf_4 = L\eta = \eta \text{ where } f_4 = \tfrac{1}{2}(1-e_1)\tfrac{1}{2}(1-e_{24}) \tag{191}$$

It is a very essential result of this rigor that a fermion is always the sum of a non-relativistic "Euclidean primitive idempotent" and an isospin component. While the Lorentz transformation contains at least one isospin generator of the constitutive Lie group, the pure state under consideration contains a second isospin component as a partner. The colour space, Ch_1 in our case, is of course expanded by the Lorentz transformation. Its number of dimensions is raised from four to six. We must next find out what that space is like into which the pure states are boosted. We investigate the holomovement of the u–quark induced by all three isospin components.

HOLOMOTION OF A $|u\rangle$-QUARK

We repeat the Lorentz transformation imposed on the u-quark or state f_3 by the isospin generating element λ_1.

$$L_1(\phi)f_3 = \tfrac{1}{2}\eta + L_1(\phi)\lambda_3 \qquad \text{with detailed summand} \tag{192}$$

$$L_1(\phi)\lambda_3 = -\tfrac{1}{2}AB(e_{23} - e_{123}) - \tfrac{1}{4}(A^2 + B^2)(e_{24} - e_{124}) \tag{193}$$

But we are also interested in knowing the exact terms for $L_2 f_3$. Therefore let us consider conjugation $\exp(2\theta\,\lambda_2)\, e_{(1)} \exp(-2\theta\,\lambda_2)$:

(i) $\qquad e_1' = e_1$

(ii) $e_2' = (\cos\theta) e_2 + (\sin\theta) e_3$

(iii) $e_3' = -(\sin\theta) e_2 + (\cos\theta) e_3$

(iv) $e_4' = (\cos\theta) e_4 + (\sin\theta) j$

(v) $j' = -(\sin\theta) e_4 + (\cos\theta) j$ (194)

We stop here, since those are the most important transformations of the 16 base units. They show us that the second component is a pure rotation of the plane $\{e_2, e_3\}$ plus a rotation like conversion of time into space-time volume. That is, while the field rotates its space-time volume oscillates. This is a natural consequence of the existence of the constitutive Lie group in the space-time algebra. Putting a = sin θ and b = cos θ we obtain the Lorentz transformation imposed on the u-quark by the second isospin component $L_2(\theta) f_3$

$$L_2(\theta) f_3 = \tfrac{1}{2}\eta + \tfrac{1}{4}(b^2 - a^2)(e_{24} - e_{124}) + \tfrac{1}{2}ab(e_{34} - e_{134}) =$$
$$= \tfrac{1}{2}\eta + \tfrac{1}{4}(\cos 2\theta)(e_{24} - e_{124}) + \tfrac{1}{4}(\sin 2\theta)(e_{34} - e_{134}) \quad (195)$$

Investigating this expression we again find out that $L_2(\theta) f_3$ consists of the Euclidean idempotent and an element from the manifold of order 3 idempotents. All the equations from (78) to (82) are also valid for the $L_2(\theta) f_3$. What has been said about the structural features of the fermion in the last paragraph of the previous section is also valid now. Testing for $L_3(\theta)$ does not disconfirm this because $L_3(\theta)$ preserves f_3. We obtain the list

(i) $L_1 f_3 = \tfrac{1}{2}\eta + L_1 \lambda_3$ in accord with (83), (84)

(ii) $L_2 f_3 = \tfrac{1}{2}\eta + L_2 \lambda_3$ as in (86)

(iii) $L_3 f_3 = f_3$

(iv) $L_4 f_3 = f_3$

(v) $L_5 f_3 = f_3$

(vi) $L_6 f_3 = \tfrac{1}{4}\big((Id - e_{124}) - (A_2 + B_2)(e_1 + e_{24}) + 2AB(e_4 + e_{12})\big)$

(vii) $L_7 f_3 = \tfrac{1}{4}\big((Id - e_{124}) - (b^2 - a^2)(e_1 + e_{24}) - 2ab(e_2 + e_{14})\big)$

(viii) $L_8 f_3 = f_3$ (196)

With the idempotent $\iota = \frac{1}{2}(Id - e_{124}) \in Ch_1$ we obtain the split

$$\text{(i)} \quad L_6(\varphi) f_3 = \tfrac{1}{2}\iota + L_6(\varphi)\lambda_6^{(\tau)} \quad \text{with } \lambda_6^{(\tau)} = w_{12}\lambda_6 w_{12}$$

$$\text{(ii)} \quad L_7(\varphi) f_3 = \tfrac{1}{2}\iota + L_7(\varphi)\lambda_6^{(\tau)} \tag{197}$$

Here we have respected that f_3 can also be written in the form

$$f_3 = \tfrac{1}{2}\iota + \lambda_6^{(\tau)} \tag{198}$$

analogous to the split (185). Again we obtain the result that a Lorentz transformation by v-spin λ_6, λ_7 of the $|u\rangle$–quark preserves the split into an invariant idempotent and a Lorentz-variant component which is capable to "see" λ_6. So what is a boost of a fermion field in this system?

An application of the generator $\lambda_1 \in \mathcal{L}^{(2)}$ causes a classical boost of $\{e_3, e_4\}$ while preserving e_1 and boosting e_2 in correlation with the space-time volume j. In addition it boosts the fermion field according to (196) (i).

$$\text{(i)} \quad e_1' = L_1(\phi)e_1 = e_1$$

$$\text{(ii)} \quad e_2' = L_1(\phi)e_2 = \left(1 - \tfrac{v^2}{c^2}\right)^{-1/2} e_2 - \tfrac{v}{c}\left(1 - \tfrac{v^2}{c^2}\right)^{-1/2} j$$

$$\text{(iii)} \quad e_3' = L_1(\phi)e_3 = \left(1 - \tfrac{v^2}{c^2}\right)^{-1/2} e_3 - \tfrac{v}{c}\left(1 - \tfrac{v^2}{c^2}\right)^{-1/2} e_4$$

$$\text{(iv)} \quad e_4' = -\tfrac{v}{c}\left(1 - \tfrac{v^2}{c^2}\right)^{-1/2} e_3 + \left(1 - \tfrac{v^2}{c^2}\right)^{-1/2} e_4$$

$$\text{(v)} \quad f_3' = L_1(\phi) f_3 = \tfrac{1}{2}\eta - \tfrac{vc}{2(c^2 - v^2)}\left(e_{23} - e_{123}\right) - \tfrac{c^2 + v^2}{4(c^2 - v^2)}\left(e_{24} - e_{124}\right) \tag{199}$$

Notice, if we can measure velocity by a real number v as before, all quantities on the right hand side undergo a boost. All the coefficients including those of the terms of the u-quark diverge as v approaches the velocity of light. The scales of the terms of the fermion pure state go to infinity in accordance with

$$\lambda_1: \qquad \text{as} \qquad v \longrightarrow c$$

$$f_3' \xrightarrow{grad Lorentz} \tfrac{1}{2}\eta + \infty(-e_{23}, +e_{123}, -e_{24}, +e_{124}) \tag{200}$$

but the idempotent $\eta = \frac{1}{2}(1 - e_1)$ is untouched. It is interesting to realize that this quantity brings forth the well known spinor space in the Pauli algebra by forming the minimal left ideal $Cl_{3,0}\,\eta$.

If we recall that the elements $\lambda_i \in Cl_{3,1}$ represent fields, generators of isospin and Lorentz transformations, we obtain a certain unifying interpretation of a u-quark. Namely, if we turn from a standard representation of the quark

$$|u\rangle = f_3 = \tfrac{1}{2}(1-e_1)\tfrac{1}{2}(1+e_{24}) = \tfrac{1}{2}\eta + \lambda_3 \tag{201}$$

in the original standard inertial system 1 to another 2 which moves with velocity v relative to 1, the Lorentz transformed fermion has to be represented by the

Relativistic isospin decomposition of a fermion

$$L_1(v,c)|u\rangle = \tfrac{1}{2}\lambda_1\lambda_1 + \tfrac{2vc}{c^2-v^2}\lambda_2 - \tfrac{c^2+v^2}{c^2-v^2}\lambda_3 \tag{202}$$

Analogous formulas exist for the other quarks. The beauty of such a design is in the fact that each fermion sees the isospin as it is essentially built up by isospin fields. For very small velocities $v \ll c$ we can write

$$|u'\rangle = \tfrac{1}{2}\lambda_1\lambda_1 + 2\beta\lambda_2 - (1+\beta^2)\lambda_3 \quad \text{with} \quad \beta = \tfrac{v}{c} \tag{203}$$

When we denote the field as resting, that is, $\beta=0$, the second term vanishes and the state is preserved $|u'\rangle \equiv |u\rangle$. The advantage of such a decomposition is obvious. We can legitimately hope to obtain eigenvalue spectra for the rest mass and at the same time a reasonable relativistic energy.

An isospin transformation between $|u\rangle$ and $|d\rangle$ is achieved by

$$L_2(\pi)|u\rangle = |d\rangle \quad \text{and} \quad L_2(\pi)|d\rangle = |u\rangle \tag{204}$$

It is clear we have the eigenvalue equations

$$\lambda_3|u\rangle = \tfrac{1}{2}|u\rangle \quad \text{and} \quad \lambda_3|d\rangle = -\tfrac{1}{2}|d\rangle \quad \text{isospin} \tag{205}$$

$$Y|u\rangle = \tfrac{1}{3}|u\rangle \text{ and } \quad Y|d\rangle = \tfrac{1}{3}|d\rangle \quad \text{hypercharge} \tag{206}$$

with $\quad Y = \frac{1}{\sqrt{3}}\lambda_8$

Chapter 28

TOPOLOGICAL DEFECTS AND CONFORMAL LORENTZ TRANSFORMATIONS

GEOMETRY AND TOPOLOGY

In the history of science there has appeared an obvious difference between a geometric point of view and a topological attitude. I shall only highlight that difference by mentioning the geometric calculus as favoured by Hestenes (2005) and the topological approach to evolutionary processes forwarded by Kiehn. It is clear that this apparent antagonism of 'geometry versus topology' is more durable than those recent contributions to physics in general and fields in particular. But I would like to compare the works of Kiehn and Hestenes since those two have also cultivated some reconciling kind of thought.

Kiehn begins his fifth volume on »Non-Equilibrium Systems and Irreversible Processes«, - *Topological Torsion and Macroscopic Spinors* -, by dismantling the failure of diffeomorphic analysis. He totally disagrees with the old idea of Eddington and Einstein that all physical phenomena can be described in terms of tensor analysis. The same says our friend Jose Vargas (2009). Rather he realizes that with Dirac's spinors something has slipped through the net". Kiehn explores the relationship between topological torsion, isotropic macroscopic spinors and topological evolution, and he denotes this relation as most important. He holds close contact with Cartan's work on differential forms and spinors, his, as he calls it, *magic formula* which he could use to connect dynamic systems with thermodynamics, but, luckily, without referring to statistics. In all his contributions to HEPhy and cosmology, as he sees it, Kiehn investigates the interplay between spinors, topological torsion and differential forms. He tries to explain "how these topics relate to the theory that quantization is a topological, not a geometrical concept" (Kiehn 2008, p. 23).

The other author has developed nowadays physics within the context of geometric Clifford algebra. His numerous, thoroughly worked out contributions to the space-time algebra, and space-time calculus have become fundamental for quantum mechanics and are based on geometric calculations. Hestenes shows a quite tolerant attitude where he quotes Henri Poincaré who concluded that "*one geometry cannot be more true than another, it can only be more convenient*" (Hestenes 2005, p. 903).

For me it is quite clear that almost everything that can be said about thermodynamic instabilities, topological defects, fractals and wakes can also be said in terms of geometric algebra. Hestenes, especially in his work on »Gauge Theory Gravity (GTG) with Geometric Calculus« identifies the chief innovation of GTG in a displacement gauge principle that asserts global homogeneity of space-time. Hestenes speaks about the new contribution in most beautiful tones: "With rotation gauge equivalence and displacement gauge equivalence combined, GTG synthesizes Einstein's principles of equivalence and general relativity into a new general principle of gauge equivalence. [...] GTG is also unique in its use of geometric calculus".

Obviously that is not quite the request of Kiehn. It seems somewhat old-fashioned, - in thermodynamic equilibrium, on the mass shell, displaced from wakes and Helmholtz instabilities. He mentioned Cartan's formula, but didn't really make use of it. Hestenes, it seems, paid his tribute to Riemann and Einstein, but thereby implicitly preferred equilibrium systems before systems involving torsion and Pfaff topological dimension (PTD) ≥ 3. In the sense of Cartan's spinor theory and Kiehn's topological model he devaluated the physical vacuum. But Kiehn is not unforgiving. He writes: "In more recent times, a number of mathematicians used concepts that do not require the assumptions of relativity and/or quantum mechanics, but then they often use the popularity and notoriety of the 1921-1828 physics experiments and analysis to justify the utility of their research in pure mathematics. A very few voices in the wilderness, such as that of D. Hestenes, have lashed out against the dogmatic physical basis of the concept of spinors as artefacts of the microscopic relativistic domain. It is of interest to realize that even E. Cartan recognized that there was a difference between Spinors as ordered arrays of functions with zero norm, and vectors as tensors: "*These classical techniques (of Riemannian geometry) are applicable to vectors and ordinary tensors, which besides their metric character, posses a purely affine character; but they cannot be applied to spinors, which have a metric but not affine characteristics.*" It appears that Cartan (unconventionally) uses the word 'affine' to describe all unconstrained matrices that represent linear transformations". (Kiehn 2008, p. 22).

As we have seen in chapter "escape from chaos" the constitution of flavour- and colour-dynamics involves extended Lorentz transformations and nonlinear trigonometric functions of rotation angles. Indeed, in the last section somewhere between equations (199) and (206) there emerged various mysterious relations: an extended Lorentz transformation which brought in a transformation of space-time areas and volumes, an unexpected energy term, a decomposition of pure states into isospin components and a variety of properties that we have not even analyzed yet. But while going through those rather fascinating calculations we overlooked some very simple, but extremely important thing, namely the topological phenomena that were addressed at the same time. We missed the macroscopic spinors we were perpetually using. Fermions and bosons need not be microscopic appearances. Majorana particles as we constructed can take the form of macroscopic topological defect structures beyond instability. Unfortunately, most engineers are acquainted with symmetric matrices having real eigenvalues and eigenvectors. Quantum mechanics has indeed enriched those objects by Hermitean matrices with complex entries which also have real eigenvalues. But what about antisymmetric matrices with real valued entries? Those may disclose enantiomorphic pairs of eigenvectors with complex eigenvalues. Kiehn has denoted them as *eigendirection fields* which, in the Cartan sense, can very well be called spinors. The 'eigendirection fields' have a null quadratic form. They may be looked at as macroscopic spinors, since their emergence does not depend on scale, they may represent particles or other double-valued dynamic systems with zero diameter but finite surface. Interesting enough, the isospin element λ_2, we used so many times, is such a macro-spinor. To show this consider the real representation of $Cl_{3,1}$ by matrices

$$e_1 = \begin{bmatrix} +1 & 0 & 0 & 0 \\ 0 & -1 & 0 & 0 \\ 0 & 0 & -1 & 0 \\ 0 & 0 & 0 & +1 \end{bmatrix} \qquad e_2 = \begin{bmatrix} 0 & +1 & 0 & 0 \\ +1 & 0 & 0 & 0 \\ 0 & 0 & 0 & +1 \\ 0 & 0 & +1 & 0 \end{bmatrix}$$

$$e_3 = \begin{bmatrix} 0 & 0 & +1 & 0 \\ 0 & 0 & 0 & -1 \\ +1 & 0 & 0 & 0 \\ 0 & -1 & 0 & 0 \end{bmatrix} \qquad e_4 = \begin{bmatrix} 0 & -1 & 0 & 0 \\ +1 & 0 & 0 & 0 \\ 0 & 0 & 0 & -1 \\ 0 & 0 & +1 & 0 \end{bmatrix} \tag{207}$$

(Note, even time base unit e_4 represents an eigendirection field.) We obtain for the "real valued second Gell-Mann matrix"

$$2\lambda_2 = \tfrac{1}{2}(-e_{23} + e_{123}) = \begin{bmatrix} 0 & 0 & 0 & 0 \\ 0 & 0 & -1 & 0 \\ 0 & +1 & 0 & 0 \\ 0 & 0 & 0 & 0 \end{bmatrix} \tag{208}$$

This has two enantiomorphic pairs of eigenvectors

$$s_1 = \begin{bmatrix} 0 \\ +i \\ 1 \\ 0 \end{bmatrix} \quad \text{and} \quad s_2 = \begin{bmatrix} 0 \\ -i \\ 1 \\ 0 \end{bmatrix} \tag{209}$$

have a quadratic form equal to $1+i^2$ = zero. Thus the s_1, s_2 are eigenfields of the second isospin component[1]. Kiehn associates complex spheres with zero radii with point particles. Macroscopic spinors disclose null congruence while they allow us to derive concepts of 'length with no length', 'area without area' and 'volume without volume'. These objects have minimal surfaces with zero mean curvature. The minimal surfaces come in conjugate pairs like the s_1 and s_2. Macro-spinors, as physical artefacts are associated with phenomena of double-valuedness polarization and chirality. The antisymmetries are represented by enantiomorphic pairs of mirror reflected images up to total space-inversion. They are not connected by congruent maps. We can therefore go further and ask if the equations give us enough information concerning the transition between the enantiomorphic pairs of isospin

[1] I am adapting to the matrix calculus Kiehn is using, in order that the readers con follow the relationships in the straightest way. I also believe that this a good exercise to follow the topological argumentation.

states. To go deeper into the phenomenology, consider the Lorentz transformed state $|u'\rangle$ of equation (199)(v) and apply from the left the $2\lambda_1$, $2\lambda_2$ and $2\lambda_3$:

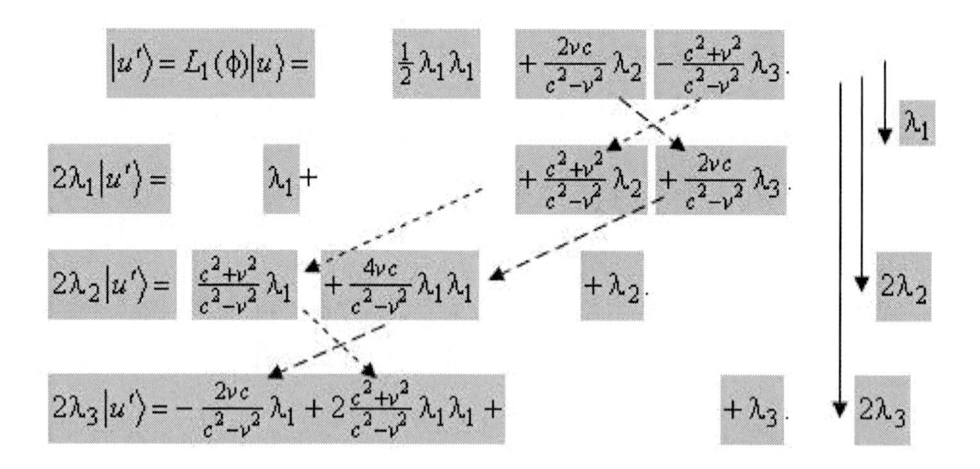

Array 1. Enantiomorphism in isospin decomposition.

See how the isospin elements shift the relativistic energy contributions from component to component. This means a transposition of topological states from space-time constituents such as $\frac{1}{2}(e_{24}-e_{124})$ to $\frac{1}{2}(-e_{23}+e_{123})$ or to $\frac{1}{2}(e_{34}-e_{134})$ or finally to the paravector $\frac{1}{2}(Id-e_1)$. This shift of energy among the constitutive geometric subspaces of fractal motion is typically accompanied by macroscopic isospinors. Notice that matrices λ_5, λ_7 also have eigenvectors with zero radii like λ_2. If we multiply those by the unit imaginary to obtain the original Gell-Mann matrices, obviously, that doesn't alter the pairing of eigenvalues and zero quadratic form. That confirms that these matrices can be comprehended as real forms of $SU(3)$ matrices.

Chapter 29

CONFORMAL TIME-SPACE

TIME-SPACE EXPANSION SHEARS

Pierre de Fermat postulated that a light ray connecting two points in space minimizes the material light path length. Following this idea, Heinrich Bruns (1885) published the Eikonal function determined by optical conductors. In 1959 Wladimir Alexandrowitsch Fock formulated the general law in arbitrary coordinates „*to which any disturbance travelling with limiting velocity is subject*" (Fock 1959, chapter iv, p. 146). According to this view there exists an equivalence class of diffeomorphic reference frames in which a propagating discontinuity of an electromagnetic wave can likewise be seen. The propagating disturbance appears to be a solution to the Null Eikonal equation. This is the nonlinear first order partial differential equation

$$(\partial_i \varphi)^2 = 0 \tag{210}$$

Fock established the diffeomorphic equivalence by a further postulate, namely that the only linear transformation group that would preserve the Null Eikonal equation, while relating one observer to another, was a Lorentz transformation. Consider the diagonal matrix

$$(\partial_i \varphi) f_i \qquad \text{with the four idempotents } f_i \text{ as in (30, 31)} \tag{211}$$

and define the ket vector of 1-forms

$$|d\varphi\rangle = (\partial_i \varphi) f_i \left| dx^k \right\rangle \text{ with } \qquad \partial_4 \varphi = \partial\varphi / c\partial t \tag{212}$$

Then the Eikonal equation $\langle d\varphi | \eta | d\varphi \rangle = 0$ transforms as

$$\langle d\varphi | \eta | d\varphi \rangle = 0 \Rightarrow \langle \partial\varphi \left| L^\dagger \right| \eta | L | \partial\varphi \rangle \tag{213}$$

with the *reversal dagger conjugation* † of L being equal to the inverse L^{-1} and η the metric tensor in signature (+ + + –). Kiehn (2004, 2008 p. 222) pointed out there is a much larger class of nonlinear transformations that preserve the perception of the discontinuity, namely the class of *conformal extensions of the Lorentz transformation*. A sufficient condition for the discontinuity of the field amplitude to be maintained is that the null line element must be preserved.

$$(k\,L)^{\dagger}\,\eta\,(kL) = k^2\eta \tag{214}$$

As Kiehn said, such groups of extended Conformal Lorentz Transformations (CLT) have already been found by Fock who demonstrated these preserve the propagating discontinuity.

Investigating the holomovement of fermions, we have seen how a generalized graded Lorentz transformation imposed by $L^{(2)}$ on the u–quark brings forth a nonlinear relativistic isospin decomposition of the fermion. We will regard this contribution proportional to $((v^2+c^2)/(c^2-v^2))\lambda_3$ in equation (93) as an expression of the thermodynamic physical environment of nuclear power. Yet, those graded Lorentz transformations induced in the ground space which generates the Clifford algebra are still classically linear. Therefore, to understand the nonlinearity in cosmic conformal evolution, we have to go deeper into the non-equilibrium thermodynamics as is connected with conformal nonlinear Lorentz transformations, that is, fluid dynamics of wakes together with SU(3) transformations of time-space. Let us go slowly into this. We first have to set the frame.

Consider as a standard basis of time-space the small three-dimensional subspace of the geometry $Cl_{3,1}$

$$\mathrm{U} \overset{def}{=} span\{e_4,\, e_{123},\, j\} \qquad \text{having signature } \{-\,-\,-\} \tag{215}$$

measured by time as ct, space volume v and space-time volume u. With the angular function $\theta(x)$ depending at least on none and at most on 16 real numbers, we expect a transformation by λ_2 as given by

$$\begin{aligned}
L_2(\theta)(ct\,e_4) &= (\cos\theta/2)(ct\,e_4) + (\sin\theta/2)(u\,j) \\
L_2(\theta)(v\,e_{123}) &= v\,e_{123} \\
L_2(\theta)(u\,j) &= -(\sin\theta/2)(ct\,e_4) + (\cos\theta/2)(u\,j)
\end{aligned} \tag{216}$$

and this can easily be verified by using the explicit geometric algebra representation of $L^{(2)}$. Thus, group actions induced by the isospin generator λ_2 establish a harmonic relation among the time-like triple e_4, e_{123} and j and therefore act like planar Euclidean rotations about the fixed directed volume $v\,e_{123}$. The rotation appears in a quasi Euclidean plane spanned by e_4 and j. As we have eight generators and 16 base elements on which the group elements can act, it may seem reasonable to construct a classification scheme in order to understand those actions.

First we differ between harmonic and hyperbolic relations as induced by the isospin generators. Second we list up n-ary relations the base units must be members of. For example, HY^2 denotes a binary relation or a pair, and $\mathrm{HY}^2(\lambda_1/\,e_3,\,e_4)$ shall tell us that the isospin component λ_1 induces a hyperbolic binary relation – that is, a classic boost - between the space-like unit vector e_3 and the time-like unit vector e_4, and $HY^2(\lambda_3/\ e_2,\ e_4)$ shall express that the isospin component λ_3 establishes a classic boost connection between the space-like unit vector e_2 and the time-like unit vector e_4. We also have boosts that relate the directed time unit e_4 with a space-like space-time area e_{34} and a ternary boost relation that connects e_4 with e_1, space-like space-time area e_{24} and time-like spatial bivector e_{12}. We denote this by the word $HY^4(\lambda_6/\ e_1,\ e_4,\ e_{12},\ e_{24})$.

We concentrate on the time-space U generated by time like base units $\{e_4,\ e_{123},\ j\}$ because they are connected by trigonal $SU(3)$ rotations that are purely harmonic. We can give a list of transformational relations containing boosts, rotations and fixations:

$HY^2(\lambda_1/\ e_4,\ e_3)$, $HY^2(\lambda_3/\ e_4,\ e_2)$, $HY^2(\lambda_4/\ e_4,\ e_{34})$,
$HY^4(\lambda_6/\ e_4,\ e_1,\ e_{12},\ e_{24})$; $HA^2(\lambda_2/\ e_4, j)$, $HA^2(\lambda_5/\ e_4,\ e_{123})$, $HA^1(\lambda_7/\ e_4)$,

$HY^4(\lambda_1/\ e_{123},\ e_{23},\ e_{24},\ e_{124})$, $HY^4(\lambda_3/\ e_{123},\ e_{34},\ e_{23},\ e_{134})$,
$HY^2(\lambda_4/\ e_{123},\ e_{14})$, $HY^2(\lambda_6/\ e_{123},\ e_{234})$, $HA^2(\lambda_7/\ e_{123}, j)$; $HA^1(\lambda_2/\ e_{123})$,

$HY^2(\lambda_1/\ j,\ e_2)$, $HY^2(\lambda_3/\ j,\ e_3)$, $HY^4(\lambda_4/\ j,\ e_1,\ e_{13},\ e_{124})$,
$HY^2(\lambda_6/\ j,\ e_{134})$; $HA^1(\lambda_5/\ j)$. (217)

From those we deduce triple relations with fixed volume e_{123}.

$HA^3(\lambda_2/\ e_4,\ e_{123}, j)$, $HA^3(\lambda_5/\ e_4,\ e_{123}, j)$, $HA^3(\lambda_7/\ e_4,\ e_{123}, j)$. (218)

We realize that such triple relations might also be induced within the Cartan subalgebra $\{e_1,\ e_{24},\ e_{124}\}$ correlated with space ch_1 since they are connected by flavour rotations too.

False: $HA^3(\lambda_2/\ e_1,\ e_{24},\ e_{124})$ or $HA^3(\lambda_5/\ e_1,\ e_{24},\ e_{124})$.

But this is false! Since the elements are transformed by nonlinear trigonometric coefficients. They take a form which brings about a complete penta-relation with fixed e_1

$$\begin{aligned}
L_2(\theta)e_1 &= e_1\\
L_2(\theta)e_{24} &= B^2e_{24} + A^2e_{124} + ABe_{34} - ABe_{134}\\
L_2(\theta)e_{124} &= A^2e_{24} + B^2e_{124} - ABe_{34} + ABe_{134}\\
L_2(\theta)e_{34} &= -ABe_{24} + ABe_{124} + B^2e_{34} + A^2e_{134}\\
L_2(\theta)e_{134} &= ABe_{24} - ABe_{124} + A^2e_{34} + B^2e_{134}
\end{aligned} \tag{219}$$

Thus we classify $HA^5(\lambda_2/\ e_1,\ e_{24},\ e_{124},\ e_{34},\ e_{134})$. This relation is extremely important when we elaborate conformal Lorentz transforms of the pure states in ch_1. From this classification we can learn two things. First, the isospin components λ_2, λ_5, λ_7 which constitute a Clifford algebra-form of the $su(2)$ put together in a triple relation those units of the Cartan algebra and those pure states from the colour space ch_1 that are transformed into each other, classically, by trigonal flavour rotations of the *su*(3). Therefore we can classify those by denoting

$$\mathrm{HA}^3(su_{\mathrm{Cl}}(2)/\, e_1, e_{24}, e_{124}) \text{ and } \mathrm{HA}^3(sl_{\mathrm{Cl}}(3)/\, e_1, e_{24}, e_{124}) \qquad (220)$$

without specifying the trigonal rotations in $su_{\mathrm{Cl}}(2)$ and $sl_{\mathrm{Cl}}(3)$.

Second, we learn that the time-space triple $\{e_4, e_{123}, j\}$ belongs to the ternary relation too, that is, we have

$$\mathrm{HA}^3(su_{\mathrm{Cl}}(2)/\, e_4, e_{123}, j) \text{ and } \mathrm{HA}^3(sl_{\mathrm{Cl}}(3)/\, e_4, e_{123}, j). \qquad (221)$$

Therefore the three-dimensional time-space is correlated with the Cartan subalgebra by the strong force. It is dynamically equalized with the dynamics of the pure states. It partakes in isospin-, generalized Lorentz transformations and flavour-rotations with equal rights. We can also expect some ternary relation which solely concerns the ground space $\mathbb{R}^{3,1}$ and essentially classifies the classically known Lorentz boost plus one Euclidean rotation as depicted by

$$\mathrm{HY}^2(\lambda_1/\, e_4, e_3),\ \mathrm{HY}^2(\lambda_3/\, e_4, e_2),\ \mathrm{HA}^2(\lambda_2/\, e_2, e_3) \qquad (222)$$

Terms [111) denote a typical $sl_{\mathrm{Cl}}(2)$ triple $\mathrm{HY}^3(\mathrm{sl}_{\mathrm{Cl}}(2)/\, e_2, e_3, e_4)$ correlating with form

$$\begin{aligned} L_2(\theta)e_2 &= (\sin\theta/2)e_2 + \cos(\theta/2)e_3 \\ L_1(\phi)e_3 &= (\cosh\phi/2)e_3 - (\sinh\phi/2)e_4 \\ L_3(\varphi)e_4 &= (\cosh\varphi/2)e_4 - (\sinh\varphi/2)e_2 \end{aligned} \qquad (223)$$

and base units with signature $\{+ + -\}$. Obviously, there exist three such standard structures of the algebra $sl_{\mathrm{Cl}}(2)$ and three isomorphic images of the positive spingroup $\mathrm{Spin}^+(2,1) \simeq SL(2)$.

Now, in topological thermodynamics we regard the Lorentz transformations as given by (223) as *basis frames* acting on exact differential 1-forms of 'differential position' vectors. We select some set of base variables, here $\{x_3, x_4, x_5\}$, which are transformed in a linear collineation by a Lorentz matrix. More generally we probe the infinitesimal collineation. For the above variety, the collineal system in $Cl_{3,1}$ (Maple Clifford) appears as

$$L\left|dx^k\right\rangle = \left|\sigma^k\right\rangle \qquad k = 3, 4, 5 \qquad (224)$$

The cosmological field space is a vector space defined by the invertible basis frames that map a vector of exact differentials $|dx^k\rangle$ into a vector of exterior differential 1-forms $|\sigma^k\rangle$. The vector of 1-forms $|\,\sigma^k\rangle$, in some cases like in (223), can be regarded as an array of exact differentials and L is integrable. But this need not always be the case. Let us slowly apply these considerations to the equation (215). We begin with

$$L_2(\theta)\begin{vmatrix} cdt \\ dv \\ du \end{vmatrix}\Big\rangle = \begin{vmatrix} (\cos\theta/2)cdt + (\sin\theta/2)du \\ dv \\ -(\sin\theta/2)cdt + (\cos\theta/2)du \end{vmatrix}\Big\rangle = \left|\sigma^k\right\rangle \tag{225}$$

where du denotes the infinitesimal space-time volume to the director j and $dv\ e_{123}$ the differential directed space volume. Extending the Lorentz transformation, the group requirement is fixed by the conformal constraint (214). The new 1-forms brought about by the extended L may be integrable only over a limited domain.

We obtain a *'shears of expansion'* conformal transformation compatible with a flavour rotation dynamics by introducing the extended Lorentz matrix with an arbitrary dimensionless function $\theta(x_i)$ of all the independent variables x_i, $i = 1, \ldots, 16$. It shall be represented by some arc length up to scaling.

$$L(\theta) = \begin{bmatrix} \dfrac{1}{\cos\theta/2} & 0 & \dfrac{\sin\theta/2}{\cos\theta/2} \\ 0 & 1 & 0 \\ -\dfrac{\sin\theta/2}{\cos\theta/2} & 0 & \dfrac{1}{\cos\theta/2} \end{bmatrix} \tag{226}$$

This matrix also preserves the Minkowski metric, as there exists a Frobenius normal form with a matrix D such that

$$L(\theta)^{-1}\eta\, L(\theta) = \eta' \quad \text{and} \quad D^{-1}\eta' D = \eta \tag{227}$$

providing a linear similar basis. Calculating the differential 1-forms $|\sigma^k\rangle$ by $L(\theta)$ we obtain

$$\left|\sigma^k\right\rangle = \begin{bmatrix} \frac{1}{\cos\theta/2} & 0 & \frac{\sin\theta/2}{\cos\theta/2} \\ 0 & 1 & 0 \\ -\frac{\sin\theta/2}{\cos\theta/2} & 0 & \frac{1}{\cos\theta/2} \end{bmatrix}\begin{vmatrix} cdt \\ dv \\ du \end{vmatrix}\Big\rangle = \begin{vmatrix} \frac{1}{\cos\theta/2}cdt + \frac{\sin\theta/2}{\cos\theta/2}du \\ dv \\ -\frac{\sin\theta/2}{\cos\theta/2}cdt + \frac{1}{\cos\theta/2}du \end{vmatrix}\Big\rangle \tag{228}$$

The shears expansion is an important example of a Lorentz transformation with non exact differentials, as we have for example $\sigma^1 = \frac{1}{\cos\theta/2}cdt + \frac{\sin\theta/2}{\cos\theta/2}du$. If the angular function would be a constant, the new time-coordinate were well defined. Integration would yield

$$ct' = \frac{1}{\cos\theta/2}ct + \frac{\sin\theta/2}{\cos\theta/2}u \tag{229}$$

and $d\sigma^1 = 0$. [1] It is a distinguished situation when the angular function is a constant, but it is generally no reasonable assumption under primordial thermodynamic conditions of strong interaction. If it is a constant, the vector of 2-forms of field intensities vanishes, of course. However, in primordial events far from thermodynamic equilibrium the function θ(x) is not a constant but rather a function that depends on arc length. We still have an excellent situation with inexact differential 1-forms in a closed algebra, but non-vanishing 2-forms, and 4-forms. Congruously Kiehn (2008, p. 231) has denoted the closure 2-forms as $d|\,\sigma^k\rangle$. Here they take a quite analogous shape. The difference consists in the two facts that now we establish a direct relation between time and space-time volume, and further that relation conforms to the strong interaction dynamics:

$$d\left|\sigma^k\right\rangle = \left|\begin{matrix} -\frac{\sin\theta/2}{\cos^2\theta/2}\,cdt\wedge d\theta - \frac{1}{\cos^2\theta/2}\,du\wedge d\theta \\ dv \\ +\frac{\sin\theta/2}{\cos^2\theta/2}\,cdt\wedge d\theta - \frac{1}{\cos^2\theta/2}\,du\wedge d\theta \end{matrix}\right\rangle \neq 0 \qquad (230)$$

The differential 4-forms are exterior products $d\sigma^k \wedge d\sigma^k$ and do not vanish either. Kiehn has found a surprising similarity of such 2-forms (intensities) of a Lorentz transformation bringing forth shear rotation/expansion with the thermodynamics of fluids depending on some basic classes of infinitesimal arc length.

The fundamental rigor is known since long. But it has been developed much further by Kiehn. We shall repeat here its main arguments and calculations by using Maple Clifford, in order that the readers get a feeling for the phenomenology. As was said, Fock had demonstrated that conformal Lorentz transformations also preserve the propagating discontinuity. Kiehn vitalized this discovery by some very important explorations of non-equilibrium systems and irreversible processes of topological evolution. He showed that thermodynamic instabilities were absolutely consistent with principles of relativity. To understand those research denouements we have to expand our view towards fluid dynamics and similar things and do some step away from high energy physics. Though, we must confess, this step leads us rather deeper into the domain of subnuclear matter. In our approach to primordial space we introduce manifolds of conformal Lorentz transformations as special classes of nonlinear basis frames as Kiehn has done. This has the advantage that such frames are suitable to describe a pregeometric environment where fields underlie the symmetries of the strong force and at the same time allow for constitution of tangential discontinuities of the directed space-time volume and a directed time measure which is both wrapped and deconvoluted like in a harmonic wake.

Once in a science bookstore in Thessaloniki 1978, I acquired the book on »Physical Fluid Dynamics« by D. J. Tritton (1977) and saw in it photographs of various phenomena of convective flow and the repeating images of the Kelvin-Helmholtz instability. Then I was convinced of the lively meaning of those nonlinear phenomena for quantum physics. But I could not yet tell anyone why. Because I started to comprehend the outstanding importance of macroscopic spinors, only after I had begun to study Clifford algebra in the 90s. Then I

---

1 It is legitimate and has some advantages here to restrict calculations to matrices Mat(3, ℝ) within the Majorana algebra Cl3,1 ~ Mat(4, ℝ) since the base elements {Id, e4, e123, j} form a closed subalgebra in Cl3,1.

found, many of us were still astonished about the recurrence of some types of wake patterns and detected that there existed no satisfactory explanation for their appearance (Browand 1986, 117, quoted after Kiehn 2008). But R. M. Kiehn notices that "*from a topological viewpoint it was remarkable how often flow instabilities and wakes take on one or another of two basic scroll patterns*". Those were depicted by the images of the Kelvin-Helmholtz - and the Raleigh-Taylor instability. The first geometric analyses of the Navier Stokes equations had not given satisfying results where the explanation of those phenomena is concerned. But in 1993 Kiehn had developed a theory so far that he could explain the repeated occurrence of those patterns. Surprisingly the most important contributions to the solution came from Frenet's theory of space curves and Cartans method of moving frames. Let us repeat the main arguments that led to mathematical families for two types of instabilities.

The Rayleigh–Taylor-Instability is a hydrodynamic instability. It gives rise to an exponentially growing disturbance at the boundary layer between two fluids with different mass. It is a two-phase instability and occurs when there is acceleration between two layers of fluids having different densities. There exist many appearances of this phenomenon reaching from hydrodynamic wakes to the gestalt of exploding supernovae. Its most prominent form is that of a mushroom. The Kelvin-Helmholtz-Instability then again appears in the shear force at the boundary between two fluid layers with different velocity. It discloses the forms of shear rolls, wakes and turbulent clouds. We identify it on the gas planets Jupiter and Saturn just as well as in interstellar clouds. To obtain a reasonable model of the dynamics one had to answer the question under what conditions the fluid boundary layer is metastable towards small disturbances. The most important contribution to the answer was given by differential geometry. Only the minimal surfaces associated with a harmonic vector field will turn out robust enough to build up a wake without decaying. Those will dissipate minimal energy.

Consider Frenet space curves in the plane in parametric form

$$\frac{dx}{ds} = \sin Q(s)$$
$$\frac{dy}{ds} = \cos Q(s) \tag{231}$$

constituting a unit vector t(s) with positive signature for all s:

$$\text{t(s)} = \left| \begin{matrix} d_x s \\ d_y s \end{matrix} \right\rangle \tag{232}$$

Differentiation of this unit vector with respect to arclength s leads to the classic expression for the Frenet curvature:

$$d_s\,\text{t(s)} = \kappa n(s) = (d_s Q)\left| \begin{matrix} \cos Q(s) \\ -\sin Q(s) \end{matrix} \right\rangle \tag{233}$$

with main normal unitvector[2] $n(s)$ and curvature $\kappa = d_s Q$. The curvature classifies some fundamental space curves. For $\kappa = 1$ integration yields a circle, for $\kappa = 1/s$ we obtain a logarithmic spiral, $\kappa = s$ results in a clothoid or Cornu-Fresnel spiral. What if the curvature takes the form $\kappa = s^n$ for integer n so that the argument function is given by

$$Q = \frac{s^{n+1}}{n+1} \text{ with curvature } \quad \kappa = d_s Q = s^n \tag{234}$$

This question could be answered by the assistance of the PC

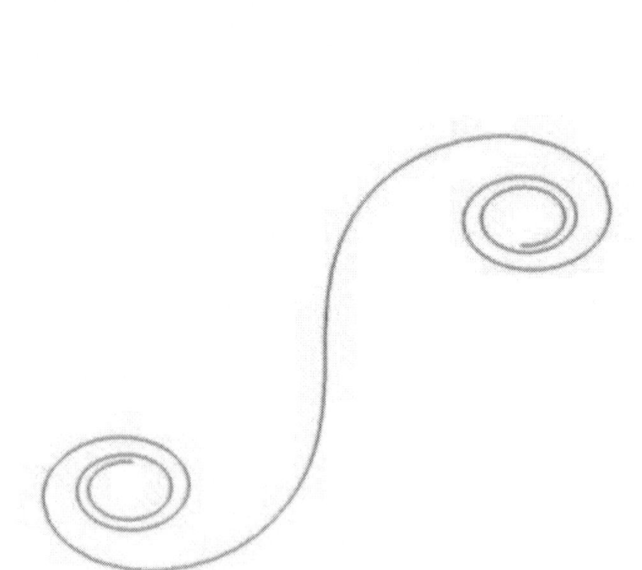

Figure 22. The Cornu-Fresnel spiral.

and respectively Maple. We essentially get two different families of space curves the integrals of which are built up by Fresnel functions for the odd integers and by Lommel functions for even integers. The first family brings forth spirals that look like the waves of the Kelvin-Taylor instability. The second gives rise to mushrooms in the shape of the Kelvin-Helmholtz instability. It is important to mention that already the Cornu spiral turned out to have great explanatory power in phenomena of wave diffraction (Figure 22).

A realistic image of a Kelvin Helmholtz instability can be obtained by the choice of

$$Q = \frac{1}{\cos^2 s} \tag{235}$$

2 'Hauptnormaleneinheitsvektor'

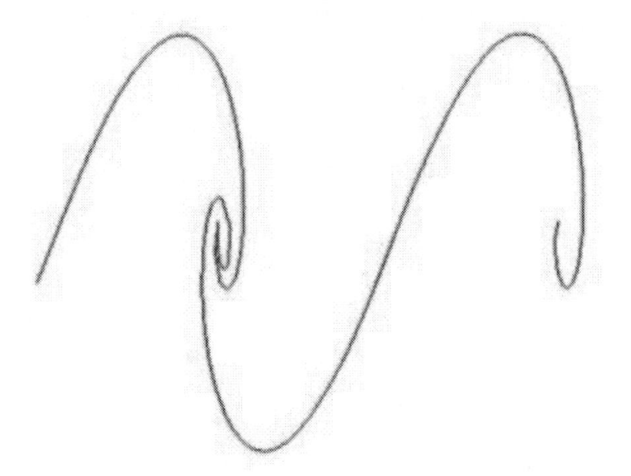

Figure 23. The Kelvin-Helmholtz instability.

$$Q = \frac{\tan s}{\cos s} \text{ generates solutions [3] as below} \tag{236}$$

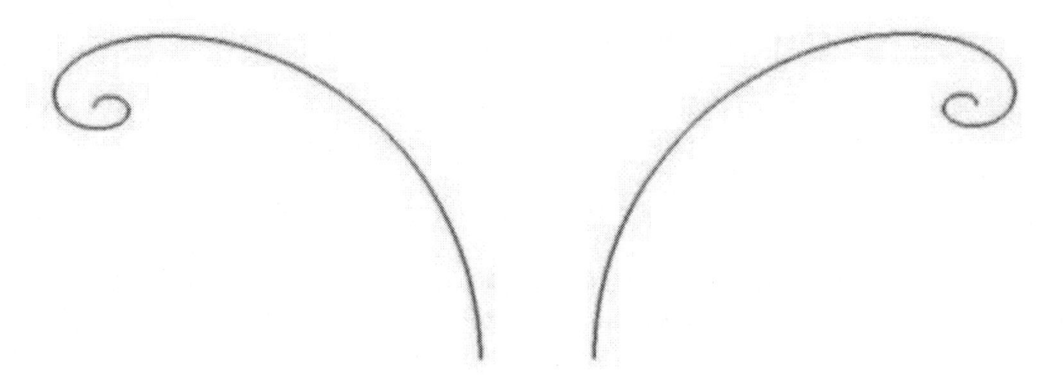

Figur 24. Rayleigh-Taylor instability, Mushroom.

Following the consideration of Kiehn, we construct from (226) the extended Lorentz transformation

$$P(\theta) = \frac{1}{\cos\theta/2} L(\theta) = \begin{bmatrix} \frac{1}{\cos^2\theta/2} & 0 & \frac{\sin\theta/2}{\cos^2\theta/2} \\ 0 & 1 & 0 \\ -\frac{\sin\theta/2}{\cos^2\theta/2} & 0 & \frac{1}{\cos^2\theta/2} \end{bmatrix}. \tag{237}$$

[3] Each figure consists of several branches. Those are special solutions to the ODE for different integration constants. The Rigor was carried out by Maple.

Surprisingly we find these functions which generate the fundamental instabilities as matrix entries of those nonlinear conformal Lorentz transformations that generate rotations and expansions. For consider the angular function $\theta \simeq s$ being proportional to arclenght. It yields the extended Lorentz transformation $L(\theta)$ with a conformality factor $k^2 = 1/\cos^2 \theta/2$ so that equation (214) should be holding: $P^{\dagger}\eta P = k^2 \eta$. Clearly, Kiehn has developed this conformal example following not only his own experience in nuclear power and hydrodynamics, but also Fock's work on »Space, time and gravitation«. He had not yet seen a necessity to apply his conformal Lorentz frame directly to the relation between observed time ct and space-time volume u. In the approach presented here, there occurs for the first time a direct correlation between the structure of space-time and the structure of the elementary particles. The instability of primordial space-time as a relation between time and space-time volume is directly reflected in the instability of the confined fermions. Since any globally occurring phenomena of, say, Rayleigh-Taylor instabilities in the triangle {time, volume, spactime volume} is to be correlated with the same instability in the Cartan-spaces which finally determines upon the structure and energy content of quarks. It is not surprising that the physicists working in the Manhattan project, in reactor experiments and other departments developing nuclear energy projects made the experience and were always convinced that cosmology and high energy physics had an immediate connection. This relation can be found confirmed even in user-friendly books like »Das Schicksal des Universums« (*The Fate of the Universe*) by Günter Hasinger (2007, chapters 2 and 3). Also Kiehn who had the opportunity to do nuclear experiments using Fermi's original reactor CP1 and who is one of the few people still alive who have witnessed atmospheric nuclear explosions, obviously knows about the indispensableness of such a mathematical connection between cosmology and subnuclear phenomenology (Kiehn 2008, 431).

In the present writing there is, for the first time, proven the direct correspondence and transferability between the dynamic, conformal space-time structure and the likewise metastable structure of fermions within the standard model. For consider a conformal Lorentz transformation of the directed space-time-volume having form

$$\left| \begin{matrix} cdt' \\ v' \\ u' \end{matrix} \right\rangle = P(\theta) \left| \begin{matrix} cdt \\ v \\ u \end{matrix} \right\rangle \tag{238}$$

This can and must immediately be applied to the pure states in ch_1. Under an $L_2(\theta)$-Lorentz transformation the fermion f_3 transforms like in equation (195) with equal factor ¼ $(b^2\text{-}a^2)$ = ¼(cos 2θ) for e_{24}, e_{124} and ½ ab = ¼ (sin 2θ) for e_{34}, e_{134}. Recall, those four are space-like areas and volumes with positive signature constitutive for states $|u\rangle$, $|d\rangle$, $|s\rangle \in Cl_{3,1}$. Therefore we could use just two symbols $dx = dx_{10} = dx_{13}$, $dy = dx_{11} = dx_{14}$ for infinitesimal coefficients. The differential 1-form of a pure state $|u\rangle = f_3$ is transformed under Lorentz non-conformal conditions by a moving Frenet-Cartan Frame in accordance with

$$L_2(2\theta)d|f_3\rangle \Rightarrow |\sigma^k\rangle = \left.\begin{bmatrix} b^2 & a^2 & ab & -ab \\ a^2 & b^2 & -ab & ab \\ -ab & ab & b^2 & a^2 \\ ab & -ab & a^2 & b^2 \end{bmatrix}\begin{vmatrix} dx_{10} \\ dx_{13} \\ dx_{11} \\ dx_{14} \end{vmatrix}\right\rangle \tag{239}$$

where we disregarded the fixated idempotent $\eta = ½\,(1 - e_1)$. Verify that the determinant of the matrix $L_2(2\theta)$ equals $a^2 + b^2$ which is unity. Be aware also that equation (239) does not show a Mat(4, $\mathbb{R}$)-representation of the Clifford algebra $Cl_{3,1}$ but a representation of the subspace spanned by $\{e_{24}, e_{124}, e_{34}, e_{134}\}$ in the fourfold real ring which is also by matrices Mat(4, $\mathbb{R}$). This colour-subspace has its own metric tensor which is the unit matrix

$$\chi = Id_4 = \begin{bmatrix} 1 & 0 & 0 & 0 \\ 0 & 1 & 0 & 0 \\ 0 & 0 & 1 & 0 \\ 0 & 0 & 0 & 1 \end{bmatrix}$$

A Lorentz transformation $L_2(\theta)$ preserves this quasi Euclidean norm. That special symmetric frame (239) is a consequence of the fact that a flavour rotation can be transposed from the triple of flavours to the triple of directed space-time-areas and -volumes. Now, this frame is not aware of discontinuities as long as the quantity a is equal to sin θ and so on. However, as soon as we take into account the conformal Lorentz transformation (237), (238), the moving frame for fermions, - omitting the fixed idempotent, - has to be

$$|\sigma^k\rangle = \left.\begin{bmatrix} \frac{1}{\cos^4\theta} & \frac{\tan^2\theta}{\cos^2\theta} & \frac{\tan\theta}{\cos^3\theta} & -\frac{\tan\theta}{\cos^3\theta} \\ \frac{\tan^2\theta}{\cos^2\theta} & \frac{1}{\cos^4\theta} & -\frac{\tan\theta}{\cos^3\theta} & \frac{\tan\theta}{\cos^3\theta} \\ -\frac{\tan\theta}{\cos^3\theta} & \frac{\tan\theta}{\cos^3\theta} & \frac{1}{\cos^4\theta} & \frac{\tan^2\theta}{\cos^2\theta} \\ \frac{\tan\theta}{\cos^3\theta} & -\frac{\tan\theta}{\cos^3\theta} & \frac{\tan^2\theta}{\cos^2\theta} & \frac{1}{\cos^4\theta} \end{bmatrix}\begin{vmatrix} dx_{10} \\ dx_{13} \\ dx_{11} \\ dx_{14} \end{vmatrix}\right\rangle \tag{240}$$

This moving frame for the fermion in an unstable space-time-field, let us denote it by R, is invertible and preserves the metric tensor χ of the colour subspace.

It would be interesting to find out if such an instability generating extended Lorentz transformation could act similar as the isospin conjugation on a fermion, namely such that its space-time structure (202) is preserved? This is the quest for an existence proof. Can R act as an affine isospin transformation of the type L_2 and preserve the properties of the fermion f_3?

Consider a deformation of the standard fermion $|u\rangle = f_3$ as is brought forth by L_2. It should have a form

$$|\upsilon\rangle = \tfrac{1}{4}(Id - e_1) + x(e_{24} - e_{124}) + y(e_{34} - e_{134}) \tag{241}$$

In order to preserve the quality of 'primitive idempotent ' it is necessary that

$$y = \pm\sqrt{\tfrac{1}{16} - x^2} \tag{242}$$

Therefore we substitute for the $|\upsilon\rangle$–state the variety:

$$|\upsilon\rangle = \left| \begin{matrix} x \\ x \\ \pm\sqrt{\frac{1}{16} - x^2} \\ \pm\sqrt{\frac{1}{16} - x^2} \end{matrix} \right\rangle \tag{243}$$

which results in a conformal Lorentz transformation by $R(\theta)$

$$R(\theta)|\upsilon\rangle = \left| \begin{matrix} \left(\frac{1}{\cos^4\theta} + \frac{\tan^2\theta}{\cos^2\theta}\right)x \\ \left(\frac{1}{\cos^4\theta} + \frac{\tan^2\theta}{\cos^2\theta}\right)x \\ \pm\left(\frac{1}{\cos^4\theta} + \frac{\tan^2\theta}{\cos^2\theta}\right)\sqrt{\frac{1}{16} - x^2} \\ \pm\left(\frac{1}{\cos^4\theta} + \frac{\tan^2\theta}{\cos^2\theta}\right)\sqrt{\frac{1}{16} - x^2} \end{matrix} \right\rangle \tag{244}$$

The function $S(\theta) = \frac{1+\sin^2\theta}{\cos^4\theta}$ (245)

discloses the singularities of the conformal Lorentz transformation. It looks like

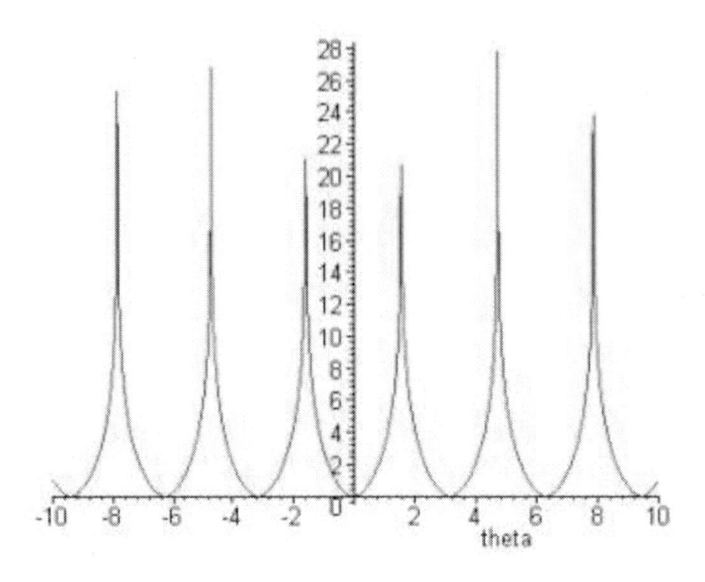

with singularities at $\theta = \pm n\pi$ with integer n

Figure 25. The Function $\log(S(\theta))$.

The variable $\varsigma(\theta)x$ is indeed the magnitude of the directed space-time area e_{24} and the directed space-time volume e_{124} after a conformal isospin transformation similar to $L_2(\theta)$. The conformal frame for the time and space-time volume during the same event of transformation is given by the matrix in (239). While the fermion space field runs through such singularities the magnitude of the space-time volume and measured time can experience an instability where the arrow of time cannot be preserved while the space-time volume oscillates.

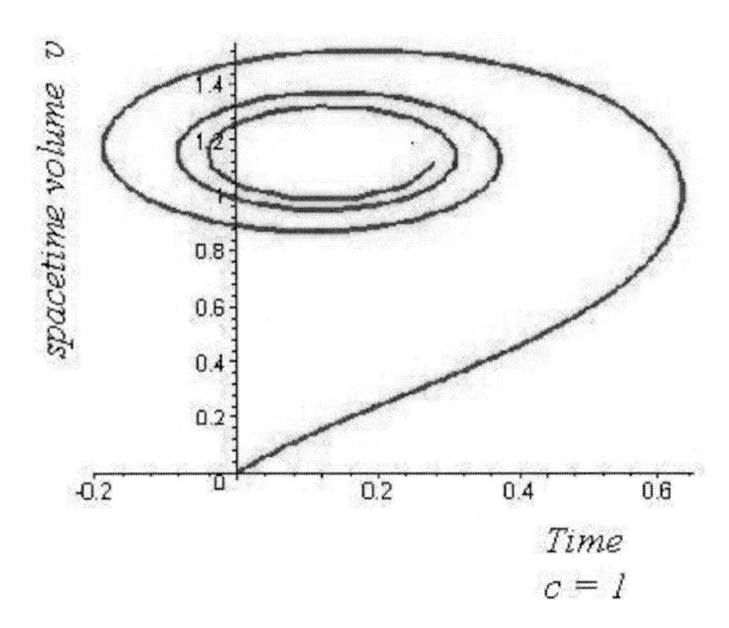

Figure 26. Time – space-time phaseplot at a singularity.

Whereas the space-time volume v appears to the observer as oscillating in the positive domain, the arrow of time can be reverted relative to its measure close to the origin.

The above result is obtained by integrating the frame equations (237), (238) under the assumptions (245) with positive arclenght going from zero to 0,43 π.

$$\theta = 2s \qquad \text{and} \qquad P(2s) \quad \text{and} \qquad Q(s) = \frac{1}{\cos^2 s}$$

and

$$\frac{du}{ds} = \sin Q(s) \qquad \frac{dt}{ds} = \cos Q(s) \tag{246}$$

with two integration constants set zero and light velocity c=1. The rigor was carried out by Maple. Starting with arclenght zero and decreasing it until to $s = -0{,}43\,\pi$:

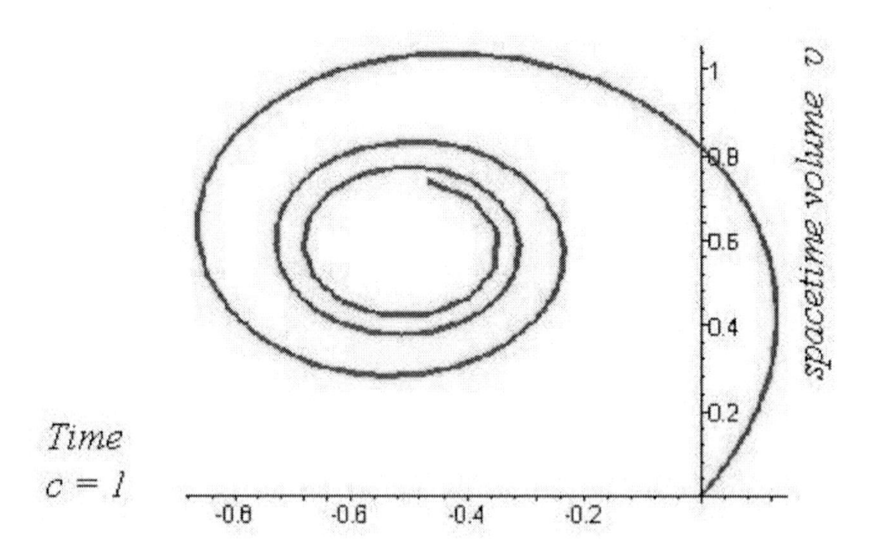

Figure 27. Time – space-time phaseplot approaching singularity with negative arclength.

Here too there exists a domain where time is reverted while space-time volume is positive.

Colour Space Analysis

A colour space is the smallest closed subalgebra with a geometric product that contains pure states. It is constituted by the scalar, a space-like unit vector, a space-like bivector and a space-time 3-volume. We take ch_1 as a standard representative of all manifolds which contain isomorphic images of the lattice of orthogonal primitive idempotents. We can regard colour spaces as fundamental vessels for fermions with baryon number 1/3. We may look at those as three different flavours, but each of the same colour plus a lepton. Or we can regard them as one flavour with three different colours plus a lepton. Which will be more appropriate, I cannot decide on mere theoretical grounds. It needs more insight into the mathematics and better empirical indicators. Perhaps those exist already and I just do not see them. I have decided intuitively that a maximal Cartan subalgebra gives rise to three flavours although there are important voices who vote against that first alternative (Vargas and Torr 2003). Whatever we may decide, the colour space contains no timelike base unit and not any algebraic operation in it can deconvolute time. The space ch_1 contains pure states and mixtures of such states which are not pure. It contains idempotents, four of which are primitive, but no more, and it encloses patterns. Whichever pattern is stored in that colour space, the dynamics stems from a dynamic process outside. The outer is the origin of the dynamics. The inner, however, holds synchronous patterns in which temporal order occurs as wrapped time. The differential equations that take effect in the colour space describe landscapes rather than evolutions. Yet they may be seen as sources or templates of evolutionary processes. They influence the outer dynamics and absorb it. It would be interesting to pose the question, since we have

$$ch_\chi = {}^4\mathbb{R} \qquad \text{with} \qquad \chi = 1, \ldots, 6 \qquad (247)$$

if the elements of a colour space can be given by components which are multiples of commuting grade-free, base-free unitals where each has equal right in the fourfold real ring. Because then we could ask if there existed, similarly as in vector spaces with homogeneous grade one, a geometric product between ∇ and a vector field $\phi \in ch_1$ which could be decomposed into an inner product $\nabla.\phi$ representing a divergence and an exterior product $\nabla\wedge\phi$ giving something like the curl of a colour field.[4]

The standard model's manifolds involve a nonlinear dynamic system in sixteen dimensions. To arrive at equations of motion implies considerable comprehension and effort. We must begin with the most fundamental, ostensibly simplest case. We have to find differentials, differential operators and forms in the smallest, closed, coloured multivector fields. We have to reconstruct calculus for the colour spaces. These are essentially constituted by maximal abelian subalgebras of the Clifford algebra.

In his habilitation treatise, Bertfried Fauser explained various ways to derive Clifford algebras by generators and relations, by factorization and cliffordization.

Table 3. Dirac group of colour space ch_1

Id	e_1	e_{24}	e_{124}
e_1	Id	e_{124}	e_{24}
e_{24}	e_{124}	Id	e_1
e_{124}	e_{24}	e_1	Id

There he pointed out that most applications actually use a Graßmann basis which is obtained by antisymmetrization of the Clifford base elements. This does not work out here in exactly the manner we are used to. Contracting the exterior product to colour spaces, we are confronted with an interesting situation, since all base units of a maximal Cartan subalgebra commute. That is, we observe a "Dirac group" (Shaw 1995) for the base elements which is isomorphic with the Klein 4-group. So we have to calculate under the rule of the group table (table 3).

REMEMBERING WEYL

By componentwise multiplication these base elements generate the Klein 4-group. Therefore we say: *the Dirac group of colour spinor spaces is the "Klein'sche Vierergruppe".* In this group each element forms its own class. It is interesting to note that Hermann Weyl established a one-one correspondence between the base units of the Klein 4-group, the Pauli matrices and the quaternion group by what he called skillful gauging (*geschickte Eichung*). Historically, the difference between an abelian basis $\{e_1, e_{24}, e_{124}\}$ and a non commutative

[4] In any case, I have determined to denote a (multi)vector field in a colour space as 'colour field'.

euclidean basis $\{\sigma_1, \sigma_2, \sigma_3\}$, where the σ_i are standard Pauli matrices, has been disclosed by Herman Weyl as what he denoted the difference between *vector*-space and *ray*-space (*Vektorkörper* and *Strahlenkörper*). Namely, when ξ is a vector, e.g. in Hilbert space, then a single quantum state must not be defined by a vector, but rather by the totality of all vectors $\{\varepsilon\xi\}$ where ε is any complex number with unit absolute value $|\varepsilon| = 1$. Two unitary transformations should have to be considered equal $U \cong \varepsilon U$ because they bring upon *the same rotation in the ray space* (Weyl 1931/1967, 161ff.). In this way, he could show that after a proper gauging, abelian groups may possess multidimensional, irreducible unitary representations which are not abelian. Anyway, the Klein-4 group provides a simple algebraic type of triality by its cyclic product rule: the product of any two base elements $\neq Id$ gives the third (table 3). Following the example of Weyl, the Pauli algebra can be conceived as a *Ray-representation* of the colour space algebra by defining $U(1) = 1$, $U(e_1) = \sigma_1$, $U(e_{24}) = \sigma_2$ and $U(e_{124}) = \sigma_3$. However, we shall not go deeper into those old questions here. But we keep in mind:

This special array of space-like blades with different grades reminds us of projective equivalence where information about signs, and in this sense orientations is lost, so that the exterior product becomes symmetric. Then the Graßmann algebra of blades appears like reduced to a lattice (Hestenes 1988). It is this forgetful operation which establishes an important relation between geometric Clifford algebra and lattice theory. Recall, we have four primitive idempotents, - interpreted as fermions, - in ch_1 which constitute a lattice. Further each element in ch_1 can be written as a linear combination of those four mutually annihilating primitive idempotents and can be represented as an element in the fourfold real ring ${}^4\mathbb{R}$.

Let us consider any two elements ξ, $\psi \in ch_1$. They may be heterogeneous multivectorial elements, that is, sums of blades with different grades. But they appear like commuting vectors of homogeneous grade 1 because of group table 3. Their grade actually does not make a difference. The exterior product can be extended over these colour fields in ch_1. Now, both, the Clifford product $\xi\psi \in ch_1$ and the wedge product $\xi \wedge \psi$ are elements of ch_1, and both commute. That is, we have

$$\xi\psi = \psi\xi \qquad \text{and} \qquad \xi \wedge \psi = \psi \wedge \xi \tag{248}$$

In this sense the information about their direction is lost. Since we have $\xi\psi \neq \xi \wedge \psi$, we may regard $\xi\psi - \xi \wedge \psi$ as the interior product which turns out as restricted to the linear space spanned by $\{Id, e_1, e_{24}\}$. The apparent loss of orientation is the origin of the phenomenon of time-wrap.

To insert differential operators into the colour spaces and to derive calculus in a proper way, we have to respect this fact and respectively the effects of table 3 on the algebra of derivatives. Consider the first order differential operator

$$\nabla \overset{def}{=} e_1 \frac{\partial}{\partial x} + e_{24} \frac{\partial}{\partial y} + e_{124} \frac{\partial}{\partial z} \tag{249}$$

also called static colour Dirac operator acting on a colour field

$$\phi = (\varphi, u, v, w) \qquad \text{or} \qquad \phi = \varphi + e_1 u + e_{24} v + e_{124} w \tag{250}$$

With constant scalar φ using table 3 we verify

$$\nabla\phi = \left(\frac{\partial u}{\partial x} + \frac{\partial v}{\partial y} + \frac{\partial w}{\partial z}\right) Id + \left(\frac{\partial v}{\partial z} + \frac{\partial w}{\partial y}\right) e_1 + \left(\frac{\partial u}{\partial z} + \frac{\partial w}{\partial x}\right) e_{24} + \left(\frac{\partial u}{\partial y} + \frac{\partial v}{\partial x}\right) e_{124} \tag{251}$$

which represents nothing else than the familiar decomposition

$$\nabla\phi = \nabla.\phi + \nabla\wedge\phi \tag{252}$$

The magnitude

$$\nabla.\phi = \frac{\partial u}{\partial x} + \frac{\partial v}{\partial y} + \frac{\partial w}{\partial z} \qquad \text{is the divergence and} \tag{253}$$

$$\nabla\wedge\phi = \left\{\left(\frac{\partial v}{\partial z} + \frac{\partial w}{\partial y}\right), \left(\frac{\partial u}{\partial z} + \frac{\partial w}{\partial x}\right), \left(\frac{\partial u}{\partial y} + \frac{\partial v}{\partial x}\right)\right\} \tag{254}$$

represents the *colour field curl.* Cartan discussed the meaning of these magnitudes in a chapter on linear representations of the group of rotations in Euclidean space (Cartan 1937, ch. IV) in a section on "special cases; Harmonic polynomials" and "applications" where those sets give the velocity field for a rigid body displacement. We should compare this expression with the classic curl in Euclidean 3-space where the + would have to be replaced by a minus sign:

$$\overset{euclid}{curl}\,\phi \overset{def}{=} \left\{\left(\frac{\partial v}{\partial z} - \frac{\partial w}{\partial y}\right), \left(\frac{\partial u}{\partial z} - \frac{\partial w}{\partial x}\right), \left(\frac{\partial u}{\partial y} - \frac{\partial v}{\partial x}\right)\right\}$$

This makes the difference between Euclidean and colour space.

BASIC LAWS OF PATTERN FORMATION

A standard colour field is given base-free by a real quadruple

(i) $\phi = \{\varphi, x, y, z\}$ or equivalently by a multivector

(ii) $\phi = \varphi + x e_1 + y e_{24} + z e_{124}$ in the Clifford algebra $Cl_{3,1}$. (255)

It can be written as a linear combination of the orthogonal primitive idempotents $\phi = a_1 f_1 + a_2 f_2 + a_3 f_3 + a_4 f_4$ with

$$a_1 = \varphi + x + y + z$$
$$a_2 = \varphi + x - y - z$$
$$a_3 = \varphi - x + y - z$$
$$a_4 = \varphi - x - y + z \qquad (256)$$

We assume that the dynamic process in the whole space of $Cl_{3,1}$, - what we may call the system's equation of motion, - causes pattern convolution in the colour field. That will result in functions $F(\phi)$ of ϕ which can be written in the form of geometric Taylor polynomials. The elementary building block will be a term like ϕ^n which is

$$\phi^n = (a_1 f_1 + a_2 f_2 + a_3 f_3 + a_4 f_4)^n \quad \text{which is equal to} \qquad (257)$$

$$a_1^n f_1 + a_2^n f_2 + a_3^n f_3 + a_4^n f_4 \quad \text{because of orthogonality}$$

relations $f_i f_j = \delta_{ij} f_i$. This gives back a multivector in colour space ch_1 with the following symmetric solution

(i) $\phi^n = (\varphi, x, y, z)$ with

(i) $\varphi = \frac{1}{4}(a_1^n + a_2^n + a_3^n + a_4^n)$

$$x = \tfrac{1}{4}(a_1^n + a_2^n - a_3^n - a_4^n)$$
$$y = \tfrac{1}{4}(a_1^n - a_2^n + a_3^n - a_4^n)$$
$$z = \tfrac{1}{4}(a_1^n - a_2^n - a_3^n + a_4^n) \qquad (258)$$

Those can be labelled as the structure equations characteristic for the colour space. They seem utterly harmless in this context, but can have dramatic consequences as soon as we consider the primitive idempotents which are without doubt constitutive for the properly arrangement of space-time in connection with uncertainty. It cannot be guaranteed that the formation of an idempotent field is sharp, in the sense that the idempotency relation does not show the slightest deviation from $ff = f$. Rather we expect an uncertainty which is essentially given by Planck's constant h.

The Unstable Almost Pure State

The central idea is here that we tolerate small conformal deformations of the primitive idempotents. The most simple though very general form is to permit deviations δ such like

$$\psi_3 = \tfrac{1}{2}(Id - (1-\delta)e_1)\tfrac{1}{2}(1 + e_{24}) \text{ for very small } \delta. \qquad (259)$$

This results in an equation

$$\psi_3^n = \left(\frac{1}{4} - \frac{n}{8}\delta + \frac{n(n-1)}{16}\delta^2 - \binom{n}{3}\frac{1}{32}\delta^3 \ldots\right) f_3 \tag{260}$$

for the n't power of the *almost primitive idempotent* evaluated up to order 3. Therefore a systematic conformal deviation will cause a primitive idempotent to shrink if $0 < \delta << 1$ and to inflate if $\delta < 0$. The same happens if we do not decrease the scale of e_1 by a factor $1 - \delta$ but instead shrink the scalar ¼ Id by some small amount as in equation

$$\psi_3 = \tfrac{1}{2}(Id - \delta Id - e_1)\tfrac{1}{2}(Id + e_{24}) = f_3 - \tfrac{1}{4}\delta(Id + e_{24}) \tag{261}$$

The deformation of the idempotent is determined by a single constant depending on the small deviation δ, namely $\psi_3 = k.f_3$.

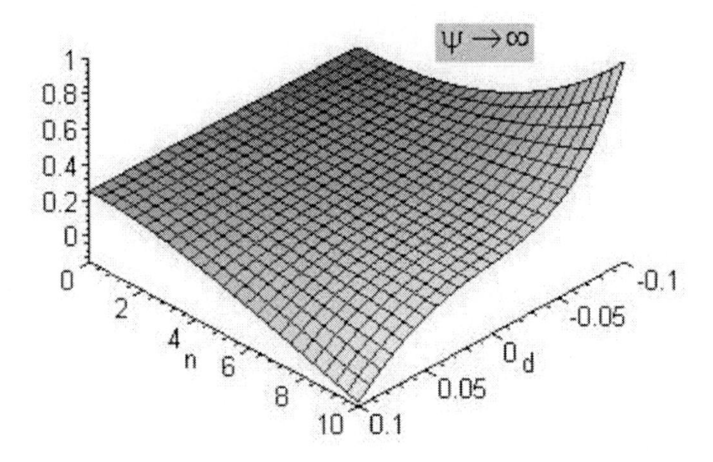

Figure 28. Iteration ψ^n of the 'almost idempotent' ψ.

Timespace Spinor- and Colour Spinor Space

Cartan had based the concept of a spinor on the existence and construction of isotropic vectors (x_1, x_2, x_3) in Euclidean 3-space. Those have zero length although their components are not trivially null. As we put

$$\begin{aligned} x_1 &= \xi_0^2 - \xi_1^2 \\ x_2 &= i(\xi_0^2 + \xi_1^2) \\ x_3 &= -2\xi_0\xi_1 \end{aligned} \tag{262}$$

we get

$$x_1^2 + x_2^2 + x_3^2 = 0 \tag{263}$$

(Cartan 1937, section 52 “Definition”).

and the pair of quantities (ξ_0, ξ_1) constitutes a spinor. Everything else we can use can be taken in nowadays terms from Pertti Lounesto’s book »Clifford Algebras and Spinors«.

A timespace is given here by triples $U = \{(c\,dt\,e_4, v\,e_{123}, u\,j)\}$, a colour space by forms $ch_1 = \{\zeta = (\frac{1}{4})(x_1 + x_2 e_1 + x_3 e_{24} + x_4 e_{124})\}$. Both spaces have something important in common: both triples of base units, - here (e_4, e_{123}, j), there (e_1, e_{24} ,e_{124}), - are connected by trigonal flavour rotations of the $SU(3)$ and respectively the $SL(3)$. But between both spaces there is also a very important difference.

We calculate the geometric product which, in analogy with (151), should vanish. It is obvious that form

$$\begin{aligned}\varsigma\varsigma &= \tfrac{1}{16}(x_1^2 + x_2^2 + x_3^2 + x_4^2)Id - \tfrac{1}{8}(x_1x_2 + x_3x_4)e_1 + \\ &+ \tfrac{1}{8}(x_1x_3 + x_2x_4)e_{24} - \tfrac{1}{8}(x_1x_4 + x_2x_3)e_{124} \equiv 0\end{aligned} \tag{264}$$

allows only for the trivial solution $x_1 = x_2 = x_3 = x_4 \equiv 0$.

However, if we consider timespace elements

$$\mathrm{Y} = x_1e_4 + x_2e_{123} + x_3 j \tag{265}$$

we obtain from the geometric product the scalar quadratic form

$$\mathrm{Y\,Y} = (-x_1^2 - x_2^2 - x_3^2)Id \tag{266}$$

which provides a spinor solution of the type (262). The timespace brings on spinors and angular momentum algebra.

A timespace spinor can be written as Cartan spinor

$$\mathrm{Y} = (\xi_0^2 - \xi_1^2)e_4 + i(\xi_0^2 + \xi_1^2)e_{123} - 2\xi_0\xi_1\, j \tag{267}$$

giving square

$$\mathrm{Y\,Y} \equiv 0 \tag{268}$$

Therefore the timespace spinor is associated with an isotropic timespace 3-multivector constituted by directed time, volume and space-time-volume. It is connected with a measure of time-volume which is so to say a time-space-time-volume without volume. This isotropic time-like 3-vector covers the time-convolution and de-convolution from time to volume and space-time-volume by the strong force.

We have found out what can give rise to spinors in the timespace component of geometric space-time algebra, we have shown how zero isotropic vectors are naturally derived from colour spaces. We solved this problem which seemed so unsolvable for a while. The answer has already been given implicitly by Chevalley (1954) and also by Lounesto. The colour spinors are derived from the pure states of colour spaces by constructing minimal left ideals with appropriate elements from the Clifford algebra.

$$S_k = Cl_{3,1} f_k \qquad k = 1,\ldots,4 \tag{269}$$

where the f_k are the familiar primitive idempotents. Thus a colour spinor will be connected with an isotropic multivector from a minimal left ideal in the space-time algebra. But the problem of timespace dynamics is more complex. It involves quaternions and Fueter's differential equation.

Chapter 30

TIMESPACE ANALYSIS

The timespace is a root structure, a natural quaternion component of the Minkowski algebra. We know another such root structure, namely the bivector space of $Cl_{3,1}$ which is giving rise to the angular momentum algebra and to Lorentz transformations. In order to understand this a little better, we have to become familiar with quaternions. For the time space provides a Fueter equation for a natural force field that has a certain similarity to the electromagnetic field, but has deeper cosmogonic significance. It is perhaps helpful to consolidate a more generous meaning of the Dirac operator.

Marmolejo-Olea and Mitrea (2004, p. 92ff.) define the differential operator D of order $k \geq 1$ as the linear map

$$D: \quad C^{\infty}(M, \mathsf{E}) \longrightarrow C^{\infty}(M, \mathsf{F}) \tag{270}$$

where E, $\mathsf{F} \longrightarrow M$ are two smooth, Hermitian vector bundles and M a smooth, compact, boundaryless, orientable Riemannian manifold. Those are so to say the ideal templates into which we can ascribe less smooth events. D is local inasmuch as it does not extend the support. The proponents use "star", "t" and "bar" to denote complex adjunction, transposition and respectively complex conjugation. So they have

$$\overline{D}u = \overline{D\bar{u}} \quad D^{*} = \left(\overline{D}\right)^{t} = \overline{\left(D^{t}\right)} \qquad \text{as expected.} \tag{271}$$

By $\langle .,. \rangle$ there is denoted the *bilinear* pointwise pairing in the fibers of a Hermitian vector bundle. In particular $|u|^2 = \langle u, \bar{u} \rangle$. The meaning of the principal symbol of D is in the map

$$\sigma(D;.): \quad T^{*}M \setminus 0 \longrightarrow Hom(\mathsf{E}, \mathsf{F}) \tag{272}$$

defined in local differential topology by

$$\left(\sigma(D;\xi)e\right)_{\alpha} = \sum_{|\gamma|=k} a_{\gamma}^{\alpha\beta}(i\xi)^{\gamma} e_{\beta} \tag{273}$$

if $\left(Du\right)_{\alpha} = \sum_{|\gamma|=k} a_{\gamma}^{\alpha\beta} \partial^{\gamma} u_{\beta}$ + lower order terms, in local coordinates.

With the first-order differential operator is associated a self-adjoint, nonpositive definite, second order differential operator, the D-Laplacian, a map $-D^*D$: E ⟶ E. Marmolejo-Olea and Mitrea denote D as a differential operator of the Dirac type if its Laplacian has the property that

$$\underset{\xi \in T^* \setminus 0}{\forall} \quad \sigma(D^* D; \xi) \text{ is scalar, real and invertible.} \tag{274}$$

This definition is non-standard, less restrictive and, in particular, does not require D to be elliptic, i.e. have an invertible symbol. This latter case allows for calculations in more general nonsmooth manifolds which we shall need when discontinuous manifolds are transposed from the time line to the directed volume and from there to the space-time volume.

The principal symbol of the quaternions is $\mathbb{H}$ given by three letters i, j, k with generating relations

$$\begin{aligned} &i^2 = j^2 = k^2 = ijk = -1 \\ &ij = -ji = k \\ &jk = -kj = i \\ &ki = -ik = j \end{aligned} \tag{275}$$

The quaternion algebra $\mathbb{H}$ is sometimes also written $\mathbb{R}(i, j, k)$ (Morales-Luna 2008) to indicate it is an algebra over the real field. It is an associative, non-commutative division algebra and a 4-dimensional real vector space. The conjugate map $\mathbb{H} \longrightarrow \mathbb{H}$ takes quaternion

$$h = x_0 + x_1 i + x_2 j + x_3 k \text{ to } \bar{h} = x_0 - x_1 i - x_2 j - x_3 k. \tag{276}$$

The norm of quaternion h is $|h| = \sqrt{h\bar{h}}$ and its inverse is

$$h^{-1} = \tfrac{1}{|h|^2}\bar{h} \text{ provided the norm is not zero.} \tag{277}$$

There are several embeddings that preserve addition and multiplication. The one we prefer to use is the embedding into the real Majorana algebra Mat(4, $\mathbb{R}$) $\hookrightarrow \mathbb{H}$

$$M\!: h = x_0 + x_1 i + x_2 j + x_3 k \mapsto \begin{bmatrix} x_0 & -x_1 & x_3 & -x_2 \\ x_1 & x_0 & -x_2 & -x_3 \\ -x_3 & x_2 & x_0 & -x_1 \\ x_2 & x_3 & x_1 & x_0 \end{bmatrix} \tag{278}$$

Clearly, as such it must be a subalgebra of $Cl_{3,1}$. Consider

$$D \overset{def}{=} \partial_{x_0} + i\partial_{x_1} + j\partial_{x_2} + k\partial_{x_3} \qquad \text{so that} \qquad (279)$$

$$D^* = -\partial_{x_0} + i\partial_{x_1} + j\partial_{x_2} + k\partial_{x_3} \qquad (280)$$

from this derive the Laplacian

$$-D^* D = DD^* = \Delta, \text{ so } D \text{ is of the Dirac type.} \qquad (281)$$

There are two representations of the principal symbol. The first is derived from the even subalgebra $Cl^+_{3,1}$ and is meanwhile well known in the science community. The second, however, the importance of which has not yet been realized, is given by the three timelike base units $\{e_4, e_{123}, j\}$ of the Clifford algebra $Cl_{3,1}$. The quaternion j can indeed be identified with the director $j = e_{1234} \in Cl_{3,1}$. Quite obviously, those three symbols represent i, j, k and generate as their finite Dirac group the smallest Hamilton group, the group of Quaternions Q_8 or Dicyclic group Dic_2 with subgroups 3 x $\mathbb{Z}_4$ and $\mathbb{Z}_2$. So they constitute a negative definite quadratic space $\mathbb{R}^{0,3}$. This is indeed isomorphic with the subspace generated by the timelike bivectors $\{e_{12}, e_{13}, e_{23}\}$ A very close, but at the moment still somewhat enigmatic, relation with classical function theory can be established. We shall have to go into that.

The graded vector $Y = x_0 + ct\, e_4 + v\, e_{123} + u\, e_{1234} \in \mathcal{U}$ is a quaternion.[1] In order to see very clearly that we are now dealing with two time measures we can multiply the timespace element Y from the right by e_4. We obtain the element

$$z = x_0\, e_4 - ct + v\, j - u e_{123}$$

In accord with the defining equations (271 to 273) we wish to represent the differential operator D in timespace by

$$D \overset{def}{=} \partial_x + e_4\, \partial_y + e_{123}\, \partial_z + j\, \partial_u \qquad j = e_{1234} \qquad (282)$$

This is of course a differential operator inhomogeneous in the dimension degrees. The definition may separate an observer from the observed timespace and thus makes a difference between a time measure x_1 that is entangled in the strong force and an outer observation time x_0. Historically, the shortened or *pure* differential operator neglecting the scalar part is due to Hamilton

$$D_p \overset{def}{=} i\partial_y + j\partial_z + k\partial_u$$

[1] Note, a bijective relation with the biparavector $y = x + y\, e_{12} + z\, e_{13} + u\, e_{23}$ can be established.

and can be represented by the bivector Dirac operator

$$\nabla_p = e_{12}\frac{\partial}{\partial y} + e_{13}\frac{\partial}{\partial z} + e_{23}\frac{\partial}{\partial u}$$

This is also in a 1-1 correspondence with

$$\nabla_p = e_4\frac{\partial}{c\partial t} + e_{123}\frac{\partial}{\partial v} + j\frac{\partial}{\partial u}$$

In the following paragraphs we shall use the correspondence between quaternions and complex pairs and respectively bicomplex algebra. The Dirac operator takes the form (279) in order to stay in the mathematics framework. But we keep in mind the special representation (282) by directed time and volumes. A quaternion denoted by a four dimensional vector

$$h = x + yi + zj + uk \in \mathbb{H}$$

can be identified with the complex pair

$$(c_1, c_2) = (x + iy, z + iu) \qquad \text{since we have} \tag{283}$$

$$c_1 + c_2 j = x + iy + jz + ku \tag{284}$$

Note that we also use $j = e_{1234}$ for the director in $Cl_{3,1}$. This is possible! Otherwise we would have been forced to denote the quaternions by *i, k, l* instead of *i, j, k*. Anyway, we shall run short of symbols when we need the unit imaginary which is neither e_4 nor e_{123}, but rather, as in Maple, an »I«. Consider a bounded domain Ω in $\mathbb{H}$. A quaternion function can be written in the form

$$\Psi = \Psi_1 + k\Psi_2 \in C^1(\Omega) \tag{285}$$

It is left regular in the sense of Fueter (1935) if

$$D\Psi = \frac{\partial \Psi}{\partial x} + i\frac{\partial \Psi}{\partial y} + j\frac{\partial \Psi}{\partial z} + k\frac{\partial \Psi}{\partial u} = 0 \text{ on } \Omega . \tag{286}$$

Therefore consider a function in bicomplex space

$$\begin{aligned} &\psi(q) = (\varphi(c_1, c_2), \phi(c_1, c_2)) = \\ &= (\psi_1(x, y, z, u) + i\psi_2(x, y, z, u), \psi_3(x, y, z, u) + i\psi_4(x, y, z, u)) \\ &q = (c_1, c_2), c_1 = x + iy, \qquad c_2 = z + iu \end{aligned} \tag{287}$$

The function $\psi(q)$ is supposed to solve the Laplace equation

$$\partial_x^2\psi + \partial_y^2\psi + \partial_z^2\psi + \partial_u^2\psi = 0 . \quad (288)$$

Solutions are found by factorization of the form

$$\Delta \times \psi = 0 \quad (289)$$

where $\times$ denotes the quaternionic multiplication. As the Laplacian is a scalar quaternion the multiplication merely applies to each single component of ψ. Stefan Rönn (2008) has carried out the factorization

$$\Delta = (\partial_x + i\partial_y, \partial_z + i\partial_u) \times (\partial_x - i\partial_y, -\partial_z - i\partial_u) \quad (290)$$

Introducing the new operator $L_q = (\partial_x + i\partial_y, \partial_z + i\partial_u)$ we get

$$\Delta = L_q \times L_q^* \quad (291)$$

a factorization in terms of the Cauchy-Riemann operators

$$D_1 = \partial_x + i\partial_y \qquad \text{and} \qquad D_2 = \partial_z + i\partial_u \quad (292)$$

So we have

$$L_q = (D_1, D_2) \qquad \text{and} \qquad L_q^* = (D_1^*, -D_2) \quad (293)$$

$$\Delta = (D_1, D_2) \times (D_1^*, -D_2) \quad (294)$$

Because of $L_q \times L_q^* = L_q^* \times L_q$ we deduce from (186), (188) in logic form

$$\Delta \times \psi = 0 \Leftarrow (L_q \times \psi = 0) \vee (L_q^* \times \psi = 0) \quad (295)$$

Thus we can solve the Laplace equation by solving either disjunct of the antecedent. Focusing on the first equation, we rewrite in ${}^2\mathbb{C}$

$$L_q\psi = 0 \Leftrightarrow (D_1, D_2) \times (\varphi, \phi) = 0 \Leftrightarrow$$
$$(D_1\varphi - D_2\phi^*, D_2\varphi^* + D_1\phi) = 0 \Leftrightarrow$$
$$(D_1\varphi - D_2\phi^* = 0) \,\&\, (D_2\varphi^* + D_1\phi = 0) \quad (296)$$

Therefore the first case of the disjunction (295) is equivalent to the simultaneous validity of equations

$$D_1\varphi = D_2\phi^* \text{ and } \qquad D_2\varphi^* = -D_1\phi \qquad (297)$$

Substituting the Cauchy-Riemann operators (292) and functions φ, ϕ as have been introduced in (287), Stefan Rönn obtains a representation in the Majorana algebra Mat(4, $\mathbb{R}$)

$$\begin{aligned}
&\partial_x\psi_1 - \partial_y\psi_2 - \partial_z\psi_3 - \partial_u\psi_4 = 0\\
&\partial_x\psi_2 + \partial_y\psi_1 + \partial_z\psi_4 - \partial_u\psi_3 = 0\\
&\partial_x\psi_3 - \partial_y\psi_4 + \partial_z\psi_1 + \partial_u\psi_2 = 0\\
&\partial_x\psi_4 + \partial_y\psi_3 - \partial_z\psi_2 + \partial_u\psi_1 = 0
\end{aligned} \qquad (298)$$

It was this system of partial differential equations Fueter derived in 1935 starting off with the quaternion form

$$\partial_x\psi + i\times\partial_y\psi + j\times\partial_z\psi + k\times\partial_u\psi = 0$$

Treatment of the second equation in (295) leads to the conjugate Fueter equations derived by Lanczos in 1929.

$$\begin{aligned}
&\partial_x\psi_1 + \partial_y\psi_2 + \partial_z\psi_3 + \partial_u\psi_4 = 0\\
&\partial_x\psi_2 - \partial_y\psi_1 - \partial_z\psi_4 + \partial_u\psi_3 = 0\\
&\partial_x\psi_3 + \partial_y\psi_4 - \partial_z\psi_1 - \partial_u\psi_2 = 0\\
&\partial_x\psi_4 - \partial_y\psi_3 + \partial_z\psi_2 - \partial_u\psi_1 = 0
\end{aligned} \qquad (299)$$

In his »Clifford Algebras and Spinors« (p. 74) Pertti Lounesto was still sceptical, where the utility of quaternion calculus was concerned. Among the quaternion functions of quaternion variables ψ: $\mathbb{H} \to \mathbb{H}$, as he said, "*many generalizations are uninteresting, the classes of functions are too small or too large*". Nevertheless he carried out a few calculations and gave some hints that turned out rather far sighted. Jiří Měska (1984) was more enthusiastic when he wrote "*the development of the quaternionic analysis started only recently and – in comparison with the complex analysis – the theory is certainly still underdeveloped with its best years to come. It took a long time to find a suitable generalization of the C-R equation to have a nice, distinguished class of regular functions*". Měska gave a very substantial contribution by proving that the zero sets of complex Fueter functions are null surfaces , i.e. "*roughly, the 'gradients' at the points of the surface lie in the null cone*". The number of important contributions is now on the increase. For example, in 2005 Perotti clarified some of the relation between holomorphicity and regularity in Fueters sense, and recently Rönn by his »*Bicomplex algebra and function theory*« gave the issue a very precise push foreward by deriving new classes of regular and conjugate regular functions. At present the number of papers is substantial. But Lounesto's scepticism seems to be appropriate. Better than to concentrate too much effort on quaternion analysis is it to first investigate the combinatoric properties of Clifford algebras and then work out a general function theory for them.

We shall now proceed with the analysis of timespace in Clifford algebra.

Chapter 31

PHENOMENOLOGY OF TIMEWRAP AND STRUCTURE WAVES

Once we begin to understand how the laws of the mind and the logic of quantum physics go together, we enter a real fantastic situation. After all human beings are nature and the mind functions like nature. Nature has given to thought something which is for thought only. The creative mind has incredible degrees of freedom. But it can make proper use of them only, if it faces the absolute restrictions of thought. Those are what Piaget called operational structures of thought, or what R. D. Laing denoted as morphogenetic structures of experience. The institution of orientation, both mental and material, brings logic and geometry into the centre of analysis. The Process of Nature and the human mind are bound to those root structures of geometry and logic. It is in this prehistoric bondage where we must see the origin of the anthropic principles and their implications for the validity of our theories.

We have worked out in the bygone years how the standard model of physics can be seen as a natural outcome of a process of reorientation in the geometry of Minkowski space-time. But the resulting Lie group which is guiding the process has to be graded by necessity. The consequent mathematical problems are considerable. But they are worth to be solved. For we gain a new comprehension of the whole system of motion, for the whole array of the degrees of freedom material fields can use. Understanding these structural restrictions actually widens our view of the available degrees of freedom. Realizing this, nature all by itself, provides the purposive mathematical instruments. Nature, it seems to us, once had a special freedom of energetic motion in outer space which it gradually had to hand over to the inner space of subnuclear matter. That was the freedom to interact in the strong force and to create the stars and planets, the interstellar clouds and galaxies. Today we can only try to reconstruct those archaic events by forcing nature into conditions where she reconstructs particles and shows us the laws that we believe she had to follow.

We found out that the Pauli principle offers matter an escape from chaos. In the strong force field this means that pure states can escape equality by changing colour and flavour. Such transitions in colour space rotate a fermion in a space involving three different grades. This movement takes a distinct base unit of a maximal Cartan subalgebra to another such base unit with higher grade and from there still to a third one with still one step higher, - when modelled in a standard frame, - before it takes it back to where it started off. This is a fundamental degree of freedom in strong force fields. That we observe an escape of hadrons from spherical asymmetry by exploiting a colour degree of freedom, that we can excite a change of family within the multiplets of the $SU(3)$ is logically and geometrically equivalent to a degree of freedom to move in quaternion timespace. This follows naturally from the structure of the Clifford algebra of the Minkowski space-time which is a structure of both matter and mind. The conclusion that there is a quaternion timespace with a typical $SU(3)$ symmetric freedom of motion imposed from the outer geometry, is both necessary and sufficient for the existence of colour. It would be surprising and by experience highly improbable if nature would not make use of that freedom. It is therefore that we have to investigate the fundamental features of timespace and to some degree also the quaternion- and respectively timespace analysis.

We have heard that the set of regular functions in quaternion analysis is incredibly large. We can add to this that the degrees of freedom of the quantum field, once it unfolds from the ground state are likewise almost unlimited although it is (i) bound to the orientation symmetries and (ii) creates its own boundary conditions. But many of the phenomena we can count on can already be figured out on the ground of what we know from quantum physics. We just have to eke it.

The timespace is spanned by

$$\mathcal{U} = span_F\{e_4, j, e_{123}\} \text{ with } F \text{ equal to } \mathbb{R} \text{ or } \mathbb{C} \tag{300}$$

Base units e_4, j, e_{123} are in 1-1 correspondence with base quaternions i, j, k, the unit imaginary is denoted by symbol »I«. The Dirac like operator is given by

$$D \overset{def}{=} \partial_\tau + e_4\,\partial_t + j\,\partial_u + e_{123}\,\partial_v \quad \text{as in (170)} \tag{301}$$

We have to differ between

Definition [τ]: τ representing diachronous real time (302)

Definition [t]: t representing synchronous wrap time (303)

The restricted pure quaternion differential operator

$$\nabla_p = e_4 \frac{\partial}{c\partial t} + j\frac{\partial}{\partial u} + e_{123}\frac{\partial}{\partial v}$$

incorporates synchronous wrap time in an anti-euclidean space. The 3-dimensional anti-Euclidean space $\mathbb{R}^{0,3}$ consists of vectors

$$\xi = x\,a_1 + y\,a_2 + z\,a_3 \qquad \text{with a negative scalar product} \tag{304}$$

$$\xi \cdot \xi = -(x^2 + y^2 + z^2)$$

and the standard orthonormal basis $\{a_1, a_2, a_3\}$ obeys the rules

$$a_1^2 = a_2^2 = a_3^2 = -1 \quad \text{and}$$

$$a_1 a_2 = -a_2 a_1,\ a_1 a_3 = -a_3 a_1,\ a_2 a_3 = -a_3 a_2 \tag{305}$$

which are also satisfied by the unit quaternions *i, j, k* and likewise by

$$a_1 = e_4,\ a_2 = j,\ a_3 = e_{123} \in Cl_{3,1} \tag{306}$$

The identity $i\,j\,k = -1$ equivalent to $a_1\,a_2\,a_3 = -1$ implies that the real algebra $\mathbb{H} = \mathbb{R} \oplus \mathbb{R}^{0,3}$ can be generated by a proper subspace $\mathbb{R}^{0,2}$ of $\mathbb{R}^{0,3}$. Namely, each quaternion can be represented in the form $\xi = x\,a_1 + y\,a_2 + z\,a_1\,a_2$ with $a_1\,a_2 = a_3$. This fact is responsible for the existence of the bicomplex representation, and it led to the predication of the number field $\mathbb{H}$ to be *an algebra of the negative quadratic form* $\xi \rightarrow -x^2-y^2-z^2$ although it cannot be identified with the Clifford algebra $Cl_{0,3}$ which has dimension 8 and must be associated with the direct sum $\mathbb{H} \oplus \mathbb{H}$. Pertti Lounesto had depicted this correspondence by the multiplication table for the two-fold quaternion field. He emphasized the Clifford algebra with negative definite metric generated by $\mathbb{R}^{0,3}$ is the universal object in the category of quadratic algebras with quadratic form $\xi \rightarrow -x^2-y^2-z^2$.

Table 4. Multiplication table for $Cl_{0,3} \simeq \mathbb{H} \oplus \mathbb{H}$

$Cl_{0,3}$	$\mathbb{H} \oplus \mathbb{H}$
a_1, a_2, a_3	$(i,-i),(j,-j),(k,-k)$
a_{23}, a_{31}, a_{12}	$(i,i),(j,j),(k,k)$
a_{123}	$(-1,1)$

In the category of associative algebras there are two 3 dimensional ones with negative definite metric isomorphic with each other and thus with $\mathbb{H} = \mathbb{R} \oplus \mathbb{R}^{0,3}$. They are related by the map $a_1 \rightarrow a_1$, $a_2 \rightarrow a_2$ and $a_3 \rightarrow -a_3$. When we investigate a Schrödinger like equation of motion, we have to consider the pure restricted quaternion differential operator together with the properties of space $\mathcal{U}$.

The timespace Dirac operator $D \overset{def}{=} \partial_\tau + e_4\,\partial_t + j\,\partial_u + e_{123}\,\partial_v$ can be seen as a square root of the scalar Laplacian

$$\Delta = \partial_\tau^2\psi + \partial_t^2\psi + \partial_u^2\psi + \partial_v^2\psi \tag{307}$$

The free structure wave ψ solves the homogeneous Fueter equations

$$\partial_\tau\psi_1 - \partial_t\psi_2 - \partial_u\psi_3 - \partial_v\psi_4 = 0$$

$$\partial_\tau \psi_2 + \partial_t \psi_1 + \partial_u \psi_4 - \partial_v \psi_3 = 0$$
$$\partial_\tau \psi_3 - \partial_t \psi_4 + \partial_u \psi_1 + \partial_v \psi_2 = 0$$
$$\partial_\tau \psi_4 + \partial_t \psi_3 - \partial_u \psi_2 + \partial_v \psi_1 = 0 \qquad (308)$$

Acevedo, López-Bonilla and Sánchez-Meraz have demonstrated that we need only very few suppositions to bring the source-free Maxwell equations in the above form. Essentially we need a geometric separation between electric and magnetic component and a complex vector $F = cB + iE$ that can be subjected to duality rotations. They use a Debye potential to reformulate the C-R relations as in the above bicomplex derivations.

Let us now investigate the pure quaternion differential operator ∇_p. Consider the restricted pure Laplacian

$$\Delta_p = \partial_t^2 \psi + \partial_u^2 \psi + \partial_v^2 \psi \qquad (309)$$

and define timespace impulses $|p_t, p_u, p_v\rangle$ or $p(q) = p(t, u, v)$ depending on the pure quaternion q. Extend the *correspondence principle* to the quaternion of wrap time, space-time volume and space volume. This is possible as the three components contribute likewise to the quadratic form. We postulate that the three quantities $p_t ct, p_u u, p_v v$ have the physical dimension of an action [kg m^2 s^{-1}]. It is tempting to go deeper into the question of structure constants. That would lead us to the thermic de Broglie wave length, mathematical models of the Bose-Einstein condensate, to the Debye model, to the Grüneisen parameter which correlates the change of frequency with the change of volume $\partial(\ln \omega)/\partial(\ln v)$. But all this would propose a restriction of the phenomenology. At first we should not need more than a free wave equation. Considering such a simple equation gives us a classical hint. It is clear that we are modelling now in a rather close proximity to thermodynamics. Actually this Fueter equation should show us a bridge to thermodynamics. But it is surprising enough to realize the quantization of timespace that results from the algebra alone. Consider a Schrödinger equation of the form

$$I\hbar \frac{\partial}{\partial \tau} \psi(q, \tau) = -\frac{\hbar^2}{2m} \Delta_p \psi(q, \tau) \qquad (310)$$

It is obvious that the algebra design advises us to investigate the quantization of angular momentum. For a real time first order wave equation the space U provides a template with a synchronous or wrapped time which annihilates causality by a simple mechanism. To see how this works we must first realize that unit quaternions can be gained from the Pauli matrices by multiplying them with the unit imaginary. Notice that the Pauli matrices representing standard base units for Euclidean 3-space satisfy the angular momentum algebra. We conclude that vice versa a multiplication of the quaternion timespace unit (multi)vectors by the unit imaginary I will bring forth its own angular momentum algebra.

The bijective map $\begin{matrix} \sigma_1 & \sigma_2 & \sigma_2 \\ \downarrow & \downarrow & \downarrow I \\ i & j & k \end{matrix}$ sends the Pauli matrices to quaternions.

Thus the angular momentum algebra satisfied by the base units of Euclidean 3-space is turned over into the quaternion algebra. Likewise it is natural to regain the Euclidean space from the anti-Euclidean. That is, it should be possible to send the anti-Euclidean time space back to a Euclidean 3-space. In order to prepare the desired quantization, we first notice that the quaternions can be rotated in a triangle without altering their algebraic features. This trigonal rotation degree of freedom is actually used during strong interaction. We rotate and thereby calibrate the triangle such that we obtain a Euclidean basis as follows

$$\begin{matrix} j & k & i \\ \downarrow & \downarrow & \downarrow \\ \varepsilon_1 = I\,j, & \varepsilon_2 = -I\,e_{123}, & \varepsilon_3 = I\,e_4 \end{matrix} \tag{311}$$

The ε satisfy the following commutation relations

$$[\varepsilon_1, \varepsilon_2] = -2I\,\varepsilon_3, \quad [\varepsilon_1, \varepsilon_3] = 2I\,\varepsilon_2, \quad [\varepsilon_2, \varepsilon_3] = -2I\,\varepsilon_1 \tag{312}$$

which are Euclidean. Therefore this Euclidean anti-timespace gives rise to angular momenta just like the Euclidean angular momentum algebra. We have fixed the trigonal rotation such that the time-quantity $I\ e_4$ takes in the role of the well known spin vector σ_3. We define timespace momenta

$$J_u = \tfrac{1}{2} I\,j, \quad J_v = \tfrac{1}{2} I\,e_{123}, \quad J_t = \tfrac{1}{2} I\,e_4 \tag{313}$$

with raising and lowering shift operators

$$J_+ = J_u + I\,J_v, \quad J_- = J_u - I\,J_v \qquad \text{satisfying equations} \tag{314}$$

$$[J_t, J_\pm] = \pm J_\pm \tag{315}$$

with $\mathrm{J}^2 = J_u^2 + J_v^2 + J_t^2$, the *total step momentum*, we further get

$$[\mathrm{J}^2, J_t] = [\mathrm{J}^2, J_\pm] = 0 \tag{316}$$

This is a well known design. In any problem we can find a complete set of states which are simultaneous eigenfunctions of J^2 and J_t with eigenvalues J and T. These satisfy the equations

$$\mathrm{J}^2 \left| J, \mathsf{T} \right\rangle = \hbar^2 J(J+1) \left| J, \mathsf{T} \right\rangle \tag{317}$$

$$J_t \left| J, \mathsf{T} \right\rangle = \hbar \mathsf{T} \left| J, \mathsf{T} \right\rangle \tag{318}$$

$$J_{\pm} \left| J, \mathsf{T} \right\rangle = \sqrt{J(J+1) - \mathsf{T}(\mathsf{T} \pm 1} \left| J, \mathsf{T} \pm 1 \right\rangle \tag{319}$$

The raising and lowering of the eigenvalue T is limited by the eigenvalue of step momentum *J*. These may be either integral or half integral and to any *J* there corresponds a multiplet of $2J+1$ states all having the same eigenvalue *J* and values T equal to $-J$, $-J+1$, $-J+2$, …, $J-1$, $-J$. Now pay attention that this multiplet structure is transposed by the trigonal rotation operator from unit quaternion to unit quaternion so that the strong force acts like an equalizer of the quantization in the complexified quaternion space. We obtain a limited quantum grid which equally involves the synchronous wrap time, the directed 3-volume and the space-time volume. May be that states of higher excitation can consume considerable energy. Even the minimal multiplet for the eigenvalue $J=1/2$ allows for synchronous time T = ±1/2, which means a reversal of the arrow of time or causality violation. Moving the energy through such a grid with acausal synchronous time T = ±(½)n legitimates a process which I have called *motion beyond time* as it may involve an equal distribution of negative and positive measures of time. Nevertheless, the moving field has an utterly material form of existence. However, time of the observer, the measure of τ has a definite arrow which does not violate causality.

Chapter 32

CONSEQUENCE FOR COSMOLOGY

There is a connection between the Pauli principle acting in strong force fields and the universal *SU*(3)-rotation[1] connecting the outer space. This universal trigonal transition among different fermion pure states and their direction fields operates like a real equalizer of time, volume and space-time volume. Those three, - time, volume and space-time volume, - are timelike numbers representative of the quaternion number field. It is by no means artificial to state that the triple {*time, volume, space-time volume*} generates a quaternion space just like the even bivector component of the Clifford algebra $Cl_{3,1}$. But the decomposition of the Minkowski algebra into a positive definite 10-dimensional flavour- and colour differentiating space-like subspace *Ch*, and two quaternion components with negative definite signature forms a natural 3-partition of the space-time algebra each following its own laws inscribed into the constitutive Lie algebra $\mathcal{L}^{(2)}$. But the bivector algebra and the timespace algebra are both of the type quaternion. Both give rise to a Fueter equation and respectively regular quaternionic functions. Both bring forth an angular momentum algebra, a harmonic oscillator algebra and even their own Legendre polynomials. But the harmonic transitions between different colours equalize the quantization between the three different dimensions time, volume and space-time volume. We obtain so to say a grid of quanta that is equalized among those seemingly different dimensions which are, nevertheless, quaternionic without exception. Further we must be aware that the direction field – just like the classic spinor – is not a microscopic phenomenon, but it is just as well observable in macrophysics and cosmology. So we are confronted with a most phantastic situation which can only be described in terms of a star-track block buster. The time-scale can be excited by strong interaction and all of a sudden make a quantum jump from *unit*=1 to *unit*=2 and further, just as angular momentum is doing in the old quantum mechanics. The strong force is transposing that discrete spectrum from time to volume and further to space-time volume and further to time so that directed space-time volumes are structurally wrapped into time. The velocity of light will actually disclose a discrete spectrum depending on the amount of effective energy. Motion will turn out to be compatible with forward and backward steps of time following each other. Coiling up a DNS string, - which also discloses the *SU*(3) multiplet structure, - time will indeed be able to revert its direction every turn of the double-helix. In other words, there are observer systems, both micro and macro, where there prevails motion without a

---

[1] We are using this denotation by mere historic reasons. As we know there are many more realizations of flavour and colour than just by *SU*(3). We may just as well consider any real form of *SU*(3) in Clifford algebras.

definite time, as time itself walks on maximum entropy and shows no dominant arrow. The emergence of wakes and other instabilities will make it impossible to impose a quasi historic, definitely directed time arrow to those events "beyond time", - though it might be possible to follow, to a certain extent, some kind of random walk of time. Taking into account the mixing of time with unstable directed volumes, it will be most likely that there is a considerable variety of space-time manifolds that, - close to an instability of the type big bang -, do not show the familiar regional and temporal order which allows for a history scan. In those cases, history itself turns out broken or fractal. Yet it represents a significant building block for other phenomena with a more transparent pattern of order. The timespace spinor is located at the edge between thermodynamics and geometry.

Chapter 33

THE OLD UNIFIED MOTION

Life is motion. Motion need not be unified. It is all one, verbatim: alone. Nature is alone. Only our cognitive image of it can be unified. For example we can try to find a universally valid equation of motion. Universal motion, however, goes beyond that. It is alone and includes consciousness as only one part. The physical, however, transcends beyond the conscious mind. Motion in consciousness can be called cognitive evolution. Since the earliest of times physicists have favoured the Pythagorean train of thought, harmonic analysis, squares and the like, - may be justifiably. But sometimes it was rather a hindrance. We disclosed an aesthetic fixation about classical music and wave equations. Partial differential equations began to structure the core of quantum mechanics. This can be traced to Erwin Schrödinger's talents for the arts and music, and surely brought forth enrichment for physics. But the symphony of matter turned out as much richer than our linear equations. By the time experimenters brought together an incredibly complex manifold of phenomena which showed unbelievable features of symmetry and even nonlinearity. So we had to pack up certain symmetries and recharge the equations with gauge groups. Our greatest problem was and still is that we started off with small bits of nature, subnuclear particles, fermion states with no extension, almost vanishing neutrinos and fields which could be located or globally spread out and tried to put them together by some equations of motion such that the behaviour of real matter resulted. But we should have begun with phenomena of universal interaction and derive the properties of elementary particles therefrom. We should discover the laws of motion within the universe of the observed symmetries.

Chapter 34

THE DIRAC EQUATION

In 1926 Erwin Schrödinger designed a wave-equation for the development of quantum mechanical systems. This described atomic phenomena except those which involved magnetic and relativistic motion. Next Wolfgang Pauli drew up the equation for slow fermions thus including the spin. Relativistic quantum mechanics, however, began on the mass shell

$$E^2 = m_0^2 c^4 + c^2 p^2 \text{ with energy } E \text{ and momentum } p \tag{320}$$

and obeyed the Klein Gordon equation for fields without spin

$$\hbar^2 \Box \psi = m^2 c^2 \psi \tag{321}$$

in which space and time had to have equal rights. Paul Dirac played around with the Klein-Gordon equation and linearized its square-root. So, historically, he obtained the first linear wave equation for relativistic fermions such as electrons, muons, neutrinos and quarks, all having spin ½ .

$$\gamma^\mu (\partial_\mu + iqA_\mu + im)\psi = 0 \qquad \text{with } q = \frac{e^-}{\hbar c};\ m = \frac{m_0 c}{\hbar} \tag{322}$$

where the A_μ are exterior covariant electromagnetic field components and e^- is the electron's charge. The Pauli-equation turned out to be the non-relativistic marginal case of the Dirac equation. But the Dirac representation of the Poincaré group can be reduced in two ways one of which brings on the Weyl spinor decomposition and the other a split into left and right handed chiral waves. Gamma matrices generating the Clifford algebra $Cl_{1,3}$ are

$$\gamma_0 = \gamma^0 = \begin{bmatrix} Id_2 & 0 \\ 0 & -Id_2 \end{bmatrix}, \qquad \gamma_j = -\gamma^j = \begin{bmatrix} 0 & -\sigma_j \\ \sigma_j & 0 \end{bmatrix} \tag{323}$$

with additional $\gamma_5 = -i\gamma_{0123}$ with $i = \sqrt{-1}$

The field had to meet the requirements of strong parity violation in weak interaction. Therefore such a four-component spinor decomposed into left and right handed chiral neutrino waves

$$\psi = \begin{bmatrix} \psi_1 \\ \psi_2 \\ \psi_3 \\ \psi_4 \end{bmatrix}, \quad \psi_R + \psi_L \equiv \tfrac{1}{2}\begin{bmatrix} \psi_1 + \psi_3 \\ \psi_2 + \psi_4 \\ \psi_3 + \psi_1 \\ \psi_4 + \psi_2 \end{bmatrix} + \tfrac{1}{2}\begin{bmatrix} \psi_1 - \psi_3 \\ \psi_2 - \psi_4 \\ \psi_3 - \psi_1 \\ \psi_4 - \psi_2 \end{bmatrix} \tag{324}$$

which could be projected out by two primitive idempotents

$$\eta_\pm = \tfrac{1}{2}(1 \pm \gamma_5) \qquad \text{so that} \quad \psi_R = \eta_+\psi \ \text{ and } \qquad \psi_L = \eta_-\psi. \tag{325}$$

Thus was given orientation to an elementary spatio-temporal massless fermion field in weak interaction. That was all.

The chiral components of the Dirac field should not be confused with the Weyl field components, indeed, though there are some important relations between them. The Dirac representation of the Weyl fields is obtained from ψ by a mere unitary transformation:

$$U = \tfrac{1}{\sqrt{2}}(\gamma_0 + \gamma_5) \qquad U\psi = \begin{bmatrix} \xi \\ \eta \end{bmatrix} \qquad \text{where}$$
$$\xi = \tfrac{1}{\sqrt{2}}\begin{bmatrix} \psi_1 + \psi_3 \\ \psi_2 + \psi_4 \end{bmatrix} \qquad \eta = \tfrac{1}{\sqrt{2}}\begin{bmatrix} \psi_1 - \psi_3 \\ \psi_2 - \psi_4 \end{bmatrix} \tag{326}$$

Physicists usually have used complex 4×4-matrices from the linear matrix space Mat(4, $\mathbb{C}$) isomorphic with the Clifford algebra $Cl_{4,1}$,- though the equations were designed in the space-time algebra (STA) which is $Cl_{1,3}$. The column spinor was often replaced by the algebraic spinor

$$\psi = \begin{bmatrix} \psi_1 \\ \psi_2 \\ \psi_3 \\ \psi_4 \end{bmatrix} \to \Psi = \begin{bmatrix} \Psi_{11} & 0 & 0 & 0 \\ \Psi_{21} & 0 & 0 & 0 \\ \Psi_{31} & 0 & 0 & 0 \\ \Psi_{41} & 0 & 0 & 0 \end{bmatrix} \in \mathrm{Mat}(4, \mathbb{C})f \tag{327}$$

involving a primitive idempotent from the complexified STA

$$f = \tfrac{1}{4}(1 + \gamma_0)(1 + i\gamma_1\gamma_2) \in \mathbb{C}\otimes Cl_{1,3} \tag{328}$$

In other words, the primitive idempotent which is used to design the Dirac spinor is not an element of the $Cl_{1,3}$ but of the $\mathbb{C}\otimes Cl_{1,3}$. Anyway, the spinor is now an element of a linear spinor space $S = (\mathbb{C} \otimes Cl_{1,3})\, f$. This has first been investigated by Chevalley (1954) and Riesz (1958). Later on, physicists were satisfied with the possibility to regard particles as spinors and spinors as elements of minimal left ideals of a Clifford algebra. They did not realize that not the spinors, but the primitive idempotents themselves represented pure states which could provide irreducible representations and GNS constructions in Clifford algebras. What was and still is important for physics is the richness of representation spaces and therefore the structure of the idempotent manifolds in the fundamental physical algebras. But again, the prominent lattice of idempotents that could be a reason for the emergence of strong interacting symmetries SU(3) was not a substructure of the STA, but rather of the real Majorana algebra $Cl_{3,1} \cong \mathrm{Mat}(4,\mathbb{R})$. In all this we can feel the muddle of interpretations that began to unfold from our wave equation fixation. Meanwhile we had a Schrödinger-, a Pauli-, a Dirac-, a Weyl-, a Heisenberg-, a von Neumann equation of motion and all based on Hamilton's equation, or Liouville equation, we had a pilot wave theory by Louis de Broglie and David Bohm and the former ballasted with gauge covariant elaborations, and last not least some exactly solvable equations such as those investigated by Baxter (1984). But all these waves did not provide the unity that some of us physicists were after.

Chapter 35

SELF-INTERACTING PURE STATES

FUNDAMENTAL HISTORIC APPROACHES

David Bohm (1993) was convinced there is a deeper holistic view of quantum physics. He had become friends with the philosopher Jiddu Krishnamurti. This partly explains his original search for a unity of physics.

In quantum mechanics literature, there recurs a picture where individual electrons move mathematically deterministic, though eventually on chaotic trajectories, while the statistic ensemble follows a wave. The first who communicated such an image had been Louis de Broglie who conceptualized »la theorie de l'onde pilote«, the pilot wave theory. This found little attention and fell into desuetude until David Bohm designed a new version in 1952. There were only a few followers, though rather prominent ones like John Stewart Bell, who defended Bohm's mechanics, but in the 70s even Bohm let off propagating it because, then, it seemed to him ontologically overloaded. Later experiments again supported Bohm's idea.

David Hestenes (1983) reconsidered the approach in his discourse on "*Quantum Mechanics from Self-Interaction*". There followed a number of his papers sketching the "*Zitterbewegungs - Ansatz*" in radiation theory. Bohm had maintained that the electron was a particle with a deterministic trajectory having definite locations and momenta while the wave function determined a probabilistic ensemble and provided stochastic locations and momenta of possible trajectories. Hestenes viewpoint differed from Bohms only in a few details. He preferred to use the Dirac- instead of the Schrödinger equation (Hestenes 1991, p. 13) but put Schrödinger's concept of Zitterbewegung (zbw) into the context of Clifford algebra.

If zbw were a truly objective phenomenon, Hestenes argued, it almost certainly originated from electron self-interaction. If, further, zbw was a localized helical motion of the electron with an orbital angular momentum, it could be identified with the electron spin, as was variously proposed in the literature. But it could "not be derived from the Dirac theory as it stands" (Hestenes 1983, p. 3). Hestenes could not do otherwise than to identify the Zitterbewegung with an intrinsic angular momentum or spin of magnitude $s = \hbar/2$ winding around a timelike average worldline traced out by the zbw center. But: "The self-interaction problem must be solved to derive the Dirac theory on which QED is based". Still I wonder if QED, not to say QCD, must be based on or explained by the Dirac theory. The author, then, confessed "Of course, we would need an equation of motion to deduce the curvature of the centreline. But, for the purpose of interpreting the Dirac theory, it is sufficient to take the zbw for granted, as we shall see in the next section. Then the Dirac equation can be used as the equation of motion. We are not yet in a position to derive the Dirac equation from a deeper dynamical theory" (Hestenes, p. 5).

By the time the Dirac-Hestenes equation has been worked out. In that equation, the role of the column spinors was taken over by real even multivectors. It has been pointed out that these quantities did not belong to any proper left ideal of the Clifford algebra. So the minimal ideal approach could not be applied as the appropriate tool for representation, anyway. The Hestenes equation had been criticized by various reasons and also defended by many authors (Lounesto 2001, p. 144). But, as a matter of fact, it described motion of weakly interacting particles, but sure no strong interaction. Interestingly, the Hestenes zbw -model provides not only a new interpretation of spin, but also a mechanism for the emergence of mass in fermions. The argumentation is quite fundamental:

Travelling with an orbital spin $s = \hbar/2$ the zbw frequency is identified with the de Broglie frequency

$$\omega_0 = \frac{mc^2}{\hbar} = 7{,}8 \times 10^{20} s^{-1} \tag{329}$$

The electron shows us an electric charge e but is considered massless because it moves with the speed of light. The total free self-energy of the electron is given by

$$mc^2 = E_0 + U_0 \tag{330}$$

where E_0 is the kinetic zbw energy and U_0 the potential energy in self-interaction. The zbw amplitude r is equal to the Compton wavelength

$$r = \frac{c}{\omega_0} = \frac{\hbar}{mc} = 3{,}8 \times 10^{-13} m\,. \tag{331}$$

The angular momentum has magnitude

$$s = \frac{rE_0}{c} \text{ and should be equal to } \frac{\hbar}{2} \tag{332}$$

which can be satisfied only if the inner zbw kinetic energy is

$$E_0 = s\omega_0 = \frac{mc^2}{2} \tag{333}$$

from which follows that the total energy has equal parts of kinetic and potential energy

$$E_0 = U_0 \tag{334}$$

The zbw gives rise to a magnetic moment

$$\mu = \left(\frac{ec}{2\pi r}\right)\left(\frac{\pi r^2}{c}\right) = \frac{er}{2} \tag{335}$$

Hestenes pointed out that that reproduces the result from the Dirac equation

$$\mu = \frac{e}{mc}s = \frac{eh}{2mc} \tag{336}$$

only in case the r in equation (332) is the Compton wavelength from (331). "We cannot alter E_0 without altering this result".

Like William Kingdon Clifford before him, so David Hestenes exploited the geometric meaning of a bivector e_{12} as being representative of the quantity $i = \sqrt{-1}$. Transforming the Dirac equation in this way, he discovered the importance of the unimodular even multivectors associated with a Lorentz transformation which he denoted as "unimodular spinors". His equation of motion is still bound to electromagnetic and weak interaction and does not give us satisfying results in subnuclear interaction. For instance the attempt to compose a proton by two positrons and one electron (1983, p. 14) led to no satisfactory mathematical model explaining the SU(3) symmetries in subnuclear energy. But Hestenes was aware of the incompleteness of our current geometric approaches and of the deficiencies in the equations of motion.

This situation changes significantly as soon as we become aware that the standard model can be derived from a complete investigation of the transposition involutions of Clifford algebra. The Zitterbewegung of fermions need not be identified with the orbital circulation. But the attempt to rotate a fermion in its graded state space results in a zbw which may range from chaotic over superimposed quasi-periodic to regular harmonic. This is due to the emergence of a chaotic iterator.

In some of the most harmonic scenarios, this map provides plane waves for spin and plane waves of colour to support the Pauli principle. A future challenge should be seen in the experimental material which will allow us to calibrate the energy content of such superimposed and chaotic reactions of weak and strong interacting fermions. The main structure of the full mechanics seems to contain a part which is very similar to Hestenes' model. But we neither begin with minimal left ideals, nor with unimodular spinors, but with pure states as are given by primitive idempotents. These states, by their very nature, have to follow their own symmetries which are at one time both inner and outer symmetries of motion. Before we can go into this more deeply, we have to mention a now twenty years old idea worked out by Roy Chisholm, Ruth Farwell and others.

After having clarified that spinor spaces can be regarded as minimal left ideals of geometric algebra, some of us began to investigate the lattice structure of primitive idempotents in Clifford algebras. Especially Chisholm (1986) and Farwell (1992) following some idea of Greider (1984) proposed to look at one of the tetrahedrons that was constituted by some orthogonal primitive idempotents in the lattice of $Cl_{3,1}$ as a composition of one lepton and three coloured quark states. They legitimized this approach by their correspondence principle saying "*there is no distinction between space-time and the internal interaction space*". I regarded this statement as more important today than the wave equation.

However, Chisholm stopped developing the idea further, probably because he could not figure out how the frame could be designed and what was the equation of motion for those primitive idempotents. Chisholm and Farwell have supported the elaboration of the present theory by showing me many of their results about the structure of what they denoted as physical algebras. They were aware that the loss of equations of motion had to be outbalanced by some gain in comprehending the totality of the space-time phenomena.

Chapter 36

ORIGIN OF CHAOTIC MOTION

It is true, the self-interaction problem must be solved to derive the Dirac theory. But the latter can only be a tiny shadow of the full dynamic system of subnuclear matter. We shall easily derive the Dirac equation from a deeper dynamical theory as soon as we are ready to look at the whole of our empirical findings in the context of geometric algebra. It is a timely realization that many nonlinear phenomena, - phase transitions and critical phenomena, - are connected with the observation of the organization of spin. Unfortunately, we had first cancelled out the intrinsic nonlinearity of the dynamic system by trying to linearize it. Perhaps the most serious mistake consisted in linearizing the equation of motion. We must realize that motion of a pure state in geometric Clifford algebra involves graded motion and as a consequence a trigonometric power map for the fundamental measures of space-time multivectors. That means, we cannot linearize these phenomena since the Pauli principle for coloured fermions involves an intrinsic nonlinear mechanism allowing the elementary field to escape from chaotic into metastable pure states and respectively from highly nonlocal chaotic states to localized almost linear trajectories. It is probable that even the gravitational force is a natural outcome of this graded design of motion. When Hestenes conceptualized the electronic Zitterbewegung he had to confine it to localized helical motion. But such orbital angular momentum occurs only beyond chaotic motion. The most general zbw is indeed a chaotic trajectory, and such a trajectory is not a nonlinear accident, but it is the natural appearance of elementary motion at all. As you can feel from previous chapters this whole phenomenology is somehow connected with conformal maps. Being aware that my statement may sound somewhat cryptic, let me still make it:

The extension of a certain absorption property of null vectors, some mystic mathematical property of those strange things which we call points, once continued into the spaces of graded Lie groups leads us to the absorption of the space filling constitutive Lie group by the fundamental neutrino pure state. Then we have to put up with the fact that any attempt to "apply" the elementary rotations of the corresponding orthogonal factors with considerably high probability results in a chaotic motion. The Pauli principle, however, is the result of an organization beyond chaos and thus beyond chaotic zbw. The localized helical motion as an orbital angular momentum of a fermion must be regarded as a metastable state of a freely travelling particle, a state with high degree of freedom in motion. The Pauli principle, so to say, provides stability even in strong interaction.

Chapter 37

QUARK ON TRAVEL

Consider an s-quark supposed to travel in direction e_1 acted on by one of the rotators of the isospin subgroup $\{\lambda_1, \lambda_2, \lambda_3\}$, namely λ_2 .

λ_2 e_{23}

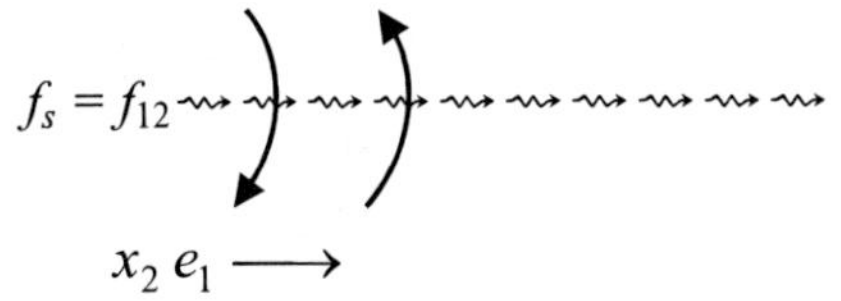

$f_s = f_{12}$

$x_2\, e_1 \longrightarrow$

Remember we had

$$\lambda_1 = \tfrac{1}{4}(e_{34} - e_{134}),\ \lambda_2 = \tfrac{1}{4}(-e_{23} + e_{123}),\ \lambda_3 = \tfrac{1}{4}(e_{24} - e_{124})$$

and the primitive idempotent $f_s = \tfrac{1}{4}(1+e_1)(1-e_{24})$. The generator λ_2 plays a remarkable role in this dynamic scenario. First it preserves x_2 and f_s , that is, it leaves the structure of the s-quark entirely unaltered and does not change the measure alongside direction e_1. The reader may easily verify these claims by calculating

$$e^{\lambda_2} f_s e^{-\lambda_2} = f_s \quad \text{as well as} \quad e^{\lambda_2} (x_2 e_1) e^{-\lambda_2} = x_2 e_1 \qquad (337)$$

Recall further that $[\lambda_2, e_{23}] = 0$. Regarding the bivector ½ e_{23} as spin in the direction of e_1, that tells us that isospin λ_2 can superimpose spin ½ e_{23}. Recall, we also had the commutators $[e_1, e_{23}] = 0$ and $[e_{23}, e_{123}] = 0$. So while preserving the f_s – pure state, by the action of λ_2 the space $Cl_{3,1}$ can also be turned about the multivectors x_2e_1, x_9e_{23}, $x_{12}e_{123}$ without changing them. Following the argumentation by Hestenes, we could now have two localized helical movements both realised by orbital angular momenta. Both can turn a vector x_3e_2 into a vector x_3e_3 but preserve the directed area x_9e_{23}. In figures 29, 30 it is shown that the action of isospin λ_2 on the state f_{13} is the same as the action of spin –½, that is, the element – ½ e_{23}. That can mislead one to believe they act the same even on more general elements of the Clifford algebra. That is not the case.

But the isospin generator λ_2 turns the plane $\{e_2, e_3\}$ twice as fast as the spin generator $s = \frac{1}{2} e_{23}$. Also the elements λ_2 and s thus calibrated turn the plane $\{e_2, e_3\}$, in opposite direction.

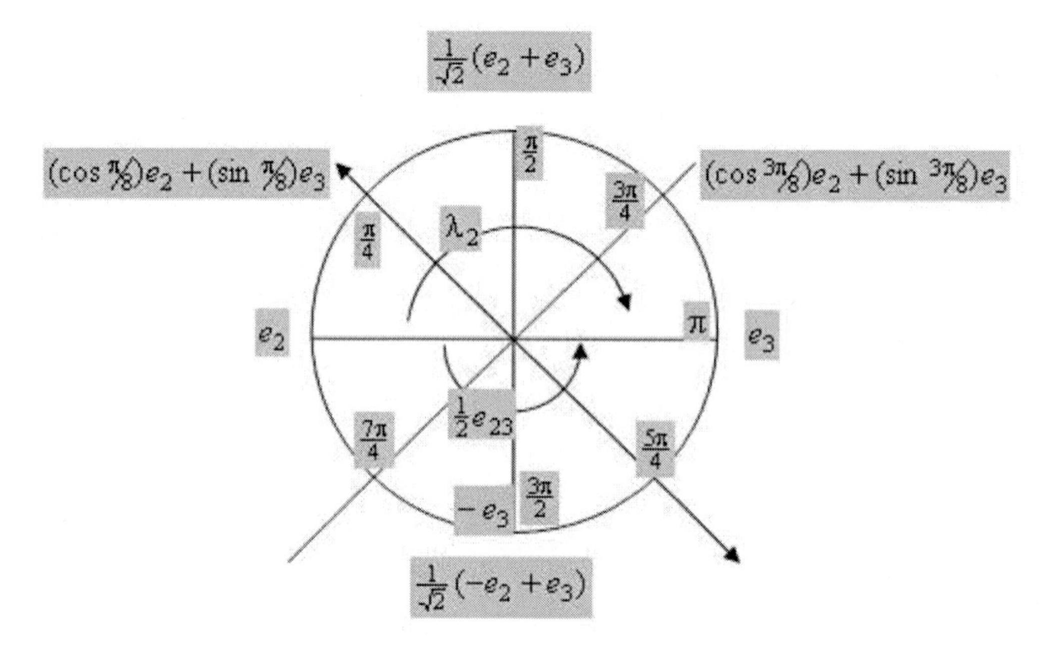

Figure 29. Action of isospin λ_2 on e_2.

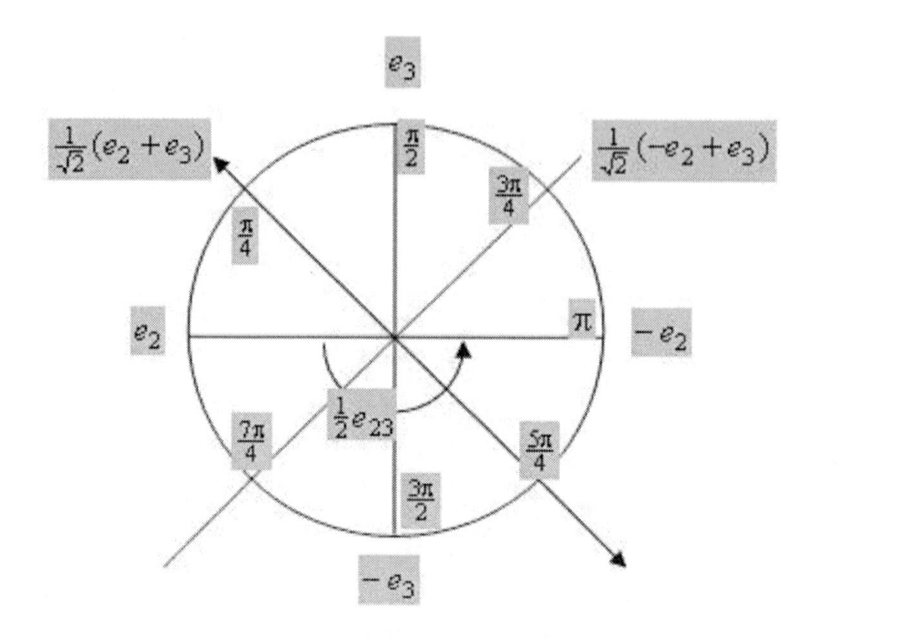

Figure 30. Action of spin $\frac{1}{2} e_{23}$ on e_2.

Surprisingly the generators λ_2 and $-\frac{1}{2}\,e_{23}$ act on the primitive idempotent f_{13} in exactly the same way.

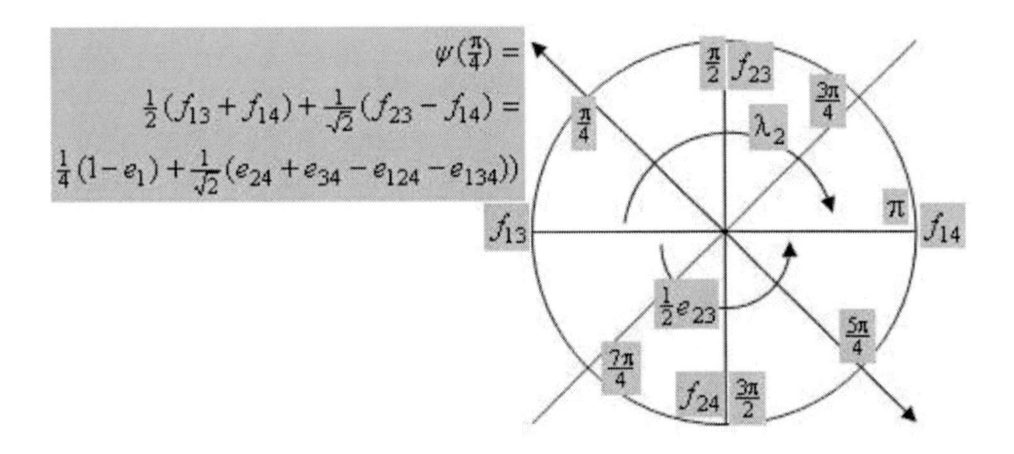

Figure 31. Action of isospin λ_2 and spin $\frac{1}{2}\,e_{23}$.

For example we transform likewise

$$f_{13} \xrightarrow{\frac{\pi}{2}\lambda_2} f_{23} \text{ and equally } f_{13} \xrightarrow{-e_{23}/2} f_{23}$$

but differently

$$e_2 \xrightarrow{\frac{\pi}{2}\lambda_2} \frac{1}{\sqrt{2}}(e_2+e_3) \text{ and } \qquad e_2 \xrightarrow{-e_{23}/2} e_3$$

Chapter 38

CONFORMAL MAPS FOR MOVING FLAVOUR

We saw the element λ_2 rotates flavour within a 6-dimensional positive subspace of the Minkowski algebra $Cl_{3,1}$ spanned by componends of the primitive idempotents $f_{13}, f_{23}, f_{14}, f_{24}$ (figure 31). Each of these is constituted by a scalar, a directed unit vector, a positive 2-blade and a positive 3-blade. Explicitely, we have

$$f_{13} = \tfrac{1}{4}(1 - e_1 + e_{24} - e_{124}) \qquad f_{23} = \tfrac{1}{4}(1 - e_1 + e_{34} - e_{134})$$
$$f_{14} = \tfrac{1}{4}(1 - e_1 - e_{24} + e_{124}) \qquad f_{24} = \tfrac{1}{4}(1 - e_1 - e_{34} + e_{134}).$$

But to comprehend strong force dynamics we have to investigate not only rotations of the SU(3) and SU(2) but also the translations alongside the directions of the Minkowski space. This seems self-evident but represents the most challenging problem of strong force dynamics. First, we are aware that in the old inhomogeneous model of the Euclidean spaces E^n the origin is a distinguished point in $\mathbb{R}^n$. This can be avoided by constructing the "conformal split" and representing points as null vectors. At the same time there is established a connection between 2-blades with positive definite signature and positive definite 3-blades which constitute the colour spaces. Therefore, one would like to first go into some peculiar conformal splits within the Clifford algebra of Minkowski space. It is obvious that the flavours in colour space ch_1 involve exactly one Euclidean plane, namely $\{e_1, e_2\}$. Therefore to understand motion one must investigate transpositions in that plane. This is one of the special problems of translational displacement we shall probably have to come back to.

But the general problem concerns translations of the whole space-time algebra and therefore a conformal split having form $Cl_{4,2} = Cl_{3,1} \otimes Cl_{1,1}$. This type of geometric algebra plays a special role in cosmology. It also houses the Anti de Sitter space »adS$_6$« and respectively a constitutive Lie group ♑●$(Cl_{4,2})$ for universal standard models. Those I described in the »*Lie group guide to the universe*« (Schmeikal 2009). Hestenes alludes its potential applications to twistor theory and cosmological models in gauge gravity.

So we begin with a Clifford algebra $Cl_{n+1,1}$ of the Minkowski type where we have associated a 4-dimensional Minkowski algebra $Cl_{1,1}$ with the Euclidean subalgebra Cl_n. The latter is mapped onto a pseudo Euclidean positive definite extension algebra which is determined by a designated 2-blade E such as e_{16}, e_{26} or e_{36}. The $Cl_{1,1}$ is generated by 2-dimensional Minkowski space $\mathbb{R}^{1,1}$. Following Hestenes we introduce a null basis

$$z_+^2 = z_-^2 = 0 \text{ such that the inner product equals } z_+.z_- = 1 \tag{338}$$

The wedge product of the nilpotents defines a positive definite plane pseudo scalar, or what I preferred to denote a 2-extension

$$z_+ \wedge z_- = E \qquad \text{with} \qquad E^2 = Id \text{ briefly=1,} \qquad \text{the identity.} \tag{339}$$

The exponent 2 refers to the geometric or Clifford product which is decomposed into an inner and an exterior product as usual:

$$\begin{aligned} &\text{(i)} \quad fg = f.g + f \wedge g \text{ so we can built up inner and exterior} \\ &\text{(ii)} \quad f.g = \tfrac{1}{2}(fg + gf) \\ &\text{(iii)} \quad f \wedge g = \tfrac{1}{2}(fg - gf) \end{aligned} \tag{340}$$

by symmetric and antisymmetric bilinear forms. We have

$$\begin{aligned} &\text{(i)} \quad z_+ z_- = 1 + E \quad \text{and} \quad z_- z_+ = 1 - E = 1 + E^\dagger \text{ and} \\ &\text{(ii)} \quad \tfrac{1}{2}\{z_+, z_-\} = 1 \qquad \tfrac{1}{2}[z_+, z_-] = E \quad \text{with the absorptions} \\ &\text{(iii)} \quad E z_+ = z_+ \qquad z_- E = z_- \end{aligned} \tag{341}$$

The conformal split was introduced by Hestenes (1991) to relate a Minkowski algebra to a maximal Euclidean subalgebra. Then he differed between an additive and a multiplicative split. The additive split is defined by a direct sum

$$\mathbb{R}^{n+1,1} = \mathbb{R}^n \oplus \mathbb{R}^{1,1} \quad \text{with the Euclidean vector space} \tag{342}$$

having an orthonormal basis $\{e_j \,/\, e_j.e_k = \delta_{jk}, \text{ j, k} = 1, 2, \ldots, \nu\}$ fulfilling the orthogonality relations

$$\begin{aligned} &z_+.e_j = z_-.e_j = 0 \qquad \text{equivalent with the anticommutation relations} \\ &\{e_k, z_+\} = \{e_k, z_-\} = 0 \end{aligned} \tag{343}$$

The multiplicative split is defined as a direct product

$$Cl_{n+1,1} = \mathcal{R}_n \otimes Cl_{1,1} \tag{344}$$

where $\mathcal{R}_n$ is generated by a space of n trivectors having a common factor E. It is isomorphic with the Euclidean n-space. We have

$$\mathcal{R}_n = Cl(\mathbb{R}^n E) \text{ which has the basis} \tag{345}$$

$\{\boldsymbol{e}_k = e_k E = E e_k\} \subset Cl^{(3)}_{n+1,1}$ with orthogonality preserved

$\boldsymbol{e}_j.\boldsymbol{e}_k = e_j.e_k = \delta_{jk}$.

However, while the vectors anticommute with the nilpotents (338), the trivectors commute with them:

$$[\boldsymbol{e}_j, z_+] = [\boldsymbol{e}_j, z_-] = 0 \quad (346)$$

To eliminate the distinguished origin in the inhomogeneous model, from now on we can represent points in Euclidean E^n by vectors x in $\mathbb{R}^{n+1,1}$ such that a point is located on the null cone

$$\mathsf{N}^{n+1} = \{x \mid x^2 = 0\} \text{ and on the hyperplane} \quad (347)$$

$$\mathsf{P}^{n+1}(z_+, z_-) = \{x \mid z_+.x = 1\} \quad (348)$$

from which there follows $z_+.(x - z_-) = 0$. Therefore the hyperplane P^{n+1} passes through the point z_-. The intersection between $\mathsf{N}^{n+1} \cap \mathsf{P}^{n+1}$ is called the horosphere and can provide a homogeneous model of Euclidean n-space. Hestenes pointed out it was first constructed by Wachter (1792-1817), and deduced the explicit algebraic terms for the conformal split:

$$\text{sp: } \boldsymbol{x} \xrightarrow{conf} x, \qquad x = \boldsymbol{x}E - \tfrac{1}{2}\boldsymbol{x}^2 z_+ + z_- \quad (349)$$

From this is derived the inner product between two homogeneous singular points. It is equal to $-½$ the Euclidean distance:

$$x.y = \boldsymbol{x}.\boldsymbol{y} - \tfrac{1}{2}(\boldsymbol{x}^2 + \boldsymbol{y}^2)z_+.z_- = -\tfrac{1}{2}(\boldsymbol{x}^2 + \boldsymbol{y}^2) \quad (350)$$

The nilpotent z_- represents the origin in $\mathbb{R}^n$ whereas z_+ a point at infinity.

It should be clear that rotating a vector in Minkowski space-time does not mean to just take this one vector and rotate it, but it means to rotate the whole geometric algebra. Quite generally, we denote by a displacement any translation or rotation of the algebra. It is therefore a legitimate question to ask what kind of effect a translation has on a colour space or on any specific flavour.

Chapter 39

Translation Split of an Inner State

In every Clifford algebra we can locate standard sets of orthogonal primitive idempotents which span subspaces with minimal dimension and positive definite signature. The whole set of mutually annihilating primitive idempotents has two important features. First it provides a discrete structure of lattices. Second it consists of manifolds as those lattices can be unfurled by the constitutive Lie groups from the minimal colour spaces into algebra filling manifolds. In the Clifford algebras generated by Minkowski spaces $\mathbb{R}^{n,1}$ we shall always find a number of n(n-1) colour spaces with positive definite signature spanned by the elements ¼ $(1 \pm e_j)(1 \pm e_{k,n+1})$ with $j \neq k=1,\ldots,n$. So we have four orthogonal primitive idempotents which by addition constitute a lattice (Lounesto 2001, p. 227) of 16 commuting idempotents, and we have n(n-1) such 16 elements-lattices. By the constitutive Lie group ♑●$(Cl_{n,1})$ which is ♑●$(Cl_{3,1}) \cong SL_{Cl}(3)$ for $\mathbb{R}^{3,1}$ those lattices can be unfolded to continuous manifolds.

The graded structure of the constitutive group (18) follows from the nonlinear design of the pure states which again is connected with the grading of transposition involutions (Schmeikal 2004, 358). This peculiarity is responsible for the difficulty we meet when we attempt to establish a conformal split for the primitive idempotents.

Consider a colour space of minimal dimension spanned by the four commuting elements $\{1, e_j, e_{k,n+2}, e_{jk,n+2}\}$. Those involve four different grades: a scalar, a space-like vector, a bivector and a 3-blade each having positive signature. Constructing a conformal split would require a space which preserves the commutation relations. In the case of $\mathbb{R}^{n+1,1}$ this is the space of trivectors $Cl^{(3)}_{n+1,1}$ having basis $\boldsymbol{e}_k = e_k E = E e_k$ and definite grade equal 3. However, a primitive idempotent which represents a pure state has a minimal number of four different grades. Can we preserve commutation relations?

Let us consider the *a-b-c-algebra* $Ch_1 \subset Cl_{4,2}$ given by the units

$$\{1, a, b, c\} \equiv \{Id, e_1, e_{25}, e_{125}\} \tag{351}$$

We know, the Clifford algebra over Ch_1

$$Cl(Ch_1) \cong {}^4\mathbb{R} \tag{352}$$

is isomorphic with the fourfold ring of real numbers and thus has no exterior product at all. Therefore it is a trivial geometric algebra. Its geometric product is contracted to an inner pseudo-scalar product since for any two vectors ξ, ψ we have

$$\xi.\psi = \tfrac{1}{2}(\xi\psi + \psi\xi) = \xi\psi \quad \text{because of } [\xi,\psi] = 0 \tag{353}$$

Any colour space is a Cartan algebra which has lost the original Euclidean structure. The exterior product of any two vectors is nil by the geometric $\xi \wedge \psi = ½ (\xi\, \psi - \psi\, \xi)$

$$\underset{\xi,\psi\in Ch_1}{\forall} \quad \xi \wedge \psi = 0 \tag{354}$$

Nevertheless, by transforming the *a-b-c-algebra* Ch_1 by a common positive pseudoscalar factor $E = e_{46}$ with $E^2 = 1$ we obtain another *a-b-c-algebra* and the commutation relations in Ch_1E are the same. We have $[\xi,\psi] = 0$ for any ξ, $\psi \in Ch_1E \cong {}^4\mathbb{R}$. Now recall vectors from $Cl_{3,1}$ should anticommute with the nilpotents z_+, z_- whereas the trivectors after the multiplication with E commute with them. This demand cannot be extended over the colour spaces. Quantities a and c fulfil the constraints, but b does not anticommute with z_+. Also the aE, cE commute with z_+ but bE does not. To put it in a few words: a conformal split of the usual type does not in general preserve the algebraic property of idempotency. This seems to be bad if we compare the invariance of idempotency and various orthogonality relations under the constitutive group. As a consequence, a conformal translation of vectors from the Minkowski space-time which make points nil may transform the whole space, but it does not retain the pure states. This situation is somewhat disturbing. Because why should we loose the primitive idempotents by a mere translation of the space? We must go deeper into this problem.

Consider a conformal translation in $Cl_{4,2}$ by some vector $x_2\, e_1$:

$$T_1 = Id + \tfrac{1}{2} x_2 e_1 z_+ \tag{355}$$

Applying it to a vector ξ in the Minkowski subspace $Cl_{3,1}$ we obtain the translational displaced homogeneous vector

$$T_1 x T_1^{-1} \qquad \text{where } x \text{ is derived from a conformal split} \tag{356}$$

$$x = \boldsymbol{x}E - \tfrac{1}{2}\boldsymbol{x}^2 z_+ + z_- \qquad \text{with trivector } \boldsymbol{x} = \xi \wedge E \text{ as in the} \tag{357}$$

equation (349). We can apply displacement (355) to our pure state f_s in order to test the effect. The result is a surprise. A conformal displacement by $x_2\, e_1$ causes an idempotent split of f_s into an idempotent and a nilpotent

$$D_1 f_s \overset{def}{=} T_1 f_s T_1^{-1} = f_s + n_s(x_2) \tag{358}$$

where

$$n_s(x_2) = \frac{x_2}{4\sqrt{2}}(-e_4 + e_6 + e_{245} + e_{256}) \qquad x_2 \text{ real} \tag{359}$$

This seems interesting because the primitive idempotent $f_s + n_s$ preserves the property of pure state while travelling on $x_2\, e_1$. We would like to interpret the nilpotent n_s as information carrier about the location of the state f_s in Minkowski space. But that is not easy since n_s can again be decomposed by addition into two nilpotents one of which is the z_+ up to a constant while the second resembles some bivector location at infinity multiplied geometrically by some finite location $x_2\, e_2$. It looks as if a conformal translation on direction e_1 is deflected onto e_2. This would refer to phenomena of transposition involution of orientation which originally led to the search after the graded Lie group SU(3) in the algebra $Cl_{3,1}$ (Schmeikal 2004). What is the origin of such strange response of a primitive idempotent in $Cl_{3,1}$ to a translation in $Cl_{4,2}$ and respectively to a conformal split of $Cl_{4,2} = Cl_{3,1} \otimes Cl_{1,1}$? In order to understand this really deeply it is helpful to investigate the conformal transformations of the spatial planes involved by the idempotents in $Cl_{3,1}$. These conformal groups C_{ik} acting on planes $\{e_i, e_k\} \subset \mathsf{E}^2 \subset Cl_{3,1}$ are isomorphic to the Lorentz group on $\mathbb{R}^{2+1,1}$. Is the deflection of a translatory displacement from a unit vector e_1 to another orthogonal unit e_2 a peculiarity of the conformal group on the Minkowski space or does it originate in the conformal group of the plane E^2 ?

Chapter 40

TRANSLATING IN THE EUCLIDEAN PLANE

We displace the vector $\vec{x} = e_1 + e_2 \in \mathsf{E}^2$ and thereby give an example which we shall use repeatedly in the coming rigor. We take the nilpotent vectors

$$z_+ = \tfrac{1}{\sqrt{2}}(e_3 - e_4) \in Cl_{3,1} \qquad z_- = \tfrac{1}{\sqrt{2}}(e_3 + e_4) \tag{360}$$

giving the pseudoscalar $E = e_{34}$ and obtain the conformal split

$$x = \text{sp}\,(e_1 + e_2) = e_{134} - e_{234} + (\sqrt{2})e_4 \tag{361}$$

First, the reader can verify we indeed have $x^2 = 0$ and $z_+.x = 1$. That is, x represents a point in the homogenous model of the Euclidean space. The trivector indicates that we actually have mapped the point $\vec{x} = e_1 + e_2$ from the domain of the inhomogeneous model onto the homogeneous model in $Cl_{3,1}$. Notice the Lorentz group on $\mathbb{R}^{n+1,1}$ is isomorphic to the conformal group on $\mathbb{R}^n$. So the Lorentz group on the Minkowski space with the Lorentz metric is isomorphic to the conformal group on Euclidean 2-space E^2.

A translation by $a \in \mathsf{E}^n$ in the conformal group has the form

$$T_a = Id + \tfrac{1}{2} a\, z_+ \tag{362}$$

and is applied to a vector $\vec{x} \in \mathsf{E}^n$ by conjugation as in (258)

$$D_a(x) = T_a x T_a^{-1} \qquad \text{displacement by transposition} \tag{363}$$

We wish to investigate a transposition in E^2 by the element e_1. In the inhomogeneous model we should obtain a sequence

$$e_1 - e_2, 2e_1 - e_2, 3e_1 - e_2 \qquad \ldots$$

By (351) we obtain the translation element $T_{e_1} = 1 + \frac{1}{\sqrt{8}}(e_{13} - e_{14})$. Applying the transposition displacement to the start vector $x_0 = e_1 - e_2 + (\sqrt{2})e_4$ we calculate the following sequence in the homogeneous model:

$$x_1 = 2e_1 - e_2 - 1{,}061e_3 + 2{,}475e_4$$
$$x_3 = 3e_1 - e_2 - 2{,}828e_3 + 4{,}243e_4$$
$$x_4 = 4e_1 - e_2 - 5{,}303e_3 + 6{,}718e_4 \quad \ldots$$

It can easily be verified those are points on the horosphere and therefore satisfy (347) and (348). So they indeed represent vectors $x_1 = 2e_1 - e_2$, $x_3 = 3e_1 - e_2$, $x_4 = 4e_1 - e_2$ . . . in the plane E^2. We document this simple dynamics by plotting coefficients from MAPLE Clifford and thus obtain diagrams

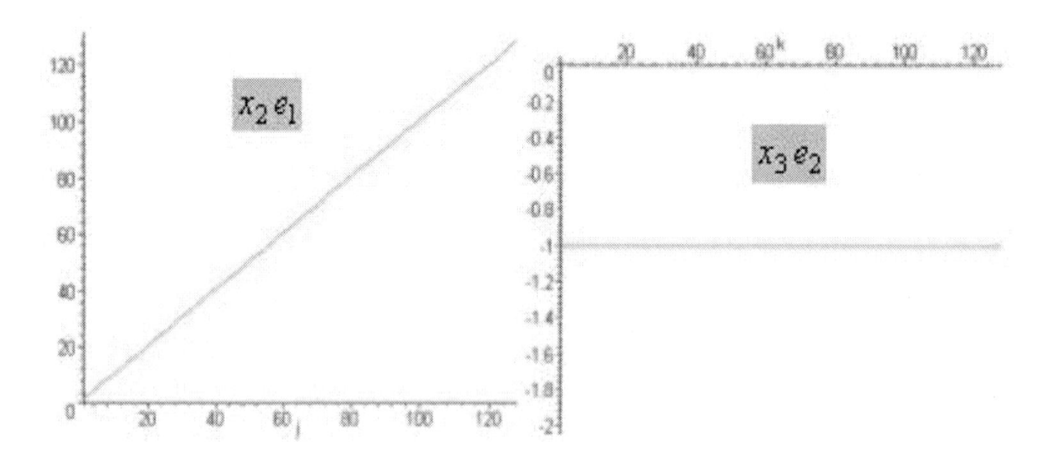

Figure 32. Coefficients at vector translation.

Chapter 41

TRANSLATING PRIMITIVE IDEMPOTENT

We would like to describe the translational mechanics of pure states by applying the method of conformal split. Referring to colour spaces (25) containing primitive idempotents of form

$$f_{(\chi)} = \tfrac{1}{2}(1 \pm e_j)\tfrac{1}{2}(1 \pm e_{k4}) \text{ with } j \neq k = 1,2,3 \tag{364}$$

we consider the additive splits

$$\mathbb{R}^{3,1} = \mathbb{R}^2 \oplus \mathbb{R}^{1,1} \text{ with the Euclidean subspaces } \mathbb{R}^2 \tag{365}$$

having orthonormal bases $\{e_j \,/\, e_j.e_k = \delta_{jk}, j\neq k=1, 2, 3\}$ fulfilling the orthogonality relations as before. Nilvectors for the split are:

$$z_+.e_j = z_-.e_j = 0 \tag{366}$$

$$z_+ = \tfrac{1}{\sqrt{2}}(e_j - e_4) \quad \text{together with the translations} \tag{367}$$

$$T_{e_k} = 1 + \tfrac{1}{2}e_k\, z_+ \overset{def}{=} D_k(x)\,. \tag{368}$$

These conformal displacements $D_k(x)$ seem to be tailor-made for the 24 standard lattice pure states $f_{\chi\eta}$ of the Clifford algebra $Cl_{3,1}$. We have three extensions $E_k = e_{k4}$ which are constitutive for both the primitive idempotents and the homogenous models of correlated Euclidean planes $\{e_j, e_k\}$. We are quite aware that a primitive idempotent is not a vector in the plane $\{e_j, e_k\}$ and is also not a point in its homogenous model. Yet it is obvious that the construction plan for the horosphere resembles that of primitive idempotents. Therefore, it should be interesting to study how an idempotent like, say, f_s responds to a conformal translation of the plane which it involves.

Playing around with any such state $f_{(\chi)} = ½\,(1+e_j)\;½\,(1+e_{k4}) \in Cl_{3,1}$ it becomes apparent that it is not at all self evident that a conformal displacement $D_j(x)$ actually results in a translation in the direction of e_j. Rather it seems the idempotent state resists the translation on the e_j and enforces a peculiar reaction on the space-time area e_{k4}. Clearly, a more complete model requires a conformal split of the whole space-time $Cl_{4,2} = Cl_{3,1} \otimes Cl_{1,1}$. But as was said, it is interesting to first investigate conformal maps of the standard Euclidean planes in the Minkowski algebra. We can legitimately do so by regarding the whole mechanics of pure states like a cybernetic system: pure states respond to translations.

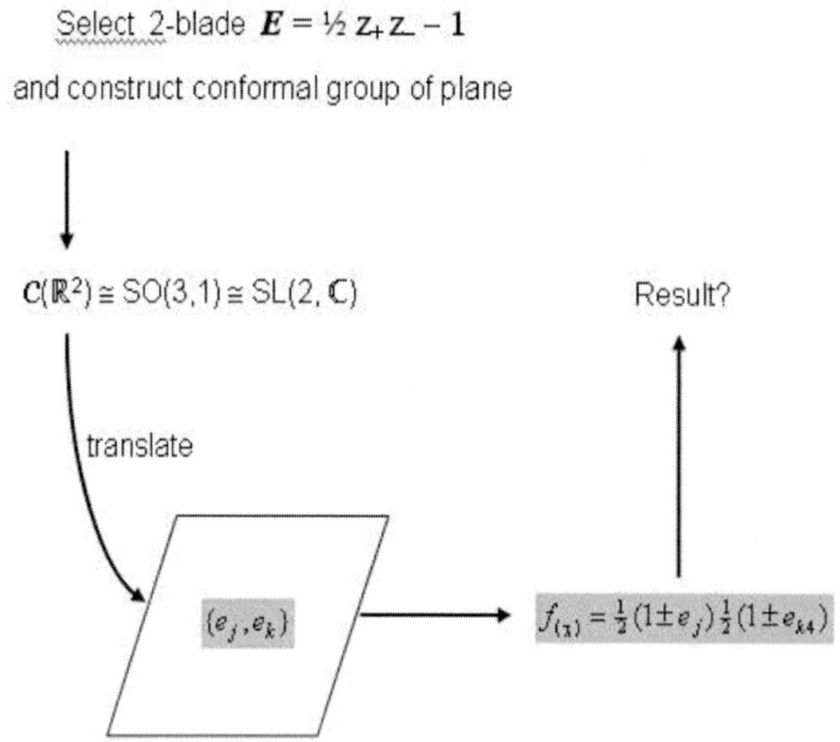

Select 2-blade $\boldsymbol{E} = \frac{1}{2} z_+ z_- - 1$

and construct conformal group of plane

$C(\mathbb{R}^2) \cong SO(3,1) \cong SL(2,\mathbb{C})$

Result?

translate

$\{e_j, e_k\}$

$f_{(\chi)} = \frac{1}{2}(1 \pm e_j)\frac{1}{2}(1 \pm e_{k4})$

Figure 33. Minimal system of conformal effects on pure states.

We call these conformal transformations minimal because they are totally contained in the Clifford algebra where the physics takes place. If they play any role in the sense of real physical interactions, is not known. If they do, this role would be a quite interesting one. If they do not, we have to carry out another rigor and construct a better conformal split for the 4-dimensional colour spaces Ch_χ.

It is impossible to effect a translation of magnitude $\Delta x\, e_j$ of any primitive idempotent $f_{(\chi)} = \frac{1}{2}(1 \pm e_j)\frac{1}{2}(1 \pm e_{k4})$ by the displacement operation $D_j(x)$. But it is possible to induce a conformal split by the space-time area $E = e_{j4}$ and thus ensure a linear translation Δx on e_k by $D_k(x)$. We choose nilpotents $z_+ = 2^{-\frac{1}{2}}(e_j - e_4)$ and $z_- = 2^{-\frac{1}{2}}(e_j + e_4)$. In that case $x\, e_k$ proceeds linearly while the coefficient of the grade 1 vector $y\, e_j$ decreases like a convex parabola. We consider a split by $E = e_{14}$ together with $D_2(x)$.

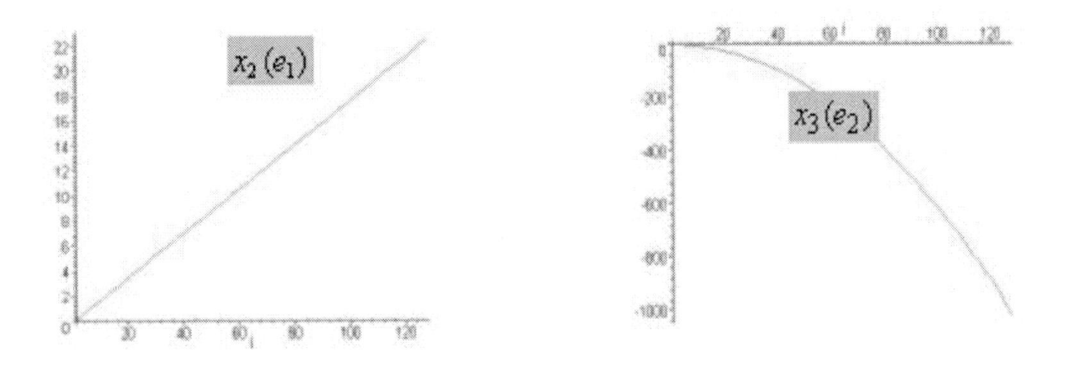

Figure 34. Coefficients at idempotent translation for f_s.

Compare with figure 35 and realize the difference. Also see the coupling between the coefficient x_2 and that of time x_5 telling us that this nonlinear motion nevertheless unfolds on the light cone, a fact characteristic for such conformal transformations. To obtain a feeling for this complex response of the pure state to a translation of its constitutive Euclidean plane, observe how the various effects of the translation compensate or respectively support each other. Namely, the resulting primitive idempotent moves in an 8-dimensional manifold consisting of the components $\{x_1Id, x_2e_1, x_3e_2, x_5e_4, x_6e_{12}, x_8e_{14}, x_{10}e_{24}, x_{13}e_{124}\}$.

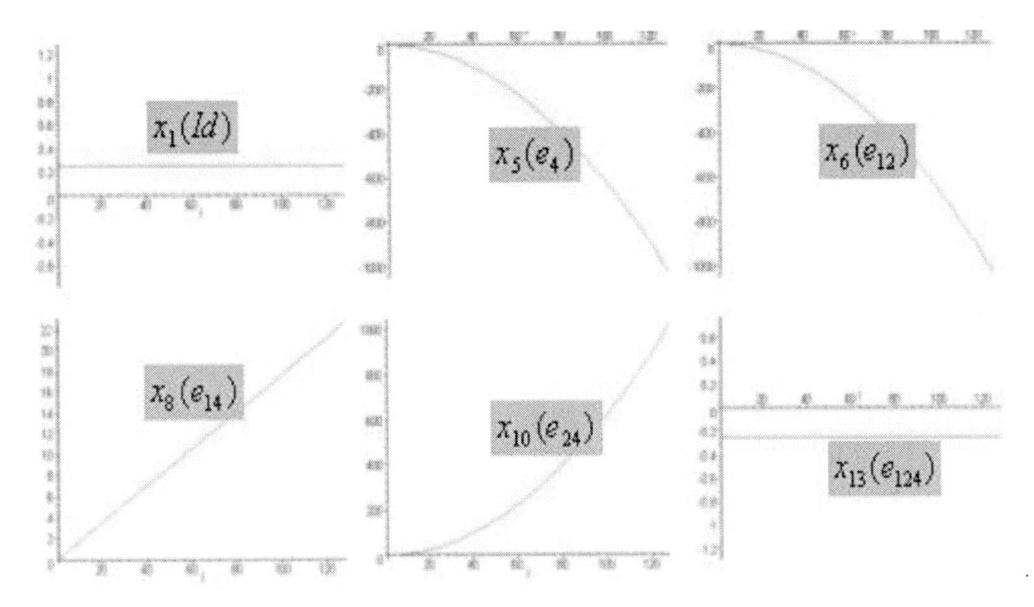

Figure 35. Further coefficients at idempotent translation for f_s.

Obviously, such a linear translation on e_2 and respectively area e_{14} of the primitive idempotent f_s preserves the algebraic property of idempotency. Because we have for any f and D_a

$$(D_a f D_a^{-1})(D_a f D_a^{-1}) = D_a f \, Id \, f \, D_a^{-1} = D_a f D_a \tag{369}$$

and this is independent of the dimension of the resulting subspace wherein the primitive idempotent moves.

We can try to enforce a displacement on e_1 by taking a translation D_1 together with the conformal split induced by the nilpotents $z_+ = 2^{-½}(e_2 - e_4)$ and $z_- = 2^{-½}(e_2 + e_4)$ so that $E = e_{24}$. But that just blocks up any dynamics on direction e_1 and instead causes a linear move on e_2 as shown in figure 36.

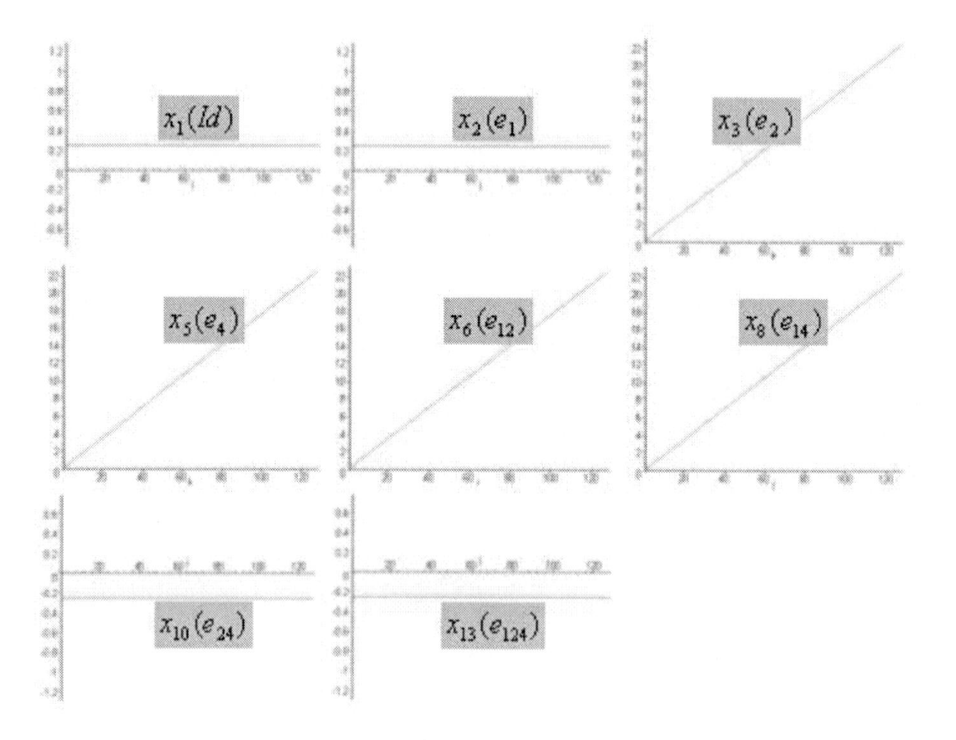

Figure 36. Nilpotent split for f_s enforced by translation T(−e1).

This discloses a linear motion of the idempotent

$$f_s(\tau) = f_s + n(\tau) \tag{370}$$

where the $n(\tau)$ represents a 1-parameter nilpotent as follows

$$n(\tau) = \tau(e_2 - e_4 - e_{12} + e_{14}) \qquad n(\tau)n(\tau) = 0 \qquad \tau \in \mathbb{R} \tag{371}$$

The four diagrams which show horizontal straight lines represent the component f_s. The raising lines belong to the components of the nilpotent. In the course of this kind of linear translation the idempotent splits off a monotonously increasing nilpotent and thus preserves its property because we obtain a multiplication table of the form:

Multiplication Table 5 for nilpotent split of idempotent

	fs	n(τ)
fs	fs	0
n(τ)	n(τ)	0

f annihilates n_τ

n_τ absorbs *f*

So we obtain

$$f_s(\tau)f_s(\tau) = (f_s + n(\tau))(f_s + n(\tau)) = f_s + 0 + n(\tau) + 0 = f_s(\tau) \qquad (372)$$

and the linear displacement preserves the idempotent quality of the $f_s(\tau)$. But this is exactly what we observed in the more general split of $Cl_{4,2} = Cl_{3,1} \otimes Cl_{1,1}$. So we can be sure that the phenomenon of idempotent split results from a conformal map of vectors in the Euclidean plane compounded with the bilinear form of the primitive idempotents.

Since the nonlinear structure of the idempotent translation of f_s involves the plane $\{e_1, e_2\}$ it should be expected that at least the displacement $D_3(x)$ = $T(e_3)$ should be carried out without a complete transposition of the dynamics onto the nonlinear constituents. This is actually the case. Coefficient x_4 can be raised linearly by $T(e_3)$. Yet, the other components result from quadratic maps. See the figure 37.

Recalling chapter 26 on inner and outer we can interpret the 1-parameter nilpotent in

$$n(\tau) = \tau(e_2 - e_4 - e_{12} + e_{14}) \quad n(\tau)n(\tau) = 0 \qquad \tau \in \mathbb{R} \qquad (371)$$

as follows: n(τ) representing the force sustained place of the pure state is an element of the outer isotropic direction field N_1 associated with the primitive idempotent of the fermion. This is indeed the solution to the representation problem. To adequately fix the state and location we need both the primitive idempotent in the inner space (colour) and the nilpotent in the outer isotropic zero space.

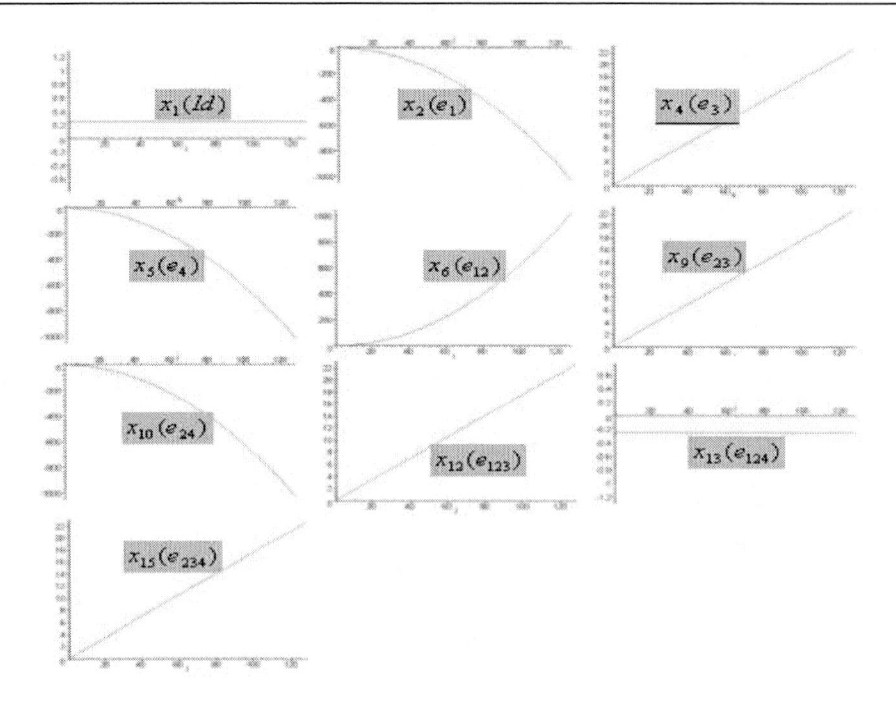

Figure 37. Displacement of f_s induced by translation $T(e_3)$.

Chapter 42

FORCE-SUSTAINED PLACES

In the new image of physical fields, primitive idempotents and nilpotents mutually complete each other. Forces sustain locations and bootstrap their structures. The idempotents react in and at the same time provide these forces while the split apart nilpotents compose their locuses and location curves. The nilpotent may seem entirely split off the pure state, but it is actually one with it, though it is mathematically strictly separable. Force and place can be decomposed although their appearance is rooted in their peculiar composition. A locus can be suspended by a force and a force determines upon the suitable degrees of freedom in local motion. In this way the space-time appears as a most vigilant interaction possessing and disclosing at the same time an intrinsic geometric nature.

Consider in $Cl_{3,1}$ a time stepping motion of a state f_s on $x_3\, e_2$ enforced by a »deflection« from e_1 to e_2.

$$D_1(\Delta x_3, \tau) \overset{def}{=} T(\Delta x_3\, \tau e_1) = 1 + \tfrac{1}{2}\Delta x_3\, \tau e_1\, z_+ \text{ with } \Delta x_3 \text{ real.} \tag{373}$$

While τ steps $\tau = 1, 2, 3, \ldots$ the pure state evolves like

$$f_s(\tau) = f_s - \frac{\Delta x_3 \tau}{4}(1 + e_{25})z_+ = f_s + n_s(\tau) = \text{force} + \text{locus} \tag{374}$$

The structure constant on the right essentially represents progress of step by Eigenzeit and does not alter the fact that $n_s(\tau)$ is a nilpotent, for all times on the trajectory, while the sum $f_s + n_s$ is still a primitive idempotent. It is important to realize that the equation can be represented in a second form:

$$f_s(\tau) = f_s + \frac{\Delta x_3 \tau}{4}(e_2\, z_{++} - z_+) \tag{375}$$

where z_{++} is the nil bivector $-e_5\, z_+$. In this form it is made explicit that $f_s(\tau)$ travels on a nilpotent, point-like trajectory in direction e_2. These facts are compiled in figure 38.

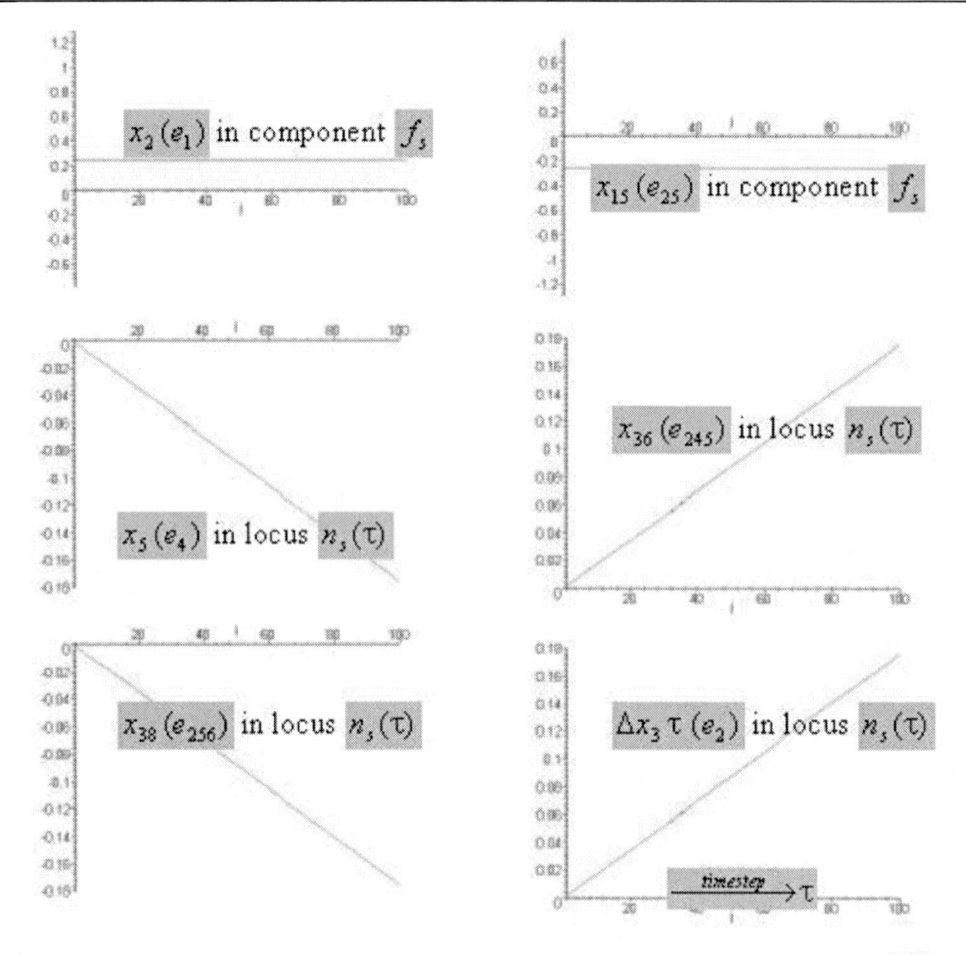

Figure 38. Translation of f_s induced by deflection $D_1(\Delta x_3,\tau)$.

The deflection $D_1(\Delta x_3,\tau)$ can be generalized for any vector in $\xi \in Cl_{3,1}$ in particular to $D_1((\Delta x_2, \Delta x_3, \Delta x_4, \Delta x_6,\tau)$ such that the resulting point $n_s(\tau)$ can represent the locus ξ of the force f_s. We locate traditionally $\xi(f_s) = x_2\, e_1 + x_3\, e_2 + x_4\, e_3 + x_6\, e_5 \in \mathbb{R}^{3,1}$.

A little algebra shows that

$$\forall_{\xi \in Cl^{(1)}_{3,1}}\ \exists_{n_S}\ f_s(\tau) = f_s + n_s(\xi,\tau) \ \text{ with } \ f_s^2(\tau) = f_s(\tau)$$

$$\text{and } n_s^2 = 0 \tag{376}$$

$$n_s(\xi,\tau) = \frac{1}{4}\big(\Delta x_2(-Id + e_{25}) - \Delta x_3(e_5 + e_{15}) - \Delta x_6(e_2 + e_{12})\big)z_+ \tag{377}$$

Factor Δx_4 cancels out.

When we started to set the pure state in motion, we began with the primitive idempotent f_s, and we overlooked that we actually began with a magnitude $f_s + 0$. This is the origin of the former difficulties that arose from my idempotent approach.

Chapter 43

QUATERNION SOLITONS AND STRINGS

A century had passed since physicists had found some important unitary elements of motion which seemed to disclose a certain universal validity. One of the most important, most of us thought was the harmonic oscillator. This could be subjected to first quantization and after that gave rise to second or field quantization. It began with a lecture in Berlin held by Max Planck in December 1900. About twenty years later Pascual Jordan laid the foundations of quantum field theory which he elaborated together with Oskar Klein, Eugene Wigner and Wolfgang Pauli. There appeared a certain problem with divergent energies of infinitely many ground states. This required some procedure of renormalization or, as it was called, regularization. Another deficiency was in its lack of topological features. Those had to be added by designing oscillating chains and other kinds of couplings defined by structure and topology. The quantized oscillator was somehow deprived of topological properties. It was a nice tool, some sort of dynamic morphogenetic type. But something was missing. As far as I can see today the missing link, the lacking dynamic element was hidden in the riches of geometric algebra or, more concretely, in the middle of quaternion analysis.

It is most surprising that some of us must have felt this or somehow were aware of it, but could not directly approach it. For instance Pertti Lounesto in »*Clifford Algebras and Spinors*« gives his students some wonderful updated versions of the Schrödinger and Schrödinger-Pauli equations (Lounesto 2001, chapter 4, p. 51) and rather quickly enters the domain of *"Quaternions"* which is indeed the title of chapter 5. The topic does not fully release him and he comes back to quaternions in his chapter 20 on *"Hypercomplex Analysis"*. Within a few pages he most creatively, almost desperately, seeks a way out of his feeling there was something missing, something sliding through the net. He writes a section *"Even fields"* and thereafter *"Irreducible fields"* (both p. 260), *"Tangential integration"* and *"The Dirac operator without coordinates"*, *"Positive and negative definite metrics"* (p. 264) and *"Monogenic homogeneous functions"*. But somehow he cannot jump over the hurdle that he himself had created in his *"Function theory of quaternion variables"* (section 5.8, p. 74). He states *"many generalizations are uninteresting, the classes of functions are too small or too large"*.

Another one, most experienced in hydrodynamics, plasma physics, cosmology and irreversible processes found the most fundamental element of motion. He named it after his "old MIT roommate, Jose Haraldo Falçao, Haraldo was always called Falaco" – the Falaco Solitons. But he too did not manage at first to embed the phenomenon into the calculus of geometric algebra. High energy physicists and string experts just did not see the enormous value of Kiehns finding, though he patiently hammered on through five volumes on *"Non equilibrium systems"*. But the element he found doesn't even require non-equilibrium. Mathematically it is a most elementary building block just like a plane wave. We shall now go into it. Because there is no proper field theory without the quaternion "Falaco solitons".

Consider two holomorphic functions

$$\phi_k : \mathbb{C}^2 \longrightarrow \mathbb{C}, \quad \mathrm{k} = 1, 2 \tag{378}$$

and bicomplex functions of the type (379) (recall (283, 284)):

$$\psi_j : \qquad \mathbb{R}^4 \longrightarrow \mathbb{R} \qquad \text{such that} \tag{379}$$

$$\begin{aligned} \psi(p) &= (\phi_1(a,b), \phi_2(a,b)) = \\ &= (\psi_1(x,y,z,u) + i\psi_2(x,y,z,u), \psi_3(x,y,z,u) + i\psi_4(x,y,z,u)) \end{aligned} \tag{380}$$

with the pair $p = (a, b)$ of complex numbers $a = x+iy$, $b = z+iu$. In agreement with Stefan Rönn (2008, p. 13) we denote $p \in \mathrm{B}$ as an element of the bicomplex number field B. The numbers p and ψ can thus be represented in B, $\mathbb{C}^2$ and $\mathbb{R}^4$. We adopt from complex analysis a statement about partial derivatives with respect to a complex variable. Consider a complex function φ holomorphic in both arguments

$$\varphi(a,b) = \xi_1(x,y,z,u) + i\xi_2(x,y,z,u). \tag{381}$$

The partial derivative $\dfrac{\partial \varphi}{\partial a}$ is given by either of the formulas

$$\begin{aligned} &\text{(i)} \qquad \frac{\partial \varphi}{\partial a} = \partial_x \xi_1 + i\partial_x \xi_2 \\ &\text{(ii)} \qquad \frac{\partial \varphi}{\partial a} = \partial_y \xi_2 - i\partial_y \xi_1 \end{aligned} \tag{382}$$

Their validity is equivalent with that of the Cauchy-Riemann equations

$$\begin{aligned} \partial_x \xi_1 &= \partial_y \xi_2 \\ \partial_x \xi_2 &= -\partial_y \xi_1 \end{aligned} \tag{383}$$

Exchanging a with b, x with z and y with u we obtain analogous equations for the partial derivative of φ with respect to b. There exists a bicomplex version of the CR-equations which guarantees that the function ψ of equation (380) is holomorphic. The holomorphism of the component functions ϕ_k taken alone is a sufficient condition for ψ to be continuous. Next we define trigonometric functions by the aid of the complex exponential:

$$\psi(p) \overset{def}{=} e^p = (e^a \cdot \cos b,\ e^a \cdot \sin b), \qquad p = (a, b) \tag{384}$$

with component functions $\phi_k(a, b)$ holomorphic in a, b.

$$\phi_1(a,b) \overset{def}{=} e^a \cdot \cos b \qquad \phi_2(a,b) \overset{def}{=} e^a \cdot \sin b \tag{385}$$

Definition (384) has the same form as that of the complex exponential function as represented by a pair of reals

$$e^a = (e^x \cdot \cos y,\ e^x \cdot \sin y), \qquad a = (x, y) \tag{386}$$

With $b = (z, u)$ the complex functions cos b and sin b take the representation by real pairs

$$\text{(i)} \qquad \cos b = (\cos z \cosh u, -\sin z \sinh u)$$
$$\text{(ii)} \qquad \sin b = (\sin z \cosh u, \cos z \sinh u) \tag{387}$$

With (386) we are able to write $\psi(p)$ as a quadruple in , that is, for $p = (a, b) = (x, y, z, u)$ we get the quaternion exponential

$$\begin{aligned} e^p &= (\psi_1, \psi_2, \psi_3, \psi_4) = \\ &= (e^x(\cos y \cos z \cosh u + \sin y \sin z \sinh u), \\ & e^x(-\cos y \sin z \sinh u + \sin y \cos z \cosh u), \\ & e^x(\cos y \sin z \cosh u - \sin y \cos z \sinh u), \\ & e^x(\cos y \cos z \sinh u + \sin y \sin z \cosh u)) \end{aligned} \tag{388}$$

The quaternion function $\psi(p)$ satisfies the differential equation

$$\frac{\partial \psi}{\partial p} = \psi \qquad \text{with} \qquad \psi = e^p \tag{389}$$

Consider the real pair $\cos b = (\cos z \cosh u, -\sin z \sinh u)$ together with u and look at the figure

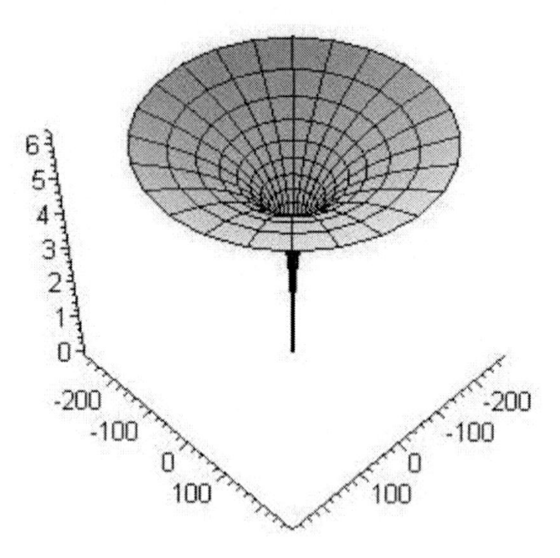

Figure 39. The real triple (cos z cosh u, - sin z sinh u, hu), z, u= 0…2π.

We have introduced the helicity $h = \pm 1$ along the real measure u to indicate that we can reverse the direction of the singularity from downside up to upside down. The triple accommodates still another chirality indicator, namely the sign of the second component of cos b. This is the handedness of the surface. The influence of the sign of the second term, that is, the handedness can be realized in figures 40 and 41. The term cos b of the component function ϕ_1 can be compared with the term sin b in the component ϕ_2. This is depicted by figure 42. We can observe the typical phase shift of $\pi/2$ between the cos- and the sin-figures.

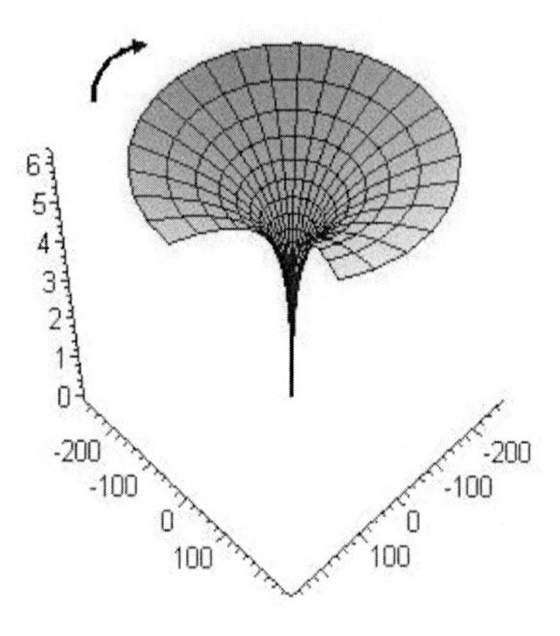

Figure 40. The real triple (cos z cosh u, - sin z sinh u, hu), z, z = 0…5π/3, u = 0…2π.

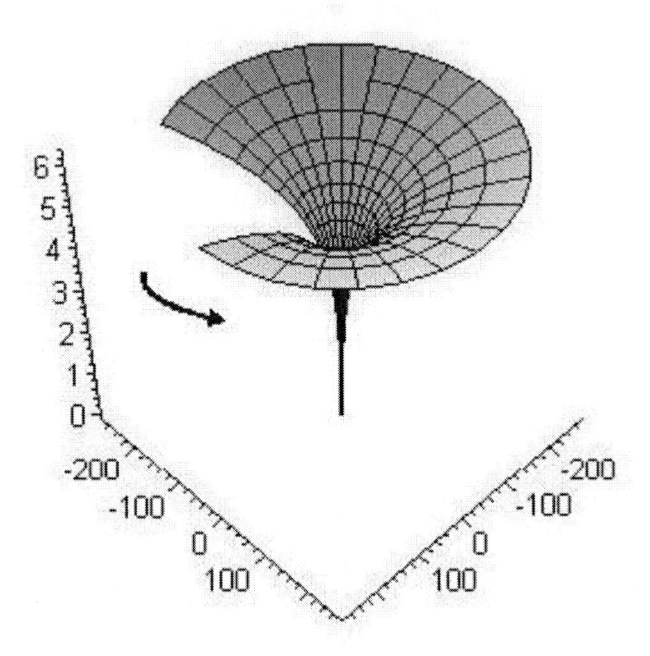

Figure 41. The real triple ($\cos z \cosh u$, $+ \sin z \sinh u$, hu), z, $z = 0 \ldots 5\pi/3$, $u = 0 \ldots 2\pi$.

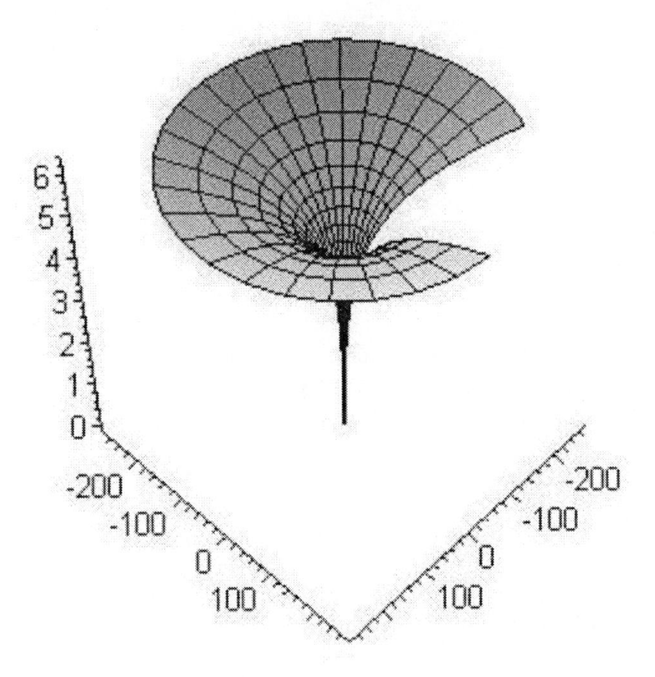

Figure 42. The real triple ($\cos z \cosh u$, $\sin z \sinh u$, hu), z, $z = 0 \ldots 5\pi/3$, $u = 0 \ldots 2\pi$.

A soliton can be coupled to another with opposite helicity.

Figure 43. Quaternion solitons with umbilic.

A soliton can be coupled to another with opposite helicity.

In his volume on »*Topological Torsion and Macroscopic Spinors*« Kiehn (2008, p. 70) defines 3 D direction fields of pure spinors as we have done by Cartan's approach in formula (109).

$$x_1 = \xi_0^2 - \xi_1^2, \qquad x_2 = i(\xi_0^2 + \xi_1^2), \qquad x_3 = -2\xi_0\xi_1$$

He also displays the perhaps more familiar Weierstrass construction in 3 D

$$f = \xi_0^2, \qquad g = \frac{\xi_1}{\xi_0},$$

$$x_1 = (1-g^2)f, \qquad x_2 = \pm i(1+g^2)f, \qquad x_3 = \mp 2fg \tag{390}$$

Still other generic forms for spinor direction fields could be deduced from the complex hyperbolic formulas:

Euclidean $\left|s_h\right\rangle = (a+ib)(-e_1 \sinh z + i e_2 \cosh z + e_3)$

Lorentz spinor $\left|s_h\right\rangle = (a+ib)(-e_1 \sinh z + i e_2 \cosh z + i e_3)$

Majorana spinor[1] $\left|s_h\right\rangle = (a+ib)(-e_1 \sinh z - i e_2 \cosh z + i e_3)$ (391)

Each complex direction field can be multiplied by any holomorphic function $F(z)$ to yield $|S\rangle = F(z)\,|s_h\rangle$. For each suitable metric we have to have

$$\langle S|\eta|S\rangle = 0 \tag{392}$$

Kiehn also displays the elliptic complex direction fields given by corresponding trigonometric formulas. Again the geometric product of the transforms of spinors by arbitrary holomorphic functions with themselves gives zero:

Euclidean spinor $\left|s_E\right\rangle = (a+ib)(e_1 \sin z + e_2 \cos z + i e_3)$

Lorentz spinor $\left|s_L\right\rangle = (a+ib)(e_1 \sin z + e_2 \cos z - e_3)$

Majorana spinor $\left|s_M\right\rangle = (a+ib)(e_1 \sin z + e_2 \cos z - e_3)$ (393)

For each appropriate choice of a metric we again obtain the identity (392). Kiehn then derives direction fields in the Lorentz metric for b=0 and z=u+iv which could lead to

$$b=0 \Rightarrow \left|s_L\right\rangle = a e_1 \sin u \cosh v + a e_2 \cos u \cosh v - a e_3$$
$$+ i a \cos u \sinh v + i a e_2 \sin u \sinh v \tag{394}$$

Finally he investigates the "*special immersion*" with a real part

$$e_1 \sinh u \cos v + e_2 \sinh u \sin v + (Au + B) e_3 \tag{395}$$

"*that can be extracted from the complex Minkowski spinor formulas leading to a surface of zero mean curvature in Lorentz, or Majorana, space (but not in Euclidean space) with a position vector*" which he writes as

R(u, v) = [sinh(u) cos (v), sinh(u) sin(v), Au+B] (396)

In figure 44 we can see the plot of the triple of real numbers [sinh(u) cos (v), sinh(u) sin(v), u] which represents Kiehn's mathematical representation of a Falaco soliton or „spinor

1 This had to be corrected. Kiehn had, $\left|s_h\right\rangle = (a + ib)(-i \sinh z, -\cosh z, 1)$ which does not satisfy the Majorana signature (– – +).

generated zero mean curvature surface with a conical singularity in a 3 D Lorentz or Majorana space". As you see from formulas (393) the elliptic terms have indeed identical representations in both signatures.

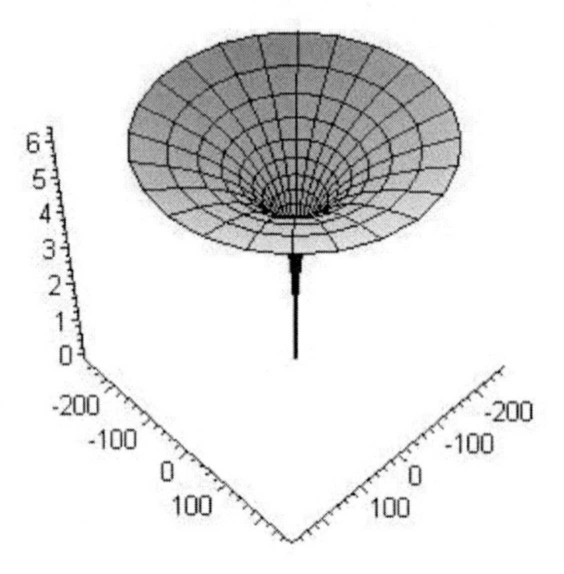

Figure 44. "Spinor generated zero mean curvature", the real triple (sinh z cos u, sinh z sin u, hu), $z = 0 \ldots 2\pi$, $u = 0 \ldots 2\pi$ - " Falaco soliton".

This does not seem to be different from figure 39. But there is a difference.

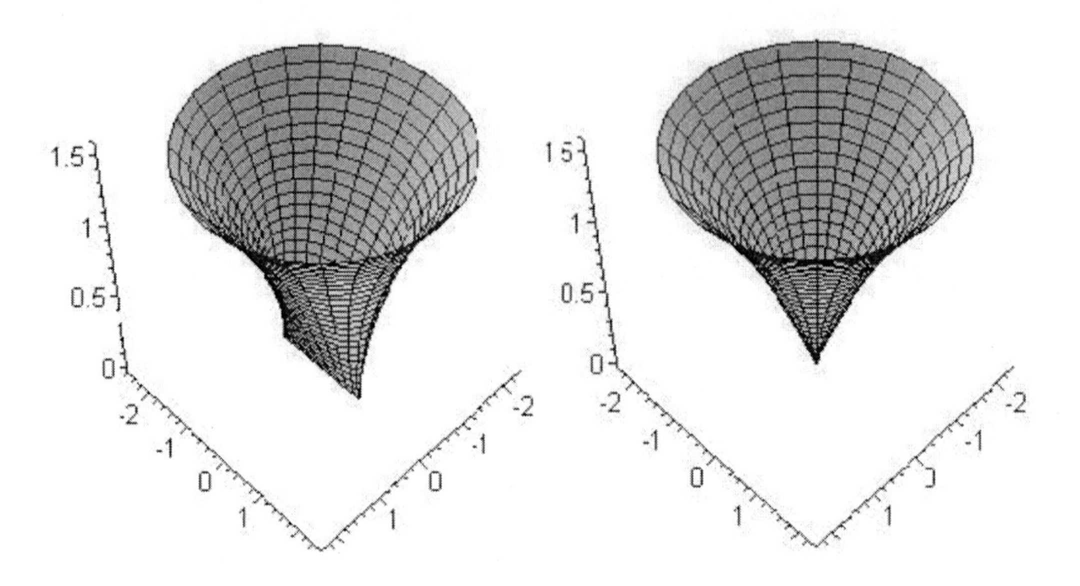

Figure 45. Quaternion soliton on the left, Falaco soliton on the right, $z = 0 \ldots 2\pi$, $u = 0 \ldots \pi/2$.

Letting $0 \leq u \leq \pi/2$ we obtain a point singularity for the Falaco soliton, but a straight line for the quaternion soliton. This result is new at the time when this chapter is written. But it is not any longer unexpected. Fernández, López and Souam (2008) have surveyed literature on minimal and maximal surface equations and related topological problems and derivated a space of complete embedded maximal surfaces with isolated subsets of singularities in the Minkowski space $\mathbb{R}^{2,1}$. What is entirely surprising, however, is the immediate relation between the quaternion soliton hypersurface with a singularity line and the Fueter equation. The connection with the work of Kiehn and the so called Falaco solitons is also evident.

Maximal immersed surfaces, as have been called maxfaces by Umehara and Yamada (to appear) share certain properties with minimal surfaces in Euclidean 3-space. They have a conformal Gauss map and admit the typical Weierstrass representation. Some of those have been investigated by Kobayashi (1984). Some new ones as the quaternion soliton in figure 45 to the left, are characterized by straight lines or complete circles of singularities located in spacelike planes parallel to the surface with zero mean curvature. The inference is that the harmonic oscillator does not represent the proper dynamic element for field quantization. A more appropriate element is provided by the soliton which is admitted by the simple quaternion equation $\partial_p \psi = \psi$. There has been published an interesting paper in honour of the 70th birthday of Professor J. A. C. Alcarás which offers some kind of celebration of the *Variational Principle*, but for *Quaternionic Quantum Mechanics*. It even provides as an application: "*The Quaternionic Harmonic Oscillator*", but does not yet realize the dynamic element of this oscillator. It is very probable that it is fully derivated at the time when this book is published.

Chapter 44

ISOSPIN DECONVOLUTION OF SOLITONS

FALACO SOLITON DYNAMICS IN CLIFFORD ALGEBRA

The layout of a quantum mechanic system in graded Lie algebra is based on the complex combinatorial structure of geometric Clifford algebra. The plot works equally well for subnuclear, nuclear, biochemic and macroscopic phenomena. Where the Lagrangian is concerned, strictly, even the trajectory of a single particle can involve all sixteen dimensions of the space-time algebra. Moving fields are thus conceived as deconvoluting entities in a high dimensional space of Clifford numbers. The old approaches cannot be launched easily.

Though we are able to distinguish precisely the isospin from graded hyperspin, the spin from Lorentz transformations and those from the dynamics of solitons, nevertheless, these are all connected within the rich representation spaces of the algebra. We can show this easily, even in the vicinity of the traditional theory in which hyperspin elements are restrictedly equivalent with Lorentz transformations. Suppose we had a particle travelling alongside the direction e_2 with relativistic velocity v. Therefore we consider the isospin component λ_3, that is, the third generating element of the real form of $SU(3)$ in $Cl_{3,1}$. By the aid of this element we carry out *transformations of places* as vectors just as well as *deconvolutions of fields* by the conjugate transformation. Recall the equations of part V !

As we chose the λ_3, this preserves all the pure states f_1, f_2, f_3, f_4 from space ch_1. Further, it induces by conjugation the Lorentz transformation

$$e_2' = e^{2\theta\lambda_3}(e_2)e^{-2\theta\lambda_3} = (\cosh\theta)e_2 - (\sinh\theta)e_4 \tag{397}$$

which is a boost transform. Here we used twice the angle θ to have it single in the hyperbolic functions. This accords with the spinor property of λ_3. In part V we wrote that as a mapping

$$\lambda_3 : e_2 \mapsto (\cosh\theta)e_2 - (\sinh\theta)e_4 \tag{398}$$

Consider further λ_2, interacting with λ_3. The second isospin component preserves the pure states f_1, f_2, but unfolds f_3, f_4 from space ch_1 into space $ch_1 \oplus ch_2$. For instance it deconvolutes the fermion states f_3, f_4 according to

$$\lambda_2 : f_3 \mapsto \tfrac{1}{4}(1-e_1)+\tfrac{1}{4}(\cos^2\theta-\sin^2\theta)(e_{24}-e_{124})+ \\ +\tfrac{1}{2}(\sin\theta\cos\theta)(e_{34}-e_{134}) \tag{399}$$

You see, the first and second terms are elements in ch_1, but the third summand extends the primitive idempotent of ch_1 into the space $ch_1 \oplus ch_2$. We obtain another transform for the forth state f_4 which is symmetric to the third

$$\lambda_2 : f_4 \mapsto \tfrac{1}{4}(1-e_1)-\tfrac{1}{4}(\cos^2\theta-\sin^2\theta)(e_{24}-e_{124})+ \\ -\tfrac{1}{2}(\sin\theta\cos\theta)(e_{34}-e_{134}) \tag{400}$$

since we have

$$f_3'+f_4'=\eta=f_3+f_4\,. \tag{401}$$

Letting the third and second isospin element act on the Minkowski space, we obtain for the direction e_2 into which the particle is travelling with relativistic velocity v the following transform

$$\lambda_2\lambda_3: \qquad e_2 \mapsto (\cos\alpha\cosh\theta)\,e_2+(\sin\alpha\cosh\theta)\,e_3 - \\ -(\cos\alpha\sinh\theta)\,e_4-(\sin\alpha\sinh\theta)\,j \tag{402}$$

which signifies a deconvolution of the direction e_2 into the 4-dimensional subspace spanned by a space unit double e_2 , e_3 and a timespace double e_4 ,j. We can make a three-dimensional smartplot using the coefficients of the spacelike e_2 , e_3 and the quantity θ = arctanh (v/c) with light velocity c. We obtain a maximal immersed surface isomorphic with the Falaco soliton

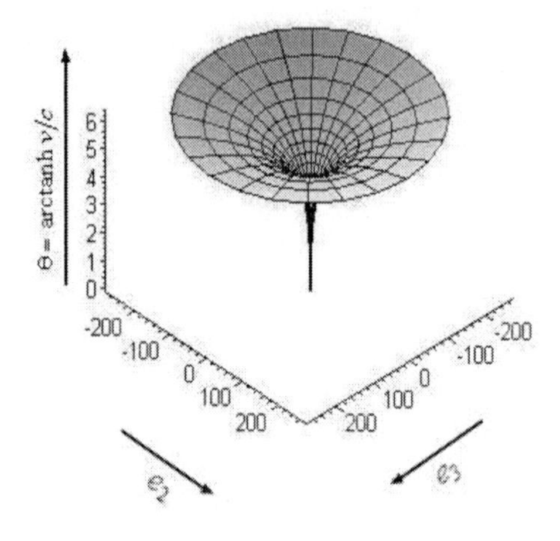

Figure 46. Isospin deconvoluted soliton, maximal immersed surface from Lorentz transformation, (cos α coshθ)e_2 + (sinα coshθ)e_3, $z = \alpha$, $u = \theta$ = arctanh v/c, $z = 0\ldots2\pi$, $u = 0\ldots2\pi$.

We can select any pair of coordinates from (402) and plot it together with θ, we always obtain the typical figure of a Falaco soliton. As in special relativity theory time and space are conceived as equivalent, we obtain symmetric solutions for that type of smart-phaseplot. For example if we select e_3, e_4 and θ we get:

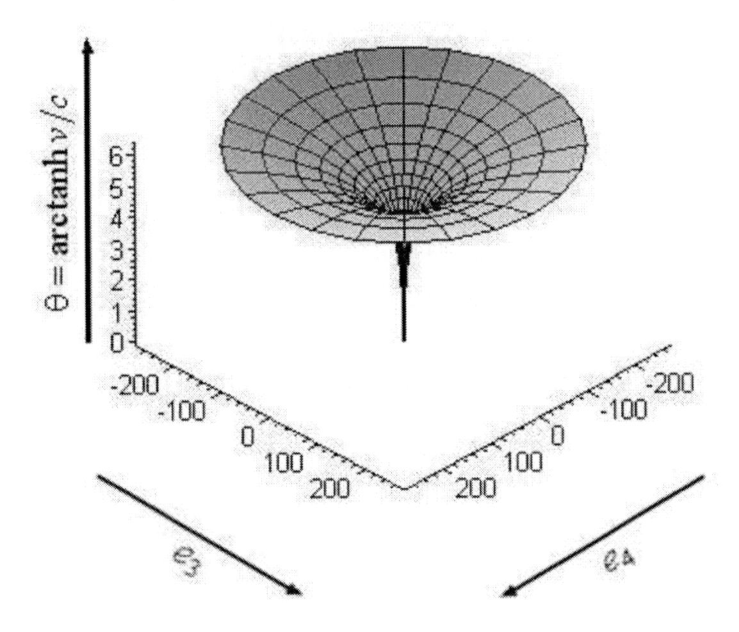

Figure 47. Isospin deconvoluted space-time soliton, maximal immersed space-time surface from Lorentz transformation, $(\sin\alpha\ \cosh\theta)e_3 - (\cos\alpha\ \sinh\theta)e_4$, $z = \alpha$, $u = \theta = \text{arctanh } v/c$, $z = 0\ldots2\pi$, $u = 0\ldots2\pi$.

Chapter 45

THE PHOTON SINGULARITY

The case $v = c$ signifies a topological defect structure. Classically it meant that the energy contribution for massive particles diverges and the measure of time converges to zero. In the new approach based on Clifford algebra the situation is not that simple. This is not surprising since, already in the classic models of special relativity, time and space are in a correlation which, strictly speaking, is a conversion relation. The equation (397) designates exactly that relation. In addition to that, the measure of time partakes in a phenomenology of solitons as shown in figure 47. That is, we obtain a maximal space-time surface.

TIME BECOMING AN ISOTROPIC DIRECTION FIELD

When Kiehn conceptualized the photon as a propagating topological singularity in the light cone, he constructed a model following the hydrodynamic idea of the Falaco soliton. He did not yet see that this type of soliton is a direct consequence of the special theory of relativity on the one hand and a quaternion type of quantum mechanics on the other. From each of the theories there follows the existence of a dynamic element which is isomorphic with the Falaco soliton.

We must be aware, the process of isospin transformation involves the classic discontinuity where the measure of time goes to zero. But this is not all. Additionally this discontinuity means a real structural transition from a directed time unit vector e_4 which squared gives -1 to a direction field which squared gives 0.

In the chapter on inner and outer states we found the three isotropic direction fields N_1, N_2, N_3 which satisfy the equations

$$N_1 \subset span_R\{e_2, e_4, e_{12}, e_{14}\} \text{ with } \qquad N_1^{\wedge 2} = ch_1$$
$$N_2 \subset span_R\{e_3, e_{13}, e_{234}, j\} \qquad N_2^{\wedge 2} = ch_1$$
$$N_3 \subset span_R\{e_{23}, e_{34}, e_{123}, e_{134}\} \qquad N_3^{\wedge 2} = ch_1$$

The N are root spaces of colour spaces. N_1 to N_3 are roots of the first colour space. They can also be denoted as root spaces of light at the singularity. To see why this must be so, let us consider the hypercharge generator λ_8 and respectively the hypercharge element Y = $(1/6)(-2e_1 + e_{24} + e_{124})$. We have

$$Y|s\rangle = Yf_2 = -\tfrac{2}{3}f_2,\ \ Y|u\rangle = Yf_3 = \tfrac{1}{3}f_3,\ \ Y|d\rangle = Yf_4 = \tfrac{1}{3}f_4 \tag{403}$$

the correct measures for hypercharge and the fact that the hyperspin-deconvolution of the three fermion pure states leaves those states unaltered. That is, we have

$$Y,\ \lambda_8:\qquad \begin{array}{l} f_2 \mapsto f_2 \\ f_3 \mapsto f_3 \\ f_4 \mapsto f_4 \end{array} \tag{404}$$

However, while preserving the pure states, the hyperspin deconvolutes the directed unit vectors as follows

$$\begin{aligned}
&\lambda_8 : e_1 \mapsto e_1 \\
&e_2 \mapsto \left(\cosh\tfrac{2\theta}{\sqrt{3}}\right)\left(2\cosh^2\tfrac{2\theta}{\sqrt{3}}-1\right)e_2 - \left(\sinh\tfrac{2\theta}{\sqrt{3}}\right)\left(2\cosh^2\tfrac{2\theta}{\sqrt{3}}-1\right)e_4 - \\
&-2\left(\sinh\tfrac{2\theta}{\sqrt{3}}\right)\left(\cosh^2\tfrac{2\theta}{\sqrt{3}}\right)e_{12} + 2\left(\sinh^2\tfrac{2\theta}{\sqrt{3}}\right)\left(\cosh\tfrac{2\theta}{\sqrt{3}}\right)e_{14} \\
&e_3 \mapsto \left(\cosh\tfrac{2\theta}{\sqrt{3}}\right)\left(2\cosh^2\tfrac{2\theta}{\sqrt{3}}-1\right)e_3 - 2\left(\sinh\tfrac{2\theta}{\sqrt{3}}\right)\left(\cosh^2\tfrac{2\theta}{\sqrt{3}}\right)e_{13} + \\
&+2\left(\sinh^2\tfrac{2\theta}{\sqrt{3}}\right)\left(\cosh\tfrac{2\theta}{\sqrt{3}}\right)e_{234} - \left(\sinh\tfrac{2\theta}{\sqrt{3}}\right)\left(2\cosh^2\tfrac{2\theta}{\sqrt{3}}-1\right)e_{1234} \\
&e_4 \mapsto -\left(\sinh\tfrac{2\theta}{\sqrt{3}}\right)\left(2\cosh^2\tfrac{2\theta}{\sqrt{3}}-1\right)e_2 + \left(\cosh\tfrac{2\theta}{\sqrt{3}}\right)\left(2\cosh^2\tfrac{2\theta}{\sqrt{3}}-1\right)e_4 + \\
&+2\left(\sinh^2\tfrac{2\theta}{\sqrt{3}}\right)\left(\cosh\tfrac{2\theta}{\sqrt{3}}\right)e_{12} - 2\left(\sinh\tfrac{2\theta}{\sqrt{3}}\right)\left(\cosh^2\tfrac{2\theta}{\sqrt{3}}\right)e_{14}
\end{aligned} \tag{405}$$

The important thing to realize is that the base units are deconvoluted into subspaces $\{e_2, e_4, e_{12}, e_{14}\}$ and $\{e_3, e_{13}, e_{234}, j\}$ which contain N_1 and N_2 as subspaces. The square of a Clifford number $x\,e_2 + x\,e_4 + y\,e_{12} + y\,e_{14}$ is equal to zero. Considering e_4 there follows that e_4 turns over into an isotropic direction field if and only if $\sinh(2\theta/\sqrt{3}) = \cosh(2\theta/\sqrt{3})$. This happens at the singularity when the velocity v equals the velocity c of light. This is a real algebraic singularity in the sense that we have

$$e_4^2 = \begin{cases} -1 & \ldots v < c \\ 0 & \ldots v = c \end{cases} \tag{406}$$

At the singularity the directed time unit is no longer a timelike unit vector. But it behaves like an isotropic direction field at infinity. This seems to be a very convincing proof of Kiehns claim that topological defects are connected with Cartans pure spinors and respectively their isotropic direction fields. In our case where the algebra is graded, we are in the lucky position that in certain subspaces the quadratic norm can be extended though the grade varies. We may wonder if this phase transition of the temporal unit from a vector to a direction field could indicate that at the singularity the bosonic field cannot be distinguished from a fermion field. Could that be an indication that the ZBW-approach to the electron as was developed by Hestenes should be extended to quarks?

ISOSPIN DECONVOLUTED TIMESPACE

The timespace $\{e_4, e_{123}, j\}$ is characterized by a number of properties, one of the most important has not yet been mentioned. It is in the fact that any base unit of the timespace turns into an isotropic direction field at the singularity and its norm changes from -1 to zero. Let us list the transforms of each timespace base unit deconvoluted by the hyperspin elements. Beginning with isospin, we obtain:

$$\begin{aligned} &\lambda_1 : e_4 \mapsto -(\sinh\theta) e_3 + (\cosh\theta) e_4 \\ &e_{123} \mapsto -(\sinh\theta)^2 e_{23} + (\sinh\theta \cosh\theta) e_{24} + \\ &+ (\cosh\theta)^2 e_{123} - (\sinh\theta \cosh\theta) e_{124} \\ &j \mapsto -(\sinh\theta) e_2 + (\cosh\theta) j \end{aligned} \tag{407}$$

Each of the fields

$$\begin{aligned} &\xi = -x e_3 + x e_4 \\ &\psi = -x e_{23} + x e_{24} + y e_{123} - y e_{124} \\ &\zeta = -x e_2 + x j \end{aligned} \tag{408}$$

is an isotropic direction field

$$\xi^2 = \psi^2 = \zeta^2 = 0 \tag{409}$$

Therefore, under a transformation by λ_1 at the singularity the norm of each timespace unit vector changes from -1 to 0. This consideration can be repeated for all eight hyperspin elements of the constitutive Lie algebra $\mathcal{L}^{(2)}$.

Four dimensional spaces involved are given by Clifford numbers having the following forms

$$\begin{aligned} &-x e_{23} + x e_{24} + y e_{123} - y e_{124} \\ &-x e_{23} + x e_{34} + y e_{123} - y e_{134} \\ &-x e_1 + x e_{13} + y e_{124} - y j \\ &-x e_1 + x e_4 + y e_{12} - y e_{24} \end{aligned} \tag{410}$$

All Lorentz transformations involving time I have decided to call time-wraps. Two of those are the well known boosts. The rest is representing proper time wraps. Those are the classic boosts

$$\begin{aligned} \lambda_1: &\quad e_4 \mapsto -(\sinh\theta)e_3 + (\cosh\theta)e_4 \\ \lambda_3: &\quad e_4 \mapsto -(\sinh\theta)e_2 + (\cosh\theta)e_4 \end{aligned} \tag{411}$$

These are the time-wraps

$$\begin{aligned} \lambda_2: &\quad e_4 \mapsto (\cos\theta)e_4 + (\sin\theta)j \\ \lambda_4: &\quad e_4 \mapsto (\cosh\theta)e_4 + (\sinh\theta)e_{34} \\ \lambda_6: &\quad e_4 \mapsto -(\sinh\theta\cosh\theta)e_1 + (\cosh^2\theta)e_4 + (\sinh^2\theta)e_{12} + (\sinh\theta\cosh\theta)e_{24} \end{aligned} \tag{412}$$

Recall, each generator of the $su(2)$- subalgebra of $\mathcal{L}^{(2)}$ preserves exactly one base unit of the timespace:

$$\lambda_7: \quad e_4 \mapsto e_4 \qquad \lambda_2: \quad e_{123} \mapsto e_{123} \qquad \lambda_5: \quad j \mapsto j \tag{413}$$

Chapter 46

PRIMORDIAL MATTER

KÄHLER-DIRAC QUARK HYDROGEN

Vargas and Torr (2004, 2008) established contact to the quark model shown in the first parts of this book. They developed new perspectives on the Kähler calculus and wave functions based thereon. Relying on the exact group theoretic solution for the real forms of the standard model in Clifford algebra $Cl_{3,1}$ and the corresponding pure states (the cursors) fixed by the constitutive Lie group (Schmeikal 2002) the authors reduced the solutions of the Kähler-Dirac equation by the aid of those symmetries to Dirac solutions. They first established the known correspondence between orthogonal idempotents of the algebra and the following constant differentials

$$
\begin{aligned}
e^{+} &\overset{def}{=} \tfrac{1}{2}(1-ie_4) \Leftrightarrow \varepsilon^{+} = \tfrac{1}{2}(1-i\,dt) \\
e^{+} &\overset{def}{=} \tfrac{1}{2}(1+ie_4) \Leftrightarrow \varepsilon^{-} = \tfrac{1}{2}(1+i\,dt)
\end{aligned}
\tag{414}
$$

$$
\begin{aligned}
t^{+} &\overset{def}{=} \tfrac{1}{2}(1+ie_{12}) \Leftrightarrow \tau^{+} = \tfrac{1}{2}(1+i\,dx_1 dx_2) \\
t^{-} &\overset{def}{=} \tfrac{1}{2}(1-ie_{12}) \Leftrightarrow \tau^{-} = \tfrac{1}{2}(1-i\,dx^1 dx^2)
\end{aligned}
\tag{415}
$$

Note that whenever we have a multiplication with no symbol this denotes a Clifford product. The product of the two forms yields zero

$$\varepsilon^{+}\varepsilon^{-} = 0 \quad \text{and} \quad \varepsilon^{+} + \varepsilon^{-} = 1 \tag{416}$$

$$\tau^{+}\tau^{-} = 0 \quad \text{and} \quad \tau^{+} + \tau^{-} = 1 \tag{417}$$

And their commutators vanish

$$[\varepsilon^{\pm}, \tau^{+}] = 0 \text{ and } \quad [\varepsilon_{\pm}, \tau_{-}] = 0 \tag{418}$$

The quantity dx^1dx^2 is related to rotations about the direction e_3 and dt can be solved in terms of the ε's and respectively in terms of the τ's. Next define wave functions by the equation

$$u = u' idt + u'' \tag{419}$$

where neither u' nor u'' contains dt as a factor. Now we write

$$u = u'(\varepsilon^- - \varepsilon^+) + u''(\varepsilon^+ + \varepsilon^-) = u^+\varepsilon^+ + u^-\varepsilon^- \tag{420}$$

which uniquely determines u^+ and u^-. Indeed, if there existed different solutions for u^+ and u^- their differences v would satisfy

$$v^+\varepsilon^+ + v^-\varepsilon^- = 0 \tag{421}$$

Multiplying from the right by any one of the two idempotents ε would annihilate one of the two terms from which we can conclude that the remaining difference is zero. So the solutions are definite. If we use cylindrical coordinates, we have

$$dx^1dx^2 = \rho\, d\rho\, d\phi \tag{422}$$

Bearing in mind that $d\rho\ d\rho = 1$ and $d\phi\ d\phi = \rho^{-2}$ one readily gets

$$d\phi = i\frac{d\rho}{\rho}(\tau^- - \tau^+) \tag{423}$$

We can decompose the u^+, u^- further with respect to $d\phi$ just the same way u could be analysed with respect to dt. Vargas and Torr obtain:

$$u = u_+^+\,\tau^+\,\varepsilon^+ + u_-^+\,\tau^+\,\varepsilon_- + u_+^-\,\tau^-\,\varepsilon^+ + u_-^-\,\tau^-\,\varepsilon^- \tag{424}$$

which uniquely determines the differential forms u_+^+, …, u_-^-.

The products $\tau^\pm\,\varepsilon^\pm$ are the orthogonal, mutually annihilating primitive idempotents we used in the early parts of this book. Multiplying u by any of the four primitive idempotents annihilates three of the four terms in (424). Therefore the Kähler-Dirac wave function can be written as

$$u = u\tau^+\varepsilon^+ + u\tau^+\varepsilon^- + u\tau^-\varepsilon^+ + u\tau^-\varepsilon^- \tag{425}$$

Recall this is a typical colour space decomposition in a space spanned by primitive idempotents. If you factor in the calculations from the chapter on *basic laws of pattern formation* you would be able to immediately write down the form of any possible solution within a colour space as in equation (145) which is the same as (425). We can confirm this result. The authors also point out that the primitive idempotents in the above differential shape are projection operators over states of proper value for the third component of angular momentum and energy. Since these are constant differentials in the Lorentz-Minkowski space-time the components $u\ \tau^{\pm}\ \varepsilon^{\pm}$ are solutions of the same Kähler equation as u is. In general, the four components may depend on all space-time coordinates in general unless there are symmetries imposed on the idempotents. Assuming that the differential form on the right hand side of the Kähler-Dirac equation as well as the metric and the torsion do not depend on t and ϕ one can show, following Kähler, that the solution u in the presence of those symmetries is the sum of four independent solutions of form

$$e^{i m \phi - iEt/\hbar} q_{\pm}^{\pm}\ \tau^{\pm}\ \varepsilon^{\pm} \tag{426}$$

where m is an integer and q depends solely on the ρ and z coordinates. Vargas and Torr point out that solutions (426) in turn yield the solutions

$$e^{i m \phi - iEt/\hbar} q_{\pm}^{\pm} \tag{427}$$

which, for the hydrogen atom, are solutions of the Dirac equation, as was explicitly shown by Kähler (1960, 1961). *Clearly the validity of the form* (426) *holds for KD wave functions of any system possessing the same symmetries.*

The way the authors establish contact to 'the Schmeikal model' is very original. Actually equation (426) is an expansion of the elementary colour space wave functions in the linear space spanned by Weyl's "erzeugende Einheiten", that is, primitive idempotents. Restricting Clifford numbers by decoupling the measures from the four orthogonal primitive idempotents we obtain the four numbers in equation (427). The idea to replace the standard primitive idempotents of $Cl_{3,1}$ by differential forms $\tau^{\pm}\ \varepsilon^{\pm}$ leads to plane wave solutions for the coordinate measures of the graded pure states. Therefore the $SU(3)$ emerges as a natural space-time symmetry as in the original model. The application tells us that the equation of motion for the hydrogen atom may not be extremely sensitive to a shift from the Dirac- to the Kähler-Dirac equation. The authors argue that *the equations of motion did not emerge just as a result of trying to fit the workings of nature in the way that the Klein-Gordon, Schrödinger and Dirac-equation were.* [...] *The Kähler-Dirac equation is the central equation in the natural evolution of the calculus of differential forms beyond the exterior calculus. Eventually, it would have been found by mathematicians if physicists had not yet had a need for quantum mechanics. Assume further that, if discovered before the Dirac equation, one would still have felt the need to explain the fine structure of the hydrogen atom, which the KD equation does. It is clear that the ad hoc Dirac equation would not have seen the light of the day, unless discovered in the process of studying the reduction of the KD wavefunction when there are symmetries. Suppose that the intimations of the last sections of this paper were correct and could be verified in part. This would raise the issue of whether so far intractable problems, like the inner working of particles, could be explained in terms of some sophisticated KD equation, the sophistication arising mainly from the structural implications of having a tensor valued input differential form, and concomitant additional degrees of freedom. There is no obvious precursor in the physics paradigm from which to start the search of such a tensor-valued input differential form, at least not a physical precursor from which Kähler would have drawn guidance as he did for the treatment of the hydrogen atom. The reason is that this equation is naturally connected with geometry in the main or tangent bundle, not with gauge geometry* (*i.e. in auxiliary bundles*).

This is a most relevant contribution to the understanding of graded systems dynamics. The inner working of particles as related to the outer symmetries, a KD-like equation of motion arising from a fine grading and additional degrees of freedom are important. First of all we must accept that a particle cannot be much less than a manifold in the whole algebra. If we pick out a small trajectory restricted to a few dimensions, this might seem reasonable and easily to solve. But it does not reflect the whole. First of all the elements of a colour space, eventually spanned by four structurally equivalent primitive idempotents instead of base units having four different grades are better seen as cursors to the correlated isotropic direction fields. Those fields point beyond the tensor calculus and involve torsion and critical transitions beyond thermodynamic equilibrium. Nevertheless, the first observation may be that the universal phenomenology of hydrogen-like spectra with a fine structure also concerns any composite of quarks.

NULLITON

In the year 2007 I reviewed some interesting paper by Robert Gordon Wallace (2008) from Australia. He coined the name of the *nulliton*, a dark matter analogue to the hydronen atom. The author based his theory on an early work by Hestenes (1981) and one of my papers on *minimal spin gauge theory* (2001). He presented Clifford algebra in some as he correctly said *ungraded matrix algebra* isomorphic with the $Cl_{4,1}$. The paper was neatly done and all derivations were most cleanly elaborated. Basic rigor was carried out by a spreadsheet. So I first had to translate everything into the languages I used. Those were Maple Clifford by Rafal Ablamowicz and Bertfried Fauser and Clical by our deceased friend Pertti Lounesto. Both programs use indices for standard base units, whereas Wallace used mere capital letters for a set of matrices. He took up my saying that at its deepest level nature does not distinguish between multivector components. I had investigated the various forms of the $SU(3)$ within the algebras

$$Cl_{1,3} \simeq Cl^{+}_{1.4} \simeq Cl^{+}_{4,1} \subset Cl_{4,1} \simeq Mat(4,\mathbb{C}) \text{ where + means } \textit{even part} \text{ of algebra} \quad (428)$$

Consequently Wallace considered subalgebras of $sl(4, \mathbb{C})$ and symmetry breakage of the latter. Verbatim in the abstract he stated: *A scheme of symmetry breakage can be imposed on orthogonal directed lineelements for the algebra sl*(4, $\mathbb{C}$) *which, for* $Cl^{+}_{3,1}$, $Cl^{+}_{4,0}$ *and* $Cl^{+}_{2,2}$ *subalgebras results in a pattern corresponding to the standard model, together with elements corresponding to fundamental particles of dark matter.* Wallace had correctly picked up some algebraic investigation of my own 2001 work and tried to lift the Clifford algebraic version of the Weinberg Salam theory of electroweak interaction worked out by David Hestenes into those $Cl_{4,1}$ spaces that incorporated the Clifform of $SU(3)$ developed by me. But he was not fully aware of the consequences of the fact that the Clifform was not a Lipschitz group, but was based on some graded algebra, the grading of which became invisible, indeed, by using the matrices from $sl(4, \mathbb{C})$. Actually we had 2-fold spinors without a group to cover. So he did his best to structurally lift Hestenes' approach into the appropriate Clifford algebra. He practically found all the essential electroweak modules, but wisely did not fix the standard model. Nevertheless he visualized a possible origin of hot dark matter and the asymmetry between light photons and heavy Higgs bosons. Properly translated, Wallace used a basis

$$Cl_{4,1} \text{ is generated by } \mathrm{W} = \{e_1, e_2, e_3, ie_{1235}, e_5\} \tag{429}$$

which has a real component

$$V = span\{e_1, e_2, e_3, e_5\} \text{ and } Cl_{3,1} = \oplus_k \wedge^k \mathrm{V} \simeq \mathrm{Mat}(4, \mathbb{R}) \tag{430}$$

and an imaginary part which is simply equal to $iCl_{3,1}$. Thus the basis vector $e_4 \in Cl_{4,1}$ is the complexified Graßmann monomial ie_{1235} and the $Cl_{4,1}$ gets a real and an imaginary part, that is, it is decomposed as

$$Cl_{4,1} \simeq Cl_{3,1} \oplus iCl_{3,1} \tag{431}$$

which is in perfect correspondence with the isomorphism

$$\mathbb{C} \otimes Cl_{3,1} \simeq \mathrm{Mat}(4, \mathbb{C}) \simeq Cl_{4,1} \tag{432}$$

If we consider any multi-index η and its complement $\overline{\eta}$ we have

$$e_\eta = \pm i e_{\overline{\eta}} \quad \text{examples:} \qquad e_{45} = -ie_{123} \qquad \text{or} \qquad e_{134} = ie_{25} \tag{433}$$

The 10-dimensional subspace $\mathrm{P}_{3,1}$ with positive definite signature turns into a 10-dimensional subspace with negative definite signature. Both taken together give a module isomorphic with the $Cl_{4,1}$. Using a basis

$$e_1 = \begin{bmatrix} 0 & 0 & -1 & 0 \\ 0 & 0 & 0 & 1 \\ -1 & 0 & 0 & 0 \\ 0 & 1 & 0 & 0 \end{bmatrix} \qquad e_2 = \begin{bmatrix} 0 & -1 & 0 & 0 \\ -1 & 0 & 0 & 0 \\ 0 & 0 & 0 & -1 \\ 0 & 0 & -1 & 0 \end{bmatrix} \tag{434}$$

$$e_3 = \begin{bmatrix} -1 & 0 & 0 & 0 \\ 0 & 1 & 0 & 0 \\ 0 & 0 & 1 & 0 \\ 0 & 0 & 0 & -1 \end{bmatrix} \qquad e_4 = \begin{bmatrix} 0 & 0 & i & 0 \\ 0 & 0 & 0 & -i \\ -i & 0 & 0 & 0 \\ 0 & i & 0 & 0 \end{bmatrix} \qquad e_5 = \begin{bmatrix} 0 & 1 & 0 & 0 \\ -1 & 0 & 0 & 0 \\ 0 & 0 & 0 & 1 \\ 0 & 0 & -1 & 0 \end{bmatrix}$$

It can easily be verified how a general element of some special electroweak group $SU(2)\times U(1)$ can be constructed for instance as

$$(aId - be_{23} + ce_{13} + de_{12}) \times (Id\cos\phi - e_{123}\sin\phi) \tag{435}$$

Wallace gives matrix representations of the electroweak special unitary symmetry for some fermions like the red u-quark. Using his spreadsheet he detected six fermionic subalgebras of the type s(u(2)×u(1)) with real pseudoscalar base units, corresponding to quark families of ordinary matter and nine such algebras with imaginary pseudoscalars for dark matter fermions. To give an example, a red up quark can be represented in Clifford space

$$\begin{array}{c} SU(2) \qquad\qquad \times \qquad\qquad U(1) \\ \begin{bmatrix} a & b & -ic & id \\ -b & a & -id & -ic \\ ic & id & a & -b \\ -id & ic & b & a \end{bmatrix} \times \begin{bmatrix} j & 0 & -k & 0 \\ 0 & j & 0 & k \\ k & 0 & j & 0 \\ 0 & -k & 0 & j \end{bmatrix} \\ (aId - be_{23} + c\, ie_{13} + d\, ie_{12}) \times (Id\cos\phi - e_{1235}\sin\phi) = \\ = (aId - be_{23} + c\, ie_{13} + d\, ie_{12})\cos\phi - (ae_{1235} + be_{15} + c\, ie_{25} - d\, ie_{35})\sin\phi \\ (a^2 + b^2) - (c^2 + d^2) = 1 \qquad j^2 + k^2 = 1 \end{array} \qquad \text{quark: up red} \tag{436}$$

which can be provided by definitions

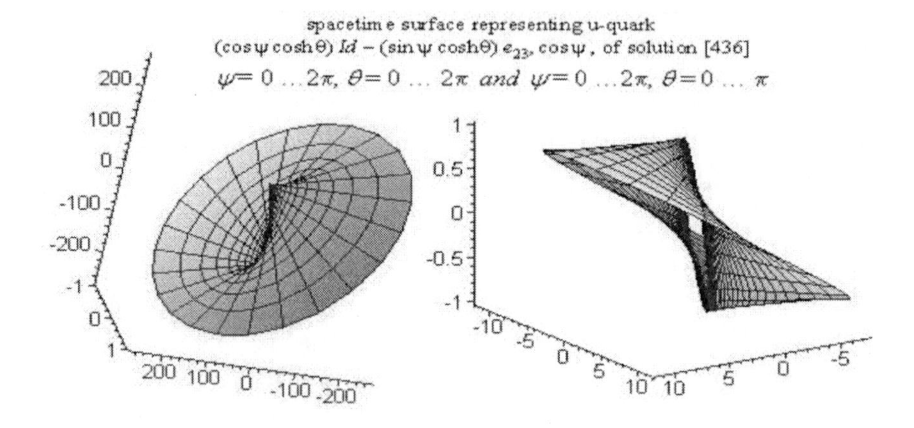

Figure 48. Clifford number 3D smartplot for Wallace SU (2) X U (1) space-time soliton.

Figure 48 represents a smartplot of the coefficients a, b plotted against j = cos ψ under the condition that cos ϕ is equal to cos ψ. So it smartplots only a small hyperplane, precisely the biparavector a(cosϕ)Id –b(cosϕ)e_{23} of the '*immersed u-quark state within the space-time–* $sl(4,\mathbb{C})$' since we use only two out of eight dimensions which we relate to the coefficient cos ϕ. Clearly we can also plot this biparavector cos ψ cosh θ Id – sin ψ cosh θ e_{23} against the quantity θ = arctanh (v/c) with light velocity c. Again we obtain a maximal immersed surface isomorphic with the Falaco soliton. In figure 49 we have plottet a high grade maximal surface of a special u-quark solution in the $SU(2)\times U(1)$ representation constructed by Wallace. Here we have used the magnitude θ, which may or may not be equal to arctanh (v/c), as the "z-axis". One can see clearly the spinor property of the Clifford number by comparing the left with the right hand figure. They differ by the interval bounds of the angles ψ and θ.

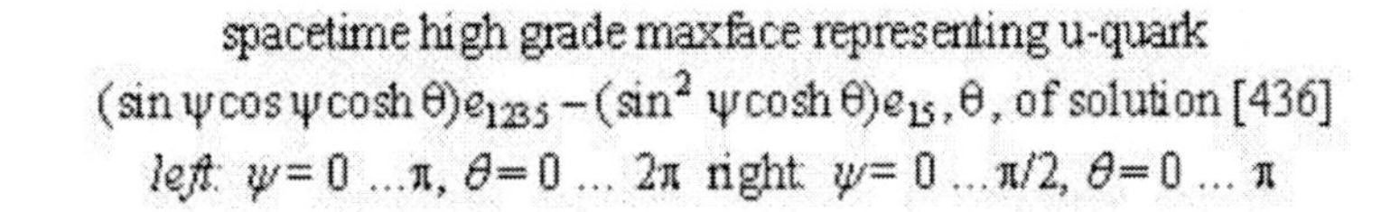

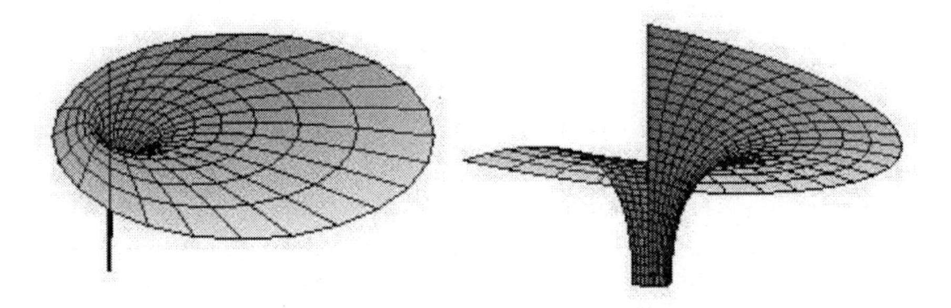

Figure 49. Clifford number 3D smartplot for Wallace SU (2) X U(1) space-time soliton.

Both figures disclose topological defect structure though the states have been constructed in order to realize the correct symmetries within the chosen algebra. Wallace did not at all aim at a model of solitons. But those drop out as a natural property of the space-time-quark model. Let us develop the model a little further. It also contains bosonic subalgebras $su(2) \oplus su(2)$ for Higgs bosons W^+, W^-, Z^0 and photons. It has space for supersymmetry by two times three special (pseudo)unitary bosonic subalgebras $su(1,1) \oplus su(1,1)$. Wallace argues that the four anti-commuting triples $\{e_{21}, e_{13}, e_{23}\}$, $\{e_{234}, e_{134}, e_{12}\}$, $\{e_{234}, e_{124}, e_{13}\}$, $\{e_{134}, e_{124}, e_{23}\}$ represent building blocks for Goldstone- and Higgs bosons. The symmetry breaking distinguishes the first triple from the other. It is indeed possible to test their behaviour in strong interacting processes by applying t, one of the universal trigonal operators. We calculate as in $Cl_{3,1}$:

$$t = (Id - 2u_1)(Id - 2u_2) \qquad \text{with} \tag{438}$$

$$u_1 = \tfrac{1}{4}(Id + e_3 - e_{25} + e_{235}) \qquad u_2 = \tfrac{1}{4}(Id + e_1 - e_{35} + e_{135}) \tag{439}$$

$$t^{-1} = t^2, \quad u_1^2 = u_1, \quad u_2^2 = u_2 \tag{440}$$

It turns out that up to direction only the first triple is preserved under a t-rotation which says nothing else than that the triple of bivectors $\{e_{21}, e_{13}, e_{23}\}$ denotes an electromagnetic triple or photon as it does not interact strongly with matter. So far the strength of the Wallace model, - it actually can describe a structure of hot dark matter.

The next important question is what happens to the red up quark from (436) under the t-operation? Actually we obtain

$$t|u\rangle t^{-1} = (a\,Id - b\,e_{12} + ic\,e_{23} + id\,e_{13})\cos\phi - -(a\,e_{123} + b\,e_3 - ic\,e_1 + id\,e_2)\sin\phi \tag{441}$$

which in Wallace' notation denotes the arrangement of complex 4×4 - matrices (S, iL, iM, N, iX, iY, Z, U) marking a green top quark $|t\rangle$. Notice, the assembly in (441) is totally analogous to (436) and thus belongs to an SU(2)×U(1)-module. If we calculate the reverted operation

$$t^{-1}|u\rangle t = (a\,Id - b\,e_{13} + ic\,e_{12} + id\,e_{23})\cos\phi + +(a\,e_5 - b\,e_{135} + ic\,e_{125} + id\,e_{235})\sin\phi \tag{442}$$

we obtain (S, M, iL, iN, Q, iP, iR, T) a blue charm quark $|c\rangle$. As the pattern of particles is not perfectly symmetrical it is supposed that the symmetry of $sl(3,\mathbb{C})$ is broken. As the symmetry breakage does not distinguish any fermion subalgebra from the others, it was suggested that there exist four fundamental dark matter fermions called Varks. Varks have six flavours like quarks do, they have colour and Texture taking two values "rough" and "smooth". Flavour can be "warm", "dull", "soft", "cool", "bright" and "hard". Therefore Wallace suggested to construct an "atom" of dark matter analogous to the hydrogen atom. He called that dark matter atom "nulliton" It looks as in

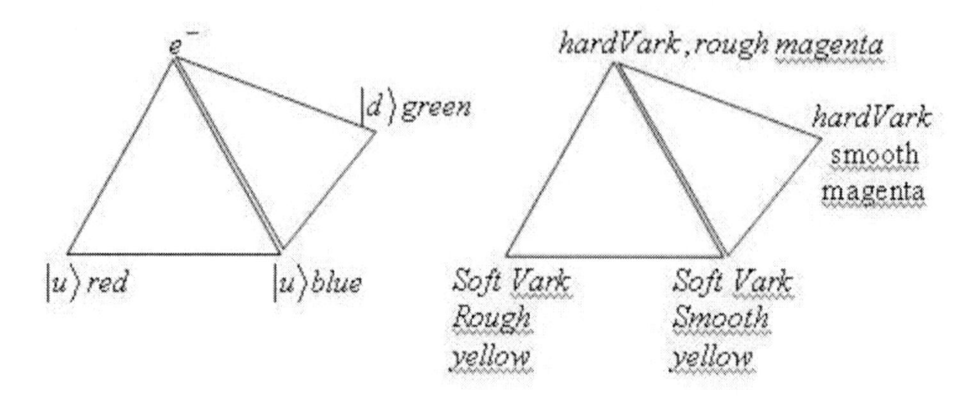

Figure 50. The Wallace Nulliton.

If the nulliton existed, it could be subjected to the Kähler-Dirac equation. Varks would have charge and spin ± ½ . Nullitons would have charge, zero spin and would be colourless.

They would interact by an equivalent of van der Waals attraction. In 2008 I thought that a neutrino model for dark matter was more probable. Let us go into it.

The Wimp

If we base the theory of dark matter on weekly interacting massive particles (wimps) which are released from the strong force, we need at least a Clifford algebra $Cl_{4,2}$ generated by a space with two time coordinates. One of the reasons is that only a neutrino built up by three factors of the type 'paravector plus two biparavectors' brings in something new. In order that it does not engage in the strong interactions it demands that it annihilates the constitutive Lie group. In the book »*Lie groups: New research*« (2009) I showed the bottom up construction of the Lie algebra $\mathcal{L}^{(6)}$ constitutive for the $SU(7)$ in the $Cl_{4,2}$.

Table 6. Bottom up construction in Lie group $\mathcal{L}^{(6)}$

$\tau_0 = \frac{1}{2}(a_d^\dagger a_d - a_u^\dagger a_u)$ $t_3 = \frac{1}{4}(e_{25} - e_{125})$	$SU(2) \subset \mathcal{L}^{(6)}$
$Y = \frac{1}{3}(-2a_s^\dagger a_s + a_d^\dagger a_d + a_u^\dagger a_u)$ $y = \frac{1}{6}(-2e_1 + e_{25} + e_{125}) = \frac{1}{\sqrt{3}}\lambda_8$	$SU(3) \subset \mathcal{L}^{(6)}$
$Z = \frac{1}{4}(a_d^\dagger a_d + a_u^\dagger a_u + a_s^\dagger a_s - 3a_c^\dagger a_c)$ $z = \frac{1}{8}(+e_1 + e_{25} + e_{36} - 3e_{125})$	$SU(4) \subset \mathcal{L}^{(6)}$
$U = \frac{1}{5}(a_d^\dagger a_d + a_u^\dagger a_u + a_s^\dagger a_s + a_c^\dagger a_c - 4a_b^\dagger a_b)$ $u = \frac{1}{10}(e_1 + e_{25} + e_{36} + e_{125} - 4e_{136})$	$SU(5) \subset \mathcal{L}^{(6)}$
$V = \frac{1}{6}(a_d^\dagger a_d + a_u^\dagger a_u + a_s^\dagger a_s + a_c^\dagger a_c + a_b^\dagger a_b - 5a_t^\dagger a_t)$ $v = \frac{1}{12}(e_1 + e_{25} + e_{36} + e_{125} + e_{136} - 5e_{2356})$	$SU(6) \subset \mathcal{L}^{(6)}$
$W = \frac{1}{7}(a_d^\dagger a_d + a_u^\dagger a_u + a_s^\dagger a_s + a_c^\dagger a_c + a_b^\dagger a_b + a_t^\dagger a_t - 6a_w^\dagger a_w)$ $w = \frac{1}{14}(e_1 + e_{25} + e_{36} + e_{125} + e_{136} + e_{2356} - 6e_{12356})$	$SU(7) \subset \mathcal{L}^{(6)}$

The Lie group $\mathcal{L}^{(6)}$ has rank 6 since we can find six non-scalar basis elements commuting with each other. Those are forming the neutrino primitive idempotent which fixes the states. The construction of such a hierarchy of fermions follows the old idea of Harry Lipkin. This allows the subgroups $SU(2)$, $SU(3)$ until to $SU(6)$ to be used in the classification of the $SU(7)$ multiplets, one after the other. Subtracting the six plus one (for B) there remains the algebra of 42 bilinear products having forms like $a_d^\dagger a_u$, $a_u^\dagger a_d$, ..., $a_d^\dagger a_w$, $a_w^\dagger a_d$ and so forth. All those operators which annihilate a *baryon* and create a *beyond state* or annihilate a *beyond* and create a *baryon*, change the eigenvalue of W, and respectively w, by ± 1. There are 12 of them; the other 30 commute with W. From the expression of W in table 6 we conclude that it takes eigenvalues

$$W\ldots\ n, n\pm\tfrac{1}{7}, n\pm\tfrac{2}{7}, n\pm\tfrac{3}{7} \qquad \text{with integer } n \tag{443}$$

With those eigenvalues there correspond seven different types of $SU(7)$-multiplets. In analogy with the $SU(3)$ decuplet in table 6 we can also classify baryons according to their beyondness YO and W. Then we combine 7 baryons and vary YO between 0 and 7 and obtain $W = 1, 0, -1, \ldots, -5$. The 7-baryon combination forms a 1716-plet. It is important to realize that the operators such as $a_d^\dagger a_w$ which change the eigenvalues of W by ± 1 not only change YO, but also change the $SU(3)$ quantum number Y by one third. Any application of a bilinear generator changing YO entails a top down amendment of quantum numbers from W down to V, U, Z and Y. For a group $\iota^{(6)}$ having rank 6 the drawing of multiplet diagrams in 2 dimensions is not feasible. But one first has to specify a specific question, and then look for a means of representation.

In such a theory, the whole dynamics of HEPhy is bound to the existence of a heavy neutrino which is constituted by the basis multivector space-time oscillators. We have

$$w_0 \overset{def}{=} \tfrac{1}{2}(1+e_1)\tfrac{1}{2}(1+e_{25})\tfrac{1}{2}(1+e_{36}) \quad \text{wimp} \tag{444}$$

analogous to the primitive idempotent f_1 in $Cl_{3,1}$. The wimp annihilates its Lie group and becomes invisible to interaction.

The numbers seven turns out as a holy number where our first statement is concerned: Orientation of space-time is the creation of matter. Namely, a period 7 -cycle plays a fundamental and very creative role when space time turns from oriented to disoriented in the sense of the Moebius-strip. So we are back at topological defects.

After studying Kiehn's work on *Falaco solitons, cosmology and the arrow of time* I began to survey my own and related works after the hidden sources of topological surprise. After a few days I was convinced that the solution to the dark matter problem is not found in the right kind of standard model alone and respectively group theory. The best sentence with which Wallace could end his paper about *the pattern of reality* was the statement "*whatever shape the dark matter will be given by us, the varks would interact by an equivalent of van der Waals attraction.*" Most probably, it is this non-equilibrium, nonlinear thermodynamic happening which brings about van der Waals attraction independent of scale and micro-macro duality. Before we investigate this topic, we look at the figures of topological discontinuities which govern the eight-dimensional state-spaces of quarks.

Chapter 47

TOPOLOGICAL DISCONTINUITIES OF QUARKS

SPACE-TIME DEFECTS $CL_{4,1}$

The Conjecture of Configuration

determines a quite general, comparatively weak demand that we allow, in order to regulate nonlinear behaviour of state functions. Namely, we assert that the geometric product of a state with itself is in the same subspace as the state. This looks simple and its meaning can best be caught by looking at the examples.

$$\Psi \in S \Rightarrow \Psi\Psi \in S \quad \text{conjecture} \tag{445}$$

A Dirac matrix spinor asserts the claim (445):

$$\Psi = \begin{bmatrix} \psi_1 & 0 & 0 & 0 \\ \psi_2 & 0 & 0 & 0 \\ \psi_3 & 0 & 0 & 0 \\ \psi_4 & 0 & 0 & 0 \end{bmatrix} \Rightarrow \Psi\Psi = \begin{bmatrix} \psi_1^2 & 0 & 0 & 0 \\ \psi_2\psi_1 & 0 & 0 & 0 \\ \psi_3\psi_1 & 0 & 0 & 0 \\ \psi_4\psi_1 & 0 & 0 & 0 \end{bmatrix} \tag{446}$$

The conjecture also holds for direction fields, for primitive idempotents, for pure states in colour spaces and, as we shall show now, for the 8-dimensional special unitary symmetry elements of quarks in $s(u(2)\times u(1))$. First consider the Dirac Kähler equation constructed by Vargas and Torr. It has form

$$\Psi = \xi_1 f_1 + \xi_2 f_2 + \xi_3 f_3 + \xi_4 f_4 \in S \text{ with } \qquad S \equiv ch_1 \tag{447}$$

The self interactive product is

$$\Psi\Psi = \tfrac{1}{4}(\xi_1^2 + \xi_2^2 + \xi_3^2 + \xi_4^2)Id + \tfrac{1}{4}(\xi_1^2 + \xi_2^2 - \xi_3^2 - \xi_4^2)e_1 + \\ + \tfrac{1}{4}(\xi_1^2 - \xi_2^2 + \xi_3^2 - \xi_4^2)e_{24} + \tfrac{1}{4}(\xi_1^2 - \xi_2^2 - \xi_3^2 + \xi_4^2)e_{124} \tag{448}$$

Clearly, there is $\Psi \in ch_1$. Thus the conjecture of configuration is satisfied by the inner states of quark-ensembles as are described by the KD-equation that has been derivated by Vargas and Torr. Notice the sign combination characteristic for a logic lattice generating geometrically the 16 binary connectives.

$$\begin{matrix} + & + & + & + \\ + & + & - & - \\ + & - & + & - \\ + & - & - & + \end{matrix}$$

Next consider the exterior state function of the red up quark $\Psi = |u_{red}\rangle$

$$\begin{aligned} &\Psi = (aId - be_{23} + c\, ie_{13} + d\, ie_{12}) \times (Id\cos\phi - e_{1235}\sin\phi) = \\ &= (aId - be_{23} + ic\, e_{13} + id\, e_{12})\cos\phi - (ae_{1235} + be_{15} + ic\, e_{25} - id\, e_{35})\sin\phi \end{aligned} \tag{449}$$

We calculate

$$\begin{aligned} &\Psi\Psi = (a^2 - b^2 + c^2 + d^2)(\cos 2\phi)Id - 2ab(\cos 2\phi)e_{23} + \\ &+ 2iac(\cos 2\phi)e_{13} + 2iad(\cos 2\phi)e_{12} - \\ &-2ab(\sin 2\phi)e_{15} - 2iac(\sin 2\phi)e_{25} + 2iad(\sin 2\phi)e_{35} - \\ &-(a^2 - b^2 + c^2 + d^2)(\sin 2\phi)e_{1235} \end{aligned} \tag{450}$$

So Ψ and $\Psi\Psi$ are both elements of a subspace S of $Cl_{4,2}$ spanned by $\{Id, e_{12}, e_{23}, e_{13}, e_{15}, e_{25}, e_{35}, e_{1235}\}$ And therefore satisfy the conjecture of configuration. The state space S of *red* u-quarks is algebraically closed with respect to the Clifford product.

Flavour rotating $|u_r\rangle$ we get the green top quark $|u_t\rangle$ located in a space $s(u(2)\times u(1))$ spanned by the Clifford monomials (Id, e_1, e_2, e_3, e_{12}, e_{23}, e_{13}, e_{123})

$$\begin{aligned} &\Psi = |t_g\rangle = (a\, Id - b\, e_{12} + ic\, e_{23} + id\, e_{13})\cos\phi - \\ &-(a\, e_{123} + b\, e_3 - ic\, e_1 + id\, e_2)\sin\phi \in S(|t_g\rangle) \end{aligned} \tag{451}$$

The squared state function is

$$\begin{aligned} &\Psi\Psi = (a^2 - b^2 + c^2 + d^2)(\cos 2\phi)Id + 2iac(\sin 2\phi)e_1 - \\ &-2iad(\sin 2\phi)e_2 - 2ab(\sin 2\phi)e_3 - 2ab(\cos 2\phi)e_{12} + \\ &+ 2iac(\cos 2\phi)e_{23} + 2ad(\cos 2\phi)e_{13} - (a^2 - b^2 + c^2 + d^2)(\sin 2\phi)e_{123} \\ &\in S(|t_g\rangle) \end{aligned} \tag{452}$$

The conjecture of configuration is fulfilled.

If we chose a blue charm quark we had to have

$$S = span_{\mathbb{C}}\{Id, e_5, e_{12}, e_{23}, e_{13}, e_{125}, e_{135}, e_{235}\} \subset Cl_{4,1} \text{ and} \tag{453}$$

$$\begin{aligned}\Psi\Psi &= (a^2 - b^2 + c^2 + d^2)(\cos 2\phi) Id + \\ &+ (a^2 - b^2 + c^2 + d^2)(\sin 2\phi) e_5 - 2iac(\cos 2\phi) e_{12} + \\ &+ 2iad(\cos 2\phi) e_{23} - 2ab(\cos 2\phi) e_{13} + 2iac(\sin 2\phi) e_{125} - \\ &- 2ab(\sin 2\phi) e_{135} + 2iad(\sin 2\phi) e_{235} \in S(|c_b\rangle)\end{aligned} \tag{454}$$

We have seen how a Lorentz boost brings on topological discontinuities in the shape of Falaco solitons. We shall now step out to show how any nonlinearity beginning with the geometric self interaction brings forth similar topological discontinuities and respectively far from equilibrium forms of action (Pfaff sequences). Recall equation (452) for the t_g. In figure 51 we show the projection of $\Psi\Psi$ for a green top quark onto the Euclidean triple $\{e_1, e_2, e_3\}$ of base units. This is given by coefficients of

$$2iac(\sin 2\phi) e_1 - 2iad(\sin 2\phi) e_2 - 2ab(\sin 2\phi) e_3:$$

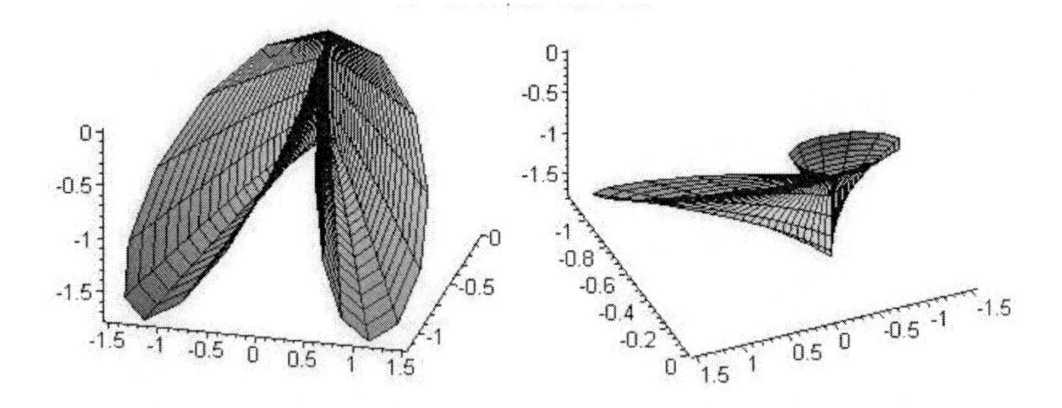

Figure 51. Topology of self interacting |tg>-soliton, (2ac sin z, - 2ad sin z, -2ab sin z), a=cos z cosh u, b=sin z cosh u, c=cos z sinh z, d=sin z sinh u, z = 0…π, u = 0… π/4.

For the top quark $|\, t_g \rangle$ the function **ΨΨ** has two leaves meeting in an edge-singularity.

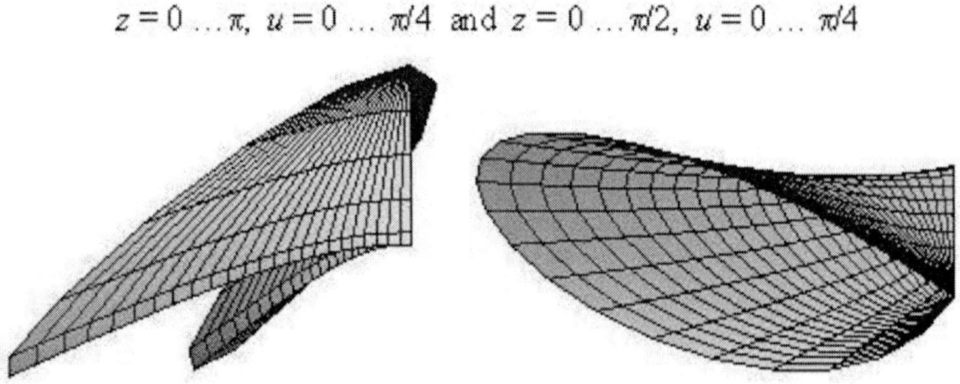

$z = 0 \ldots \pi,\ u = 0 \ldots \pi/4$ and $z = 0 \ldots \pi/2,\ u = 0 \ldots \pi/4$

Figure 52. Topology of self interacting |tg>-soliton, different views and domains of function ψ.

The object has a spinor property. With a half-turn z=0…π the figure is complete. With z=0…π/2 we plot only one leaf, that is, one half of the symmetric figure.

In figure 53 we see the absolute measures of the paravolume $\alpha Id + \beta e_{123}$ plotted against the angle u or $\theta = \text{arctanh}\,(v/c)$ of the self interacting green top quark

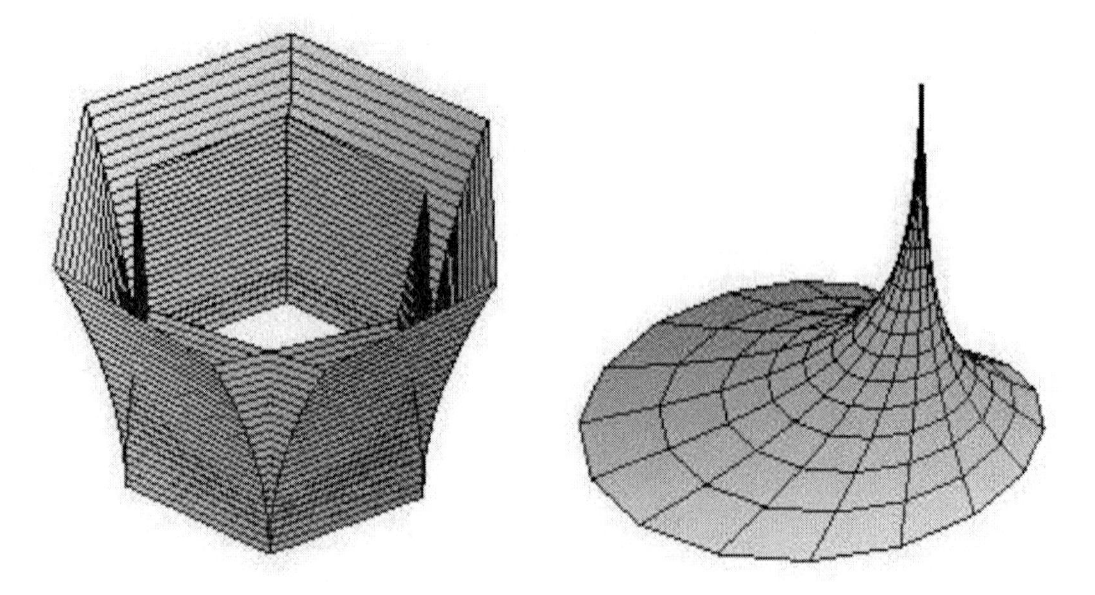

Figure 53. Paravolume of self interacting | tg>-soliton $(a^2\text{-}b^2\text{+}c^2\text{+}d^2)(\cos 2\varphi)\ Id\text{-}(a^2\text{-}b^2\text{+}c^2\text{+}d^2)(\sin 2\varphi)\ e_{123}$, plotted against u (to the left) and cos (u) (to the right), z=0…2π, u=0…π/8 and z=0…π, u=0…π.

The visualization of these topological catastrophes depend significantly on the domain of the function. If we replace the angle u by cos(u) we introduce something like a natural quantum number of helicity which behaves like the angular momentum quantum number. The figure on the left looks like a swallowtail catastrophe. But it is a strip with a loop.

Expanding the domain of u beyond π/4 we obtain the full shape of the singularity in the paravolume-component of $\Psi\ \Psi$: By self-interaction of Ψ the original umbilicus at the origin is unfolded towards a vertical line. It is interesting to see how the topology of the original state function (figure 56 on the right) namely that of the Falaco soliton is preserved and modified to bring forth figure 55.

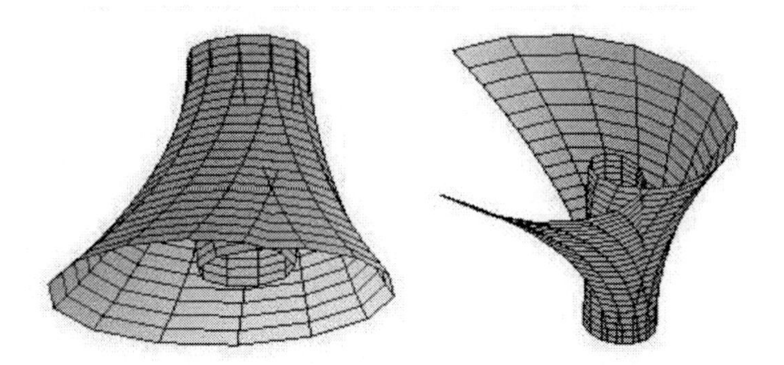

Figure 54. Paravolume of self interacting | tg>-soliton, $(a^2-b^2+c^2+d^2)(\cos 2\varphi)\ Id-(a^2-b^2+c^2+d^2)(\sin 2\varphi)\ e_{123}$, plotted against u two different perspectives $z=0\ldots\pi$, $u=0\ldots\pi/4$ and $z=0\ldots\pi$, $u=0\ldots\pi$.

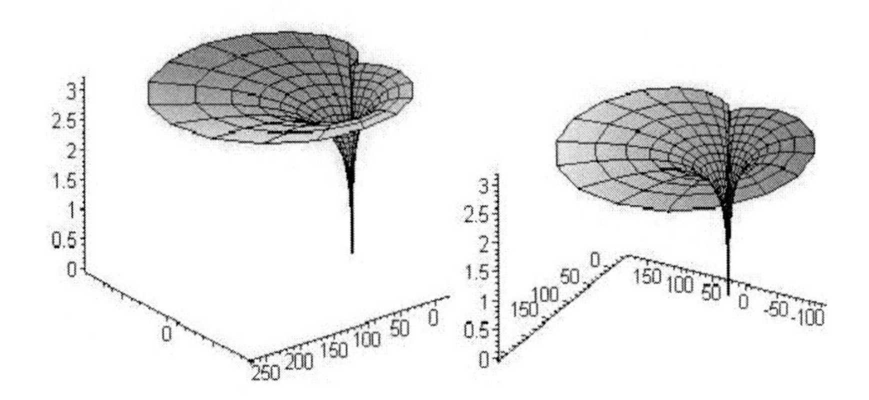

Figure 55. Paravolume of self interacting | tg>-soliton, $(a^2-b^2+c^2+d^2)(\cos 2\varphi)\ Id-(a^2-b^2+c^2+d^2)(\sin 2\varphi)\ e_{123}$, plotted against u perspective from above and below $z=0\ldots\pi$, $u=0\ldots\pi$.

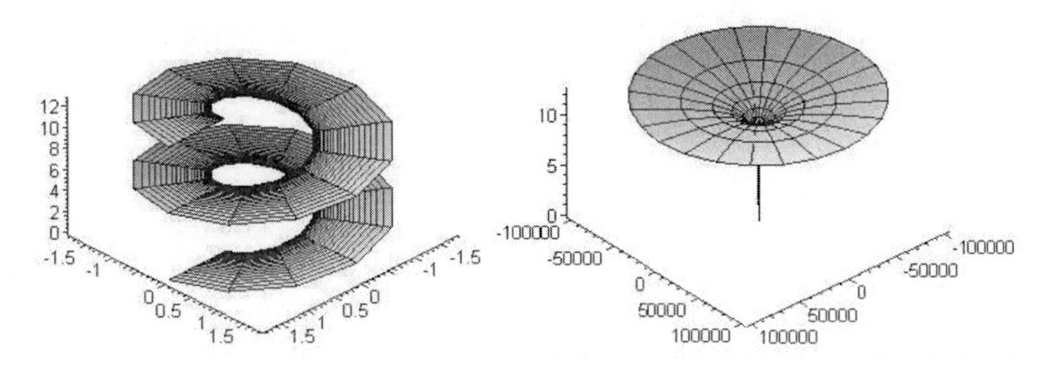

Figure 56. State function of | tg>-soliton, $-b(\cos 2\varphi)\ e_{12} - a(\sin 2\varphi)\ e_{123}$, left: plotted against z, $z=0\ldots4\pi$, $u=0\ldots\pi/2$, right: plotted against u, $z=0\ldots2\pi$, $u=0\ldots4\pi$.

Chapter 48

TOPOLOGICAL EVOLUTION OF PARTICLES

Fermions and bosons take the shape of solitons and immersion strings rather than of point particles without topology. These particles seem to have qualities which can best be derived and demonstrated by featuring their wave functions in the Clifford algebras $Cl_{3,1}$, $Cl_{4,1}$ and $Cl_{4,2}$. At least for me this was the most obvious way. Clearly, it is somewhat problematic to describe the curved and cranked multisurfaces involved without a proper concept of parallel transport and connection. The situation is somewhat similar to general relativity where we cannot easily describe a method how to transport a measurement device through a field of singularities. This becomes even more serious when the particles which constitute both the device and the space where the multi-surfaces are to be embedded, consist of dynamic topological defect structures.

A suitable measurement station for practical application in differential geometry must be a sensor so small that the curvature and geometric features become irrelevant. But as we go into it really deeply, we find out that the only instrument having that property is the fermion itself or rather a boson, best a photon. We must use the wave function as a sensor. This requires that we are able to determine all its interactions with the surrounding inner and outer spaces. Parallel transport of measured quantities u_i in coordinates x_μ is usually described by a tensor or a matrix T of a linear mapping. As long as this depends on two or more points of transport we have a non-local transport equation. If we consider transport in infinitesimal neighbourhood we obtain the covariant derivative for every direction $D_\mu = \partial_\mu u + A_\mu u$ where the field A_μ represents the geometric connection. Thus it is secured that, for every point and every direction, we obtain a matrix which describes the change of measurement values under parallel transport.

In quantum electrodynamics the covariant derivative is essentially given by the Dirac operator $D_\mu\psi = \partial_\mu\psi + ieA_\mu\psi$ which involves the electromagnetic potential. In Yang-Mills theory of strong interaction again we need covariant derivatives of a rather general nature. These describe parallel transport in the isospin spaces which at the same time classify the elementary particles.

As we have seen, it is the orientation in graded isospin space which determines upon the identity of the particle. The new situation in our theory is that, for the first time, the isospin space is not essentially different from the physical space. Just, the latter is not given by Minkowski space but by its associated algebra. Therefore we had to investigate the concept of orientation for the Clifford algebra and describe transport and identity transforms therein. Clearly, rotations in the geometric isospin spaces generally do not commute. Systems dynamics in isospin-space differs from that of electrodynamics because there are different movements that contribute to parallel transport. Those are determined by the reorientation symmetries. Clearly, the result of different isospin deconvolutions depends on the order of elementary isospin rotations applied to a particle. As a result the curvature tensor does not only contain differential one-forms, but also nonlinear functions of the connection. This means that self interacting fields cannot permeate each other without causing creation and annihilation of dynamic elements. This is only one result of the phenomena of isospin deconvolution. There are still some others that may be even more important.

In a chapter titled "*Spinor Fields in Riemannian Geometry*" Cartan wrote down a »*fundamental theorem*«

> "With the geometric sense we have given to the word "spinor" it is impossible to introduce fields of spinors into the classical Riemannian technique; that is, having chosen an arbitrary system of co-ordinates x^i for the space, it is impossible to represent a spinor by any finite number N whatsoever, of components u_α such that the u_α have covariant derivatives of the form
>
> $$u_{\alpha,i} = \frac{\partial u_\alpha}{\partial x^i} + \Lambda^{\beta}_{\alpha i}\, u_\beta$$
>
> where the $\Lambda^{\beta}_{\alpha i}$ are determinate functions of x^h."

In a letter from 5th May 2009 my friend Zbigniew Oziewicz asked "*why does the covariant derivative never never enter into thermodynamics*? *The first law of thermodynamics is the Slebodzinski-Lie derivative*". (Slebodzinski 1970, Guzmán and Z. Oziewicz 2003, Crasmareanu and Oziewicz 2002)) He defended Kiehn's revolutionary ideas that should be mentioned in my book and put the essence of some of Kiehns statements metaphoric "*Kiehn said: variational calculus is naked, it is not a god, only a top of the iceberg.*" This hits the point. Kiehn quoted Cartan's theorem and commented it. "*The problem that Cartan states above has to do with the lack of uniqueness for the covariant transplantation rule when the connection admits affine torsion of the non-integrable variety:*

$$\Lambda^{\beta}_{\alpha i} - \Lambda^{\beta}_{i \alpha} \neq 0$$

Spinors admit continuous transportation about the identity but involve the characteristic multi-valuedness. In Riemannian space, however, with a given metric the connection was uniquely determined by the Christoffel symbols." (Kiehn 2008a, p. 111f.) This feature is lost in non-equilibrium thermodynamics just as well as in quantum physics.

We cannot solve this problem in one chapter. But we can show how important it is, by giving a proof that every fermion we know from the standard model, and even some more states, bring in curved, immersed or crooked surfaces of spinor direction fields that cannot be described in terms of tensors and Riemann connection. Nevertheless those are mere space-time phenomena. As the symmetry group $SU(3)\times SU(2)\times U(1)$ is a subspace of the algebra $sl(4, \mathbb{C})$ the first we must do is to identify the state function of the known elementary particles in an abstract way. We ask, what is the algebraic form of, say, the electron, a tauon, a quark, a Higgs-boson, a photon and so on. Then we carry out the existence-proof for the associated direction field. Then we enquire its topological properties. To begin with it, consider a tauon. It can be represented by the element

$$|\tau\rangle = (aId - b\,e_{23} + c\,e_{13} + d\,e_{12})\cos\phi - (ia\,e_{45} - ib\,e_1 + ic\,e_2 + id\,e_3)\sin\phi \qquad (455)$$

In the standard Wallace representation this turns out as an element of the algebra $s(u(2)\times u(1))$ represented by 4×4-matrices with complex entries

$$\begin{bmatrix} a & b & -c & d \\ -b & a & -d & -c \\ c & d & a & -b \\ -d & c & b & a \end{bmatrix} \times \begin{bmatrix} j & 0 & 0 & k \\ 0 & j & k & 0 \\ 0 & -k & j & 0 \\ -k & 0 & 0 & j \end{bmatrix} \qquad (456)$$

$$SU(2):(a^2+d^2)+(b^2+c^2)=1 \qquad S(U(1)): j^2+k^2=1$$

In case of the leptons those conditions are realized by elliptic coefficients

$$\begin{aligned} &a=(\cos\theta)\cos\tfrac{1}{2}\psi \quad b=(\sin\theta)\cos\tfrac{1}{2}\psi \\ &c=(\sin\theta)\sin\tfrac{1}{2}\psi \quad d=-(\cos\theta)\sin\tfrac{1}{2}\psi \\ &j=\cos\phi \quad k=\sin\phi \end{aligned} \qquad (457)$$

The $|\tau\rangle$ is the Clifford product between a »torso« and its »ridge«.

$$|\tau\rangle = \underbrace{(a\,Id - b\,e_{23} + c\,e_{13} + d\,e_{12})}_{torso}\underbrace{(j\,Id - k\,i\,e_{45})}_{ridge}$$

The electron can be represented by an element

$$|e^-\rangle = (a\,Id - b\,e_{23} + c\,e_{13} + d\,e_{12})\cos\phi - \qquad (458)$$

$$+(ia\,e_4 - b\,e_{15} + c\,e_{25} + d\,e_{35})\sin\phi$$ with matrices

$$\begin{bmatrix} a & b & -c & d \\ -b & a & -d & -c \\ c & d & a & -b \\ -d & c & b & a \end{bmatrix} \times \begin{bmatrix} j & 0 & k & 0 \\ 0 & j & 0 & -k \\ -k & 0 & j & 0 \\ 0 & k & 0 & j \end{bmatrix}$$

$$SU(2):(a^2+d^2)+(b^2+c^2)=1 \qquad S(U(1)): j^2+k^2=1$$

and elliptic conditions for *a, b, c, d, j, k*. For the $|\mu\rangle$ we can also use the same *SU*(2)-»torso«, but multiply it with another »ridge«, namely the *U*(1)-element $jId - k\,e_5$.

$$\left|e^-\right\rangle = (a\,Id - b\,e_{23} - c\,e_{13} + d\,e_{12})(j\,Id + k\,e_5) =$$
$$= (a\,Id - b\,e_{23} - c\,e_{13} + d\,e_{12})\cos\phi - (ia\,e_5 - b\,e_{235} + c\,e_{135} + d\,e_{125})\sin\phi =$$
$$= (a\,Id - b\,e_{23} - c\,e_{13} + d\,e_{12})\cos\phi - (ia\,e_5 + ib\,e_{14} - ic\,e_{24} - id\,e_{34})\sin\phi \qquad (459)$$

with matrices

$$\begin{bmatrix} a & b & -c & d \\ -b & a & -d & -c \\ c & d & a & -b \\ -d & c & b & a \end{bmatrix} \times \begin{bmatrix} j & k & 0 & 0 \\ -k & j & 0 & 0 \\ 0 & 0 & j & k \\ 0 & 0 & -k & j \end{bmatrix}$$
$$SU(2):(a^2+d^2)+(b^2+c^2)=1 \qquad S(U(1)): j^2+k^2=1$$

The $|\ e^-\ \rangle$ - state function shows us quite plainly the main characteristics of the representation in $Cl_{3,1} \oplus iCl_{3,1}$. This is the mapping of the positive definite colour space $Ch \subset Cl_{3,1}$ consisting of three different grades and the scalar, onto a space with two different grades and a scalar. The mapping brings on a scalar, three vectors and 6 bivectors. The base elements are transposed as in

$$\{Id, e_1, e_2, e_3, e_{15}, e_{25}, e_{35}, e_{125}, e_{135}, e_{235}\} \mapsto$$
$$\mapsto \{Id, e_1, e_2, e_3, e_{15}, e_{25}, e_{35}, -ie_{34}, ie_{24}, -ie_{14}\} \qquad (460)$$

So it is the state of the electron which involves all six bivectors which simulate the colour space of a real component $Cl_{3,1}$. The three neutrino family members require different elements as pseudoscalars, that is, we use three different »ridges« $(j\,Id - i\,k\,e_{45})$, $(j\,Id + i\,k\,e_4)$, $(j\,Id + k\,e_5)$ for ν_e, ν_μ and ν_τ together with the torso $(a\,Id - b\,e_{23} + c\,e_{13} + d\,e_{12})$. The ridges of the leptons are eigendirection fields of antisymmetric matrices with a lucid design. We denote the ridges by the symbols $\rho(|e^-\rangle)$ $\rho(|\mu\rangle$, $\rho(|\tau\rangle)$ and obtain their matrices in $sl(4, \mathbb{C})$ with the below eigenvectors

$$\rho(\left|e^-\right\rangle) = \begin{bmatrix} j & k & 0 & 0 \\ -k & j & 0 & 0 \\ 0 & 0 & j & -k \\ 0 & 0 & k & j \end{bmatrix} \quad \rho(\left|\mu\right\rangle) = \begin{bmatrix} j & 0 & -k & 0 \\ 0 & j & 0 & -k \\ k & 0 & j & 0 \\ 0 & k & 0 & j \end{bmatrix}$$

$$\rho(\left|\tau\right\rangle) = \begin{bmatrix} j & 0 & 0 & k \\ 0 & j & k & 0 \\ 0 & -k & j & 0 \\ -k & 0 & 0 & j \end{bmatrix} \qquad (461)$$

$[-i, 1, 0, 0]$, $[0, 0, i, 1]$ and $[0, 0, -i, 1]$, $[i, 1, 0, 0]$ with eigenvalues $j + i\,k$, twofold and $j - i\,k$ twofold, for ρ_{e^-} and further

$[i, 0, 1, 0]$, $[0, i, 0, 1]$ and $[0, -i, 0, 1]$, $[-i, 0, 1, 0]$ for ρ_μ and

$[0, i, 1, 0]$, $[-i, 0, 0, 1]$ and $[i, 0, 0, 1]$, $[0, i, 1, 0]$ for ρ_τ.

The length of these vectors is zero and so distinguishes them as isotropic eigendirection-fields. E. Cartan defined these objects as *isotropic pure spinors*. They are featuring cosmic phenomena and are not just connected with microscopic quantum events as Kiehn has pointed out. A quite analogous statement applies to the *su*(2)-torso. The torso $T_\nu = (a\,Id - b\,e_{23} + c\,e_{13} + d\,e_{12})$ is essentially the same for the three neutrinos. Again we obtain an antisymmetric matrix which has double-valued eigenvectors with zero norm. When this matrix operates on a column vector, the result reminds us of the cross product of Gibbs (see also: Kiehn 2008b, p. 36).

$$T_\nu = \begin{bmatrix} a & d & -c & -b \\ -d & a & -b & c \\ c & b & a & d \\ b & -c & -d & a \end{bmatrix} \qquad (462)$$

is the antisymmetric matrix defining elliptic isotropic spinors with the typical electromagnetic structure as in a cross product operator. There is, however, an important difference, namely the trace is not zero. We shall show in a while that this is an important observation. It has to do with the existence of an associated multivector field which has the geometric square zero. Thus far the states of the known leptons. Next we turn over to quarks. We have already shown a red $|u\rangle$ can be represented in space as below

$$\begin{array}{ccc} SU(2) & \times & U(1) \end{array}$$

$$\begin{bmatrix} a & b & -ic & id \\ -b & a & -id & -ic \\ ic & id & a & -b \\ -id & ic & b & a \end{bmatrix} \times \begin{bmatrix} j & 0 & -k & 0 \\ 0 & j & 0 & k \\ k & 0 & j & 0 \\ 0 & -k & 0 & j \end{bmatrix}$$

$$(aId - be_{23} + c\,ie_{13} + d\,ie_{12}) \times (Id\cos\phi - e_{1235}\sin\phi) =$$
$$= (aId - be_{23} + c\,ie_{13} + d\,ie_{12})\cos\phi - (ae_{1235} + be_{15} + c\,ie_{25} - d\,ie_{35})\sin\phi$$
$$(a^2 + b^2) - (c^2 + d^2) = 1 \qquad j^2 + k^2 = 1$$

with hyperbolic conditions of unitarity

$$a = \cosh\theta\cos \tfrac{1}{2}\psi, \quad b = \cosh\theta\sin \tfrac{1}{2}\psi, \quad c = \sinh\theta\cos \tfrac{1}{2}\psi, \quad d = \sinh\theta\sin \tfrac{1}{2}\psi$$

$$j = \cos\phi, \quad k = \sin\phi$$

Be aware of the slight change of coordinates from elliptic to hyperbolic. The red charm quark has the same torso as the $|u_r\rangle$ and the ridge $j\ Id - k\ e_5$. The red top quark has the same torso too and the ridge $j\ Id - k\ e_5$. The matrix representations could be

$$\rho(|c_r\rangle) = \begin{bmatrix} j & -k & 0 & 0 \\ k & j & 0 & 0 \\ 0 & 0 & j & -k \\ 0 & 0 & k & j \end{bmatrix} \qquad \rho(|t_r\rangle) = \begin{bmatrix} j & 0 & 0 & k \\ 0 & j & k & 0 \\ 0 & -k & j & 0 \\ -k & 0 & 0 & j \end{bmatrix} \tag{463}$$

and satisfy the same hyperbolic conditions as the red up-quark. A trigonal $SU(3)$ rotation changes the ridge of the $|u_r\rangle$ from $j\ Id - k\ e_5$ to $j\ Id - i\ k\ e_{45}$ which is the ridge of the three top quarks. Their torso differs just by a permutation of the factors i given to the bivectors e_{12}, e_{23}, e_{13}. That is all nine quarks of the types $|\ u\ \rangle$, $|\ c\ \rangle$, $|\ t\ \rangle$ have the same torso up to factors i. Wallace used the matrices L, M, N and distributed them as follows

Table 7. The su(2) torso of fermions in matrix algebra Mat(4, $\mathbb{C}$)

	red	blue	green
$\lvert u\rangle$	$L \quad iM \quad iN$	$iL \quad M \quad iN$	$iL \quad iM \quad N$
$\lvert c\rangle$	$L \quad iM \quad iN$	$iL \quad M \quad iN$	$iL \quad iM \quad N$
$\lvert t\rangle$	$L \quad iM \quad iN$	$iL \quad M \quad iN$	$iL \quad iM \quad N$

So we have one definite torso for each colour. However, the flavour of a state is not determined by the torso, but by the ridge. In the Clifford algebra $Cl_{3,1} \oplus iCl_{3,1}$ the torso is given by the complex bivectors as below

Table 8. The *su*(2) torso of fermions in $Cl_{3,1} \oplus i\ Cl_{3,1}$

	red	blue	green
$\lvert u\rangle$	$e_{23},\ -ie_{13},\ ie_{12}$	$ie_{23},\ -e_{13},\ ie_{12}$	$ie_{23},\ -ie_{13},\ e_{12}$
$\lvert c\rangle$	$e_{23},\ -ie_{13},\ ie_{12}$	$ie_{23},\ -e_{13},\ ie_{12}$	$ie_{23},\ -ie_{13},\ e_{12}$
$\lvert t\rangle$	$e_{23},\ -ie_{13},\ ie_{12}$	$ie_{23},\ -e_{13},\ ie_{12}$	$ie_{23},\ -ie_{13},\ e_{12}$

The ridges of the $|\ u\ \rangle$, $|\ c\ \rangle$, $|\ t\ \rangle$ are given by the elements

Table 9. The $u(1)$ ridge of fermions in $Cl_{3,1} \oplus i\, Cl_{3,1}$

$\lvert u\rangle$	$Id\cos\phi + ie_4 \sin\phi$
$\lvert c\rangle$	$Id\cos\phi - e_5 \sin\phi$
$\lvert t\rangle$	$Id\cos\phi - ie_{45} \sin\phi$

Wallace has tried to account for different mass ratios for different families by varying the distribution of metric symmetry breakage. This led to the idea to use one and the same torso for $\lvert u\rangle$, $\lvert c\rangle$, $\lvert t\rangle$ but different ridges, whereas $\lvert d\rangle$, $\lvert s\rangle$, $\lvert b\rangle$ have slightly varying torsos and distinct ridges. This may not be convincing, since it is more obvious to illuminate the interval between low and high energy quarks. But as an example the idea is very useful in order to see the potentiality of representation in $sl(4, \mathbb{C})$. It is not so important to alter this model before we have better data about the interaction of constituent quarks, as the topological features of the particles remain the same. In contrast to the charged leptons, Wallace proposes to alter the torsos for the neutrinos in addition to the ridge-variation:

Table 10. The $u(1)$ ridge of neutrinos in $Cl_{3,1} \oplus i\, Cl_{3,1}$

$\lvert \nu_e\rangle$	$Id\cos\phi - e_{23} \sin\phi$
$\lvert \nu_\mu\rangle$	$Id\cos\phi + e_{13} \sin\phi$
$\lvert \nu_\tau\rangle$	$Id\cos\phi + e_{12} \sin\phi$

$$\rho(\lvert \nu_e\rangle) = \begin{bmatrix} j & 0 & 0 & k \\ 0 & j & k & 0 \\ 0 & -k & j & 0 \\ -k & 0 & 0 & j \end{bmatrix} \qquad \rho(\lvert \nu_\mu\rangle) = \begin{bmatrix} j & 0 & -k & 0 \\ 0 & j & 0 & -k \\ k & 0 & j & 0 \\ 0 & k & 0 & j \end{bmatrix}$$

$$\rho(\lvert \nu_\tau\rangle) = \begin{bmatrix} j & -k & 0 & 0 \\ k & j & 0 & 0 \\ 0 & 0 & j & k \\ 0 & 0 & -k & j \end{bmatrix} \tag{464}$$

Table 11. The *su*(2) torso of fermions | *d*⟩, | *s*⟩, | *b*⟩

d_{red}	$a\,Id + i\,be_{45} - ic\,e_5 - d\,e_{15}$
d_{blue}	$a\,Id + i\,be_{45} - ic\,e_5 + d\,e_{25}$
d_{green}	$a\,Id + i\,be_{45} - ic\,e_5 + d\,e_{35}$
s_{red}	$a\,Id + i\,be_{45} - ic\,e_5 + id\,e_{14}$
s_{blu}	$a\,Id + i\,be_{45} - ic\,e_5 + id\,e_{24}$
s_{green}	$a\,Id + i\,be_{45} - ic\,e_5 - id\,e_{34}$
b_{red}	$a\,Id - be_5 - c\,e_4 + d\,e_1$
b_{blue}	$a\,Id - be_5 - c\,e_4 + d\,e_2$
b_{green}	$a\,Id - be_5 - c\,e_4 + d\,e_3$

Though I favour the elements of colour spaces as cursors to the constituent torsos of fermions, I have to admit that the alternative constructed by Wallace is compatible with the transformations of constitutive Lie group $\mathfrak{L}^{(2)}$, in particular the *SU*(3) colour and flavour rotations. It just does not yet provide an explicit state function in *SU*(3)×*SU*(2)×*U*(1).

Let us now investigate the topological properties of the standard model under these elliptic and hyperbolic conditions of symmetric unitarity. Both involve topological discontinuities. But an electron induces a quite different topological neighbourhood than a quark. For consider the elliptic conditions to the electroweak symmetry *S*(*U*(2) ×*U*(1)).

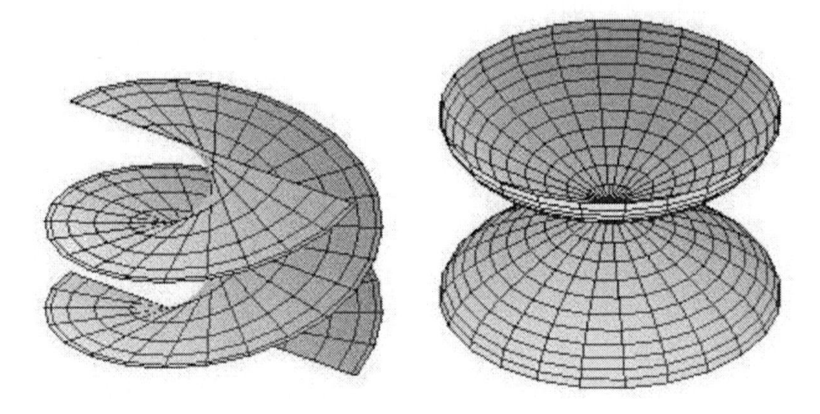

Figure 57. Torso of leptons.

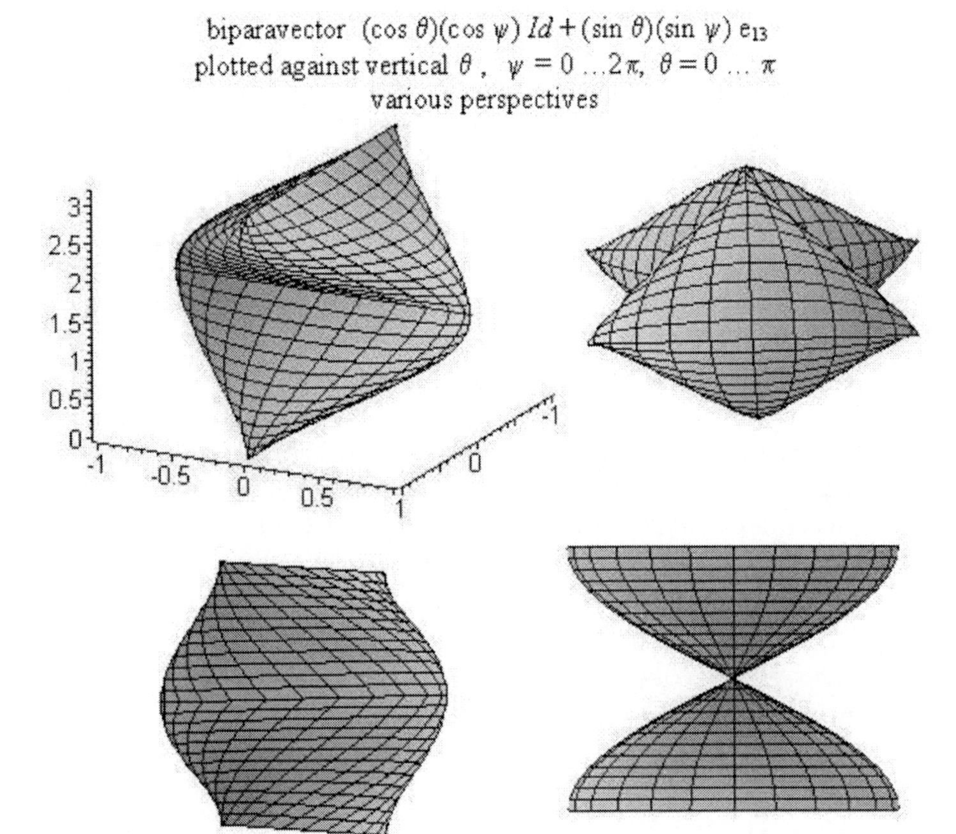

Figure 58. Torso of leptons – pillow discontinuity.

biparavector $(\cos\theta)(\cos\psi)\, Id - (\sin\theta)(\sin\psi)\, e_{23} + (\sin\theta)(\sin\psi)\, e_{13}$

$\psi = 0 \ldots 2\pi,\ \theta = 0 \ldots \pi$

2 perspectives

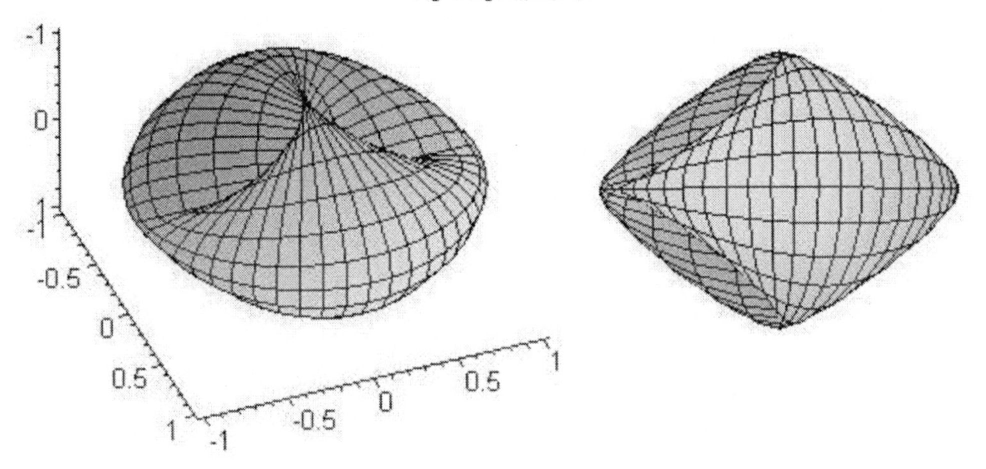

Figure 59. Pillow discontinuity.

It makes no difference if we plot the biparavectors against θ or ψ as the expression is symmetric. We can even plot a 3-dimensional biparavector.

The elliptic bivector torso of leptons is correctly returning the idea of Fock to understand propagating electromagnetic fields as dispersion of discontinuities. This needs some separate explanation, as there appears a specific relation between leptons and photons. The even subalgebras of the Clifford algebra $Cl_{4,1}$ isomorphic to $Cl_{0,3}$ have typical unitary subalgebras of the type $su(2) \oplus su(2)$. Those correspond to a boson.

The Higgs mechanism which accounts for the spontaneous symmetry breaking of the electroweak forces is responsible for the special decomposition of $Cl_{3,1} \oplus i\, Cl_{3,1}$. This allows for a partition into exactly 4 cosets of base units, let them be denoted as, $\mathsf{E}_1 = |\, e^-(½)\rangle$, $\mathsf{E}_2 = |e^+(½)\rangle$, $\mathsf{E}_3 = |e^-(-½)\rangle$ and $\mathsf{E}_4 = |e^+(-½)\rangle$. These algebras correspond to electrons and positrons with one handing, that is, E_1, E_2 and opposite spin E_3, E_4. This design which uses up all 32 base elements guarantees the Pauli exclusion principle for electroweak fields. A similar statement can be made for the other fermions. To describe the Goldstone and Higgs bosons, photons and gluons, however, we need the $su(2) \oplus su(2)$ and $su(1,1) \oplus su(1,1)$ (pseudo)-unitary algebras. For example, the Z^0 can be represented by the element:

$$\left|Z^0\right\rangle = \tfrac{1}{2}(a_1 + a_2)Id - \tfrac{i}{2}(c_1 + c_2)e_2 - \tfrac{i}{2}(d_1 - d_2)e_3 - \tfrac{1}{2}(b_1 + b_2)e_5 - \\ - \tfrac{i}{2}(a_1 - a_2)e_{14} + \tfrac{1}{2}(b_1 - b_2)e_{23} + \tfrac{i}{2}(d_1 + d_2)e_{25} - \tfrac{i}{2}(c_1 - c_2)e_{35} \qquad (465)$$

and in $sl(4, \mathbb{C})$

$$\left|Z^0\right\rangle = \begin{bmatrix} a_1 + id_1 & -b_1 + ic_1 & 0 & 0 \\ b_1 + ic_1 & a_1 - id_1 & 0 & 0 \\ 0 & 0 & a_2 + id_2 & -b_2 + ic_2 \\ 0 & 0 & b_2 + ic_2 & a_2 - id_2 \end{bmatrix}; \qquad (466)$$

$$SU(2) \oplus SU(2):\ (a_1^2 + d_1^2) + (b_1^2 + c_1^2) = 1$$

With the elliptic conditions

$$\begin{aligned} a_1 &= (\cos\theta_1)(\cos\tfrac{1}{2}\psi_1) \quad b_1 = (\sin\theta_1)(\cos\tfrac{1}{2}\varphi_1) \\ c_1 &= (\sin\theta_1)(\sin\tfrac{1}{2}\varphi_1) \quad d_1 = -(\cos\theta_1)(\sin\tfrac{1}{2}\psi_1) \\ a_2 &= (\cos\theta_2)(\cos\tfrac{1}{2}\psi_2) \quad b_2 = (\sin\theta_2)(\cos\tfrac{1}{2}\varphi_2) \\ c_2 &= (\sin\theta_2)(\sin\tfrac{1}{2}\varphi_2) \quad d_2 = -(\cos\theta_2)(\sin\tfrac{1}{2}\psi_2) \end{aligned} \qquad (467)$$

It can easily be verified that the state in (465) satisfies the conjecture of configuration. Also notice, if we satisfy the unitarity by hyperbolic conditions, we shall obtain the strong force bosons, that is, gluons. The starting point of these considerations is in the spontaneous symmetry breakage towards the peculiar decomposition of the geometry into a real and an imaginary part of the algebra. The even subalgebra can be sorted into anti-commuting triples and commuting triples. The symmetry breaking distinguishes one anti-commuting triple from the other three, suggesting identification with the photon. The other anticommuting triples combined with base units from commuting triples should form the Goldstone bosons. Wallace has proposed an arrangement of triples as below

Table 12. The *su*(2) ⊕ *su*(2) triples of bosons

Scalars, bivectors and Pseudoscalars	Even subalgebra of $Cl_{4,1}$	Special (pseudo)-unitary subalgebra
$Id, -ie_{15}, ie_{25}, -ie_{35}, e_{12}, -e_{13}, e_{23}, e_4$	$Cl^{+}_{4,0}$	$su(2) \oplus su(2)$
Anti-commuting triples	Commuting triples	particle
$e_{12}, e_{23}, -e_{13}$		photon
$(-ie_{15}, ie_{25}, e_{12}), (-ie_{15}, -ie_{35}, -e_{13}), (ie_{25}, -ie_{35}, e_{23})$	$(-ie_{15}, ie_{23}, e_4)$, $(ie_{25}, -e_{13}, e_4)$, $(-ie_{35}, e_{12}, e_4)$	W^+, W^-, Z^0

What is so special about the photonic triple of bivectors? Recall the universal trigonal rotator t from equation (438). The photon is the only anticommuting triple in the boson configuration which is preserved by this *SU*(3) rotation t. The photon disguises its participation in strong force interaction. We have

$$\begin{array}{ccccc} e_{12} & \xrightarrow{t} & e_{13} & \xrightarrow{t} & e_{23} \\ \uparrow & \longleftarrow & \underset{t}{} & \longrightarrow & \downarrow \end{array}$$

We found the photon triple contributes most to the torso of half of the known fermions, and it also contributes to the ridges of the other half. It is the essential architect of all subnuclear particles. It may very well represent the living probe not only in parallel transport, but in the whole systems dynamics of interaction. This figure need no longer surprise us:

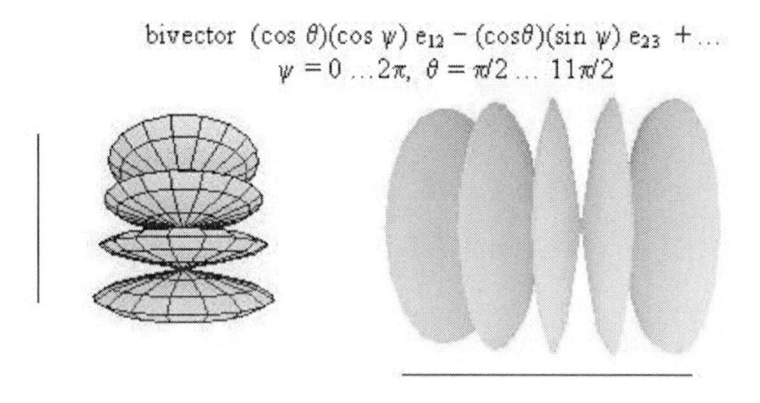

Figure 60. Photon as propagating discontinuity in su(2)+ su (2).

The corresponding pictures for quarks look different. This is due to the fact that the symmetric unitary group, for them, is brought on by hyperbolic conditions.

Left: soliton $-(\sin\psi)(\cosh\theta)\, e_{12} + (\cos\psi)(\sinh\theta)\, ie_{23}$, θ

Right: bivector $-(\sin\psi)(\cosh\theta)\, e_{12} + (\cos\psi)(\sinh\theta)\, ie_{23} + (\sin\psi)(\sinh\theta)\, ie_{13}$

Soliton and phase disc for directed areas $\psi = 0 \ldots 2\pi,\ \theta = 0 \ldots 2\pi$

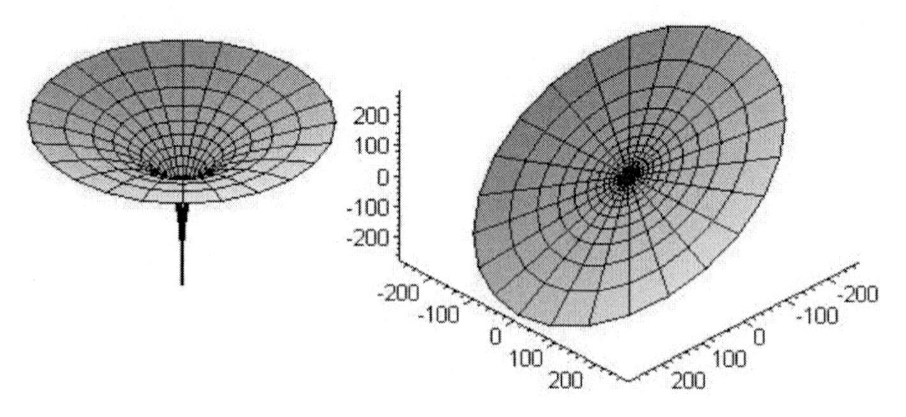

Figure 61. Green top soliton and directed areas phase disc.

The singularity of this maximal surface comes upon by the angle θ, that is, when the velocity of the photon probe, - the photon part of the torso -, reaches the light velocity, something that has to be understood verbatim. This is clear, for part of the fermion is an inner photon. Now, look to the right! As soon as we plot the magnitudes of all three directed areas, we obtain the image of a disc in 3-space. This has a definite limited size and thickness zero.

There are three magnitudes for the directed spatial areas e_{12}, e_{23}, e_{13}. They are indeed extremely small, but nevertheless have sharp boundaries which are determined by the elliptic factors in the quantities *a* to *d*. Therefore the quarks provide some special architectonic features. On the one hand the photonic singularity of the "Falaco shape" enforces extreme potential and kinetic energy. On the other hand the interplay of all the hyperbolic contributions to the symmetric unitary design balancing each other out. They keep each other in check. If we look at the areas disc from the right direction, we see there is only one dimension and one direction alongside which the fermion shows a non-vanishing small magnitude. This is a special sort of string.

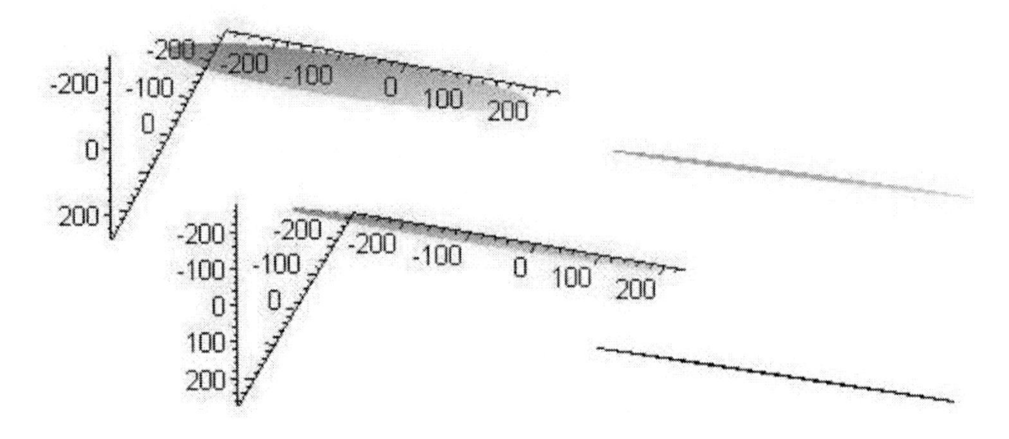

Figure 62. Green top directed areas phase disc looked at from different perspectives.

Clearly, this string appearance is not just a property of the top quark, but it involves every quark. It concerns even every other particle that directly partakes in the strong force.

Phenomenology and mathematics surrounding figure 60 confirms the claim of Fock and Kiehn that an electromagnetic signal is indeed a propagating discontinuity or topological defect. [1] It is true that all our attempts to quantize waves, field oscillations, emission of photons, de Broglie's waves, Bohr's atom, the bundle of matter waves carrying energy and momentum ultimately lead to the connection between Maxwell's equations and the concept of spinors. Spinors as eigenvectors of isotropic direction fields are not mere concepts of microphysics. But they are holding independent of scale and dimension. This is especially beautiful to be observed in the present graded theory of QCD where the spin group is so to say a pure spin group in the sense that it has no rotation group to cover, although it is quite clear that it carries out rotations in isospin space. The idea of spinor solutions to Maxwell's equations has been explained by Kiehn very eidetic by tracing it back to the topological perspective that the 2-form of field intensities $F = dA$, where A is the 1-form of action, can be represented by an anti-symmetric matrix F of functions which essentially represent the six components of the electric and magnetic field strength. The eigenvectors of F have either zero or complex eigenvalues. We shall see how this idea most harmonically can be extended further into the field of gluons. The topological argumentation can be unfolded from many various rather important perspectives. As I do not see a necessary contradiction between the geometric approach and the topological view, I prefer to work with geometric algebra and group theory. From the viewpoint of differential geometry we should rather begin with the Pfaff sequence. This begins with a 1-form of action (Kiehn 2008b, p. 255)

$$A(x,y,z,s,dx,dy,dz,ds) = -ydx + xdy - sdz + zds$$
$$F = dA = 2dx\wedge dy + 2dz\wedge ds \Rightarrow B_x dx\wedge dy + E_z dz\wedge dt$$

[1] See the internal paper by R. M. Kiehn: *Propagating Topological Singularities in the lightcone: The Photon.* (formula 9: the second somewhat heretic claim").

$$A\wedge F = 2(xdy\wedge dz\wedge ds - y\wedge dx\wedge dz\wedge ds + zdx\wedge dy\wedge ds - sdx\wedge dy\wedge dz)$$
$$F\wedge F = 8dx\wedge dy\wedge dz\wedge ds \Rightarrow 2(E\cdot B)dx\wedge dy\wedge dz\wedge ds\,. \tag{468}$$

In the second equation, the 1-form is interpreted as a Maxwell system with an E and B field both having magnitude 2 up to scale factors. The 3-form $A\wedge F$ is what we call topological torsion. In many systems just as in this one the topological torsion is not zero, but it has components proportional to a 4-vector. Like in some early works of Whittaker the E- and B-fields are parallel, so that the F may have the representation matrix

$$F = \begin{bmatrix} 0 & 2 & 0 & 0 \\ -2 & 0 & 0 & 0 \\ 0 & 0 & 0 & 2 \\ 0 & 0 & -2 & 0 \end{bmatrix} \Rightarrow \begin{bmatrix} 0 & B_z & 0 & 0 \\ -B_z & 0 & 0 & 0 \\ 0 & 0 & 0 & E_z \\ 0 & 0 & -E_z & 0 \end{bmatrix} \tag{469}$$

This matrix has four eigenvalues and four eigenvectors given by $(iE_{z,}\quad [0,0,-i,1])$; $(-iE_z,\quad [0,0,i,1])$; $(iB_{Z,}\quad [1,i,0,0])$; $(-iB_{z,}\quad [1,-i,0,0])$ and each eigenvector has zero norm in Euclidean signature. Therefore they can be interpreted as spinors in the sense of Cartan. Such matrices of field intensities can have real eigenvectors with zero eigenvalue and complex ones with non-zero eigenvalues. The first are sometimes erroneously called null vectors, the second are spinors. That a matrix of field intensities or torsion may have isotropic spinors as eigenfields, does not require that the matrix is antisymmetric. This would be a sufficient, but not a necessary condition as we shall see in the case of gluon carrier matrices. What is important is that the matrix or "the algebra" of a field that possesses an isotropic direction field is connected via the unitary symmetry to the elliptic, hyperbolic or nonlinear conditions which determine upon the topological features of the multivectors. - Wallace has studied one work by Hestenes and another one by myself, and he concluded that the gluons may be mixtures of three equivalent space-time elements composed by eight directions which have different factors, some ± 1, some $\pm\, i$. This reflects the process of symmetry breaking.

Table 13. Assembling of gluons in $Cl_{3,1} \oplus i\, Cl_{3,1}$

unitals	Id	$-e_{15}$	ie_{25}	$-e_{35}$	ie_{23}	$-e_{13}$	ie_{12}	e_4
$sl(4,\mathbb{C})$ matrix	S	D	E	F	L	M	N	iV
signature	+	+	–	+	+	–	+	+
					i		i	i
factor $\sqrt{-1}$						i	i	i
					i	i		i

These 8-dimensional elements, when properly constructed, satisfy the algebraic conditions of the special unitary group $su(2) \oplus su(2)$ or pseudounitary group $su(1,1) \oplus su(1,1)$. The symmetry breaking does not make a difference between those elements which suggests their association with the strong force.

We investigate the 2-forms of three gluons which are given by the following »*carrier 2-forms*«:

$$
\begin{aligned}
F_1 &= -idx_1 \wedge dx_2 + idx_2 \wedge dx_3 + dx_1 \wedge dx_3 - dx_1 \wedge dx_5 - idx_2 \wedge dx_5 - dx_3 \wedge dx_5 \\
F_2 &= -idx_1 \wedge dx_2 + dx_2 \wedge dx_3 + idx_1 \wedge dx_3 - idx_1 \wedge dx_5 - dx_2 \wedge dx_5 - dx_3 \wedge dx_5 \\
F_3 &= dx_1 \wedge dx_2 + idx_2 \wedge dx_3 - idx_1 dx_3 - dx_1 \wedge dx_5 + dx_2 \wedge dx_5 - idx_3 \wedge dx_5
\end{aligned} \tag{470}
$$

The *carrier 2-form* of a gluon is obtained by omitting the scalar and pseudoscalar 1-forms *Id* and e_4, that is, the matrices S and *i*V and abstracting from the constants *a*, *b*, *c*, *d*, a.s.o. As we shall find out soon that does not alter the situation. Namely, the matrix representation of F_1, F_2, and F_3 is given by

$$
[F_1] = \frac{1}{2}\begin{bmatrix} i & 1-i & 1 & 1-i \\ 1+i & i & 1+i & -1 \\ 1 & 1-i & -i & -1+i \\ 1+i & 1 & -1-i & -i \end{bmatrix} \tag{471}
$$

$$
[F_2] = \frac{1}{2}\begin{bmatrix} -1 & 0 & -i & 0 \\ 2 & 1 & 2i & -i \\ i & 0 & -1 & 0 \\ 2i & i & -2 & 1 \end{bmatrix}
$$

$$
[F_3] = \frac{1}{2}\begin{bmatrix} 1 & 0 & i & 2 \\ 2i & -1 & 0 & i \\ -i & 2 & 1 & 0 \\ 0 & -i & -2i & -1 \end{bmatrix}
$$

All three have the same set of eigenvalues (0, 0, +1, –1) indicating two valued spin 1. The eigenvectors are

$$
\begin{aligned}
&[F_1]: \; g_{11} = [i,-i,+1,+1], \quad g_{12} = [-i,-i,-1,+1], \quad g_{13} = [-1,+1,-i,-i] \\
&[F_2]: \; g_{21} = [0,+1,0,+i], \quad g_{22} = [-1,+1,+i,+i], \quad g_{23} = \{[0,+1,0,-i], g_{24} = [+1,0,+i,0]\} \\
&[F_3]: \; g_{31} = [-i,-1,+1,+i], \quad g_{32} = [+1,+i,+i,+1], \quad g_{33} = [+i,-1,+1,-i]
\end{aligned} \tag{472}
$$

Each eigenvector is isotropic as the sum of squares of coefficients is zero. But as you see, the distinction between proper *real null vectors* with eigenvalues zero and *null isotropic spinors* is not sharp. For consider for instance the isotropic spinor g_3 of F_3. It is complex and has square norm zero, but twice the eigenvalue zero. So zero eigenvalues are compatible with isotropic spinors.

From the realization of the $su(1,1) \oplus su(1,1)$ there follow the typical hyperbolic conditions for unitarity. So we obtain an ensemble of Falaco soliton plots as topological features of gluon fields.

Chapter 49

STRONG FORCE TOPOLOGICAL TORSION

Strong force fields are bound to peculiar torsion components that can only be understood in terms of topological thermodynamics. Therefore we borrow some axioms and tools from Kiehn (e.g. 2008b, p. 95) which will turn out useful (see the toolbox). The strong force is encoded into a 1-form of covariant action potentials $A_\mu(x, y, z, t \ldots)$, on a 4-dimensional variety of ordered independent variables, $\{x, y, z, t\}$ which supports a differential volume $\Omega_4 = dx\wedge dy\wedge dz\wedge dt$. According to the Sommerfeld school exterior differential forms are featuring intensities such as temperature and pressure. Kiehn upgraded this view as he realized that the Pfaff Topological Dimension (PTD) of the system 1-form of action determines whether the system is open, closed, isolated or in equilibrium. The construction of the Pfaff sequence (PS) requires an exterior differential operation and exterior product rules. Generally the PS is defined on a domain with four geometric base variables, - surprisingly this turns out to be enough for the strong force fields too. So the PS is defined as

$$PS \stackrel{\text{def}}{=} \{ A, dA, A\wedge dA, dA\wedge dA\} \qquad (473)$$

Lagrangian extremal methods are not useful now. But they have to be replaced by Cartan's "*magic formula*", that is, Cartan differentiation or "Ślebodziński-Cartan-Lie derivation" as it was called by Oziewicz. Kiehn points out "*It is usual to determine dynamical constraints by integral variational methods imposed on some Lagrange density. Such methods are not assumed herein, for typically such methods do not lead to representations of non-equilibrium systems and thermodynamically irreversible processes.*" (Kiehn 2008a, p. 175). The names of the four p-forms are *Action* A, *Vorticity* dA, *Torsion* A^dA and *Parity* dA^dA. The vanishing of forms determines upon the topological dimension (see toolbox). Recall, topological dimensions 1, 2, 3 and 4 are correlated with *equilibrium systems*, *isolated*, *closed* and *open* systems. The integer PTD determines the minimum number of independent functions required to describe the neighbourhood of the system. In a non-equilibrium thermodynamic system with action 1-form A the PTD is 3 or 4. That correlates with the thermodynamic property of systems with Cartan topology having PTD ≥ 3: they disclose a disconnected topology. (Kiehn 2008b, p. 17) Non–equilibrium systems with zero Gauss curvature but non-zero (affine) torsion led Shipov to the idea of 'absolute parallelism' and 'physical vacuum'.

In gluon fields, just as in Kiehn's "electromagnetic format", the 2-form F = dA is simply called field strength. Let us consider the action 1-form of a red u-quark

$$A = j\,(im\,x\,dy - b\,y\,dz + ic\,x\,dz) + k\,(-b\,x\,dt - ic\,y\,dt + im\,z\,dt) \qquad (474)$$

with constants b, c, m (m instead of d), j, k as in equation (436). From (474) we obtain the 2-form of field intensities

$$F = dA = j(im\,dx\wedge dy - b\,dy\wedge dz + ic\,dx\wedge dz) +$$
$$+ k(-b\,dx\wedge dt - ic\,dy\wedge dt + im\,dz\wedge dt) \qquad (475)$$

Note the exterior forms dx^dy, dy^dz, dx^dz, dx^dt, dy^dt, dz^dt are in a 1-1 correspondence with the bivectors in equation (436). For the torsion 3-form we thus obtain

$$\mathrm{A}\wedge d\mathrm{A}\,(|u_r\rangle) = -\,(2\,j^2\,c\,m\,x - j^2\,b\,m\,y)\;dx\wedge dy\wedge dz +$$
$$+ (2\,j\,k\,b\,m\,x + j\,k\,c\,m\,y - j\,k\,m^2\,z)\;dx\wedge dy\wedge dt +$$
$$+ (2\,j\,k\,b\,c\,x - j\,k\,(b^2 + c^2)\,y + j\,k\,c\,m\,z)\;dx\wedge dz\wedge dt$$
$$+ (\,j\,k(\,b^2 - m^2)\,x - j\,k\,b(2c\,y - \mathrm{m}\,z)\,)\;dy\wedge dz\wedge d \qquad (476)$$

Respecting the constants we get

$$\mathrm{A}\wedge d\mathrm{A}\,(|u_r\rangle) = \quad -\,(2\,(\cos\phi)^2\,(\sinh\theta\cos\tfrac{1}{2}\psi\,\sinh\theta\sin\tfrac{1}{2}\psi)\,x -$$
$$-\,(\cos\phi)^2\,(\cosh\theta\sin\tfrac{1}{2}\psi\,\sinh\theta\sin\tfrac{1}{2}\psi)\,y)\;dx\wedge dy\wedge dz +$$

$$+\,(\sin 2\phi)\,(\cosh\theta\sin\tfrac{1}{2}\psi\,\sinh\theta\sin\tfrac{1}{2}\psi)\,x +$$
$$+\,(\tfrac{1}{2}\sin 2\phi)\,(\sinh\theta\cos\tfrac{1}{2}\psi\,\sinh\theta\sin\tfrac{1}{2}\psi)\,y -$$
$$-\,(\tfrac{1}{2}\sin 2\phi)\,(\sinh\theta\sin\tfrac{1}{2}\psi)^2\,z)\;dx\wedge dy\wedge dt +$$

$$+\,(\sin 2\phi)\,(\cosh\theta\sin\tfrac{1}{2}\psi\,\sinh\theta\cos\tfrac{1}{2}\psi)\,x -$$
$$-\,(\tfrac{1}{2}\sin 2\phi)\,((\cosh\theta\sin\tfrac{1}{2}\psi)^2 + (\sinh\theta\cos\tfrac{1}{2}\psi)^2)\,y +$$
$$+\,(\tfrac{1}{2}\sin 2\phi)\,(\sinh\theta\cos\tfrac{1}{2}\psi\,\sinh\theta\sin\tfrac{1}{2}\psi)\,z)\;dx\wedge dz\wedge dt +$$

$$+\,(\,(\tfrac{1}{2}\sin 2\phi)\,((\cosh\theta\sin\tfrac{1}{2}\psi)^2 - (\sinh\theta\sin\tfrac{1}{2}\psi)^2)\,x -$$
$$-\,(\tfrac{1}{2}\sin 2\phi)\,(\cosh\theta\sin\tfrac{1}{2}\psi\;2(\sinh\theta\cos\tfrac{1}{2}\psi)\,y -$$
$$-\,(\sinh\theta\sin\tfrac{1}{2}\psi)\,z)\,)\;dy\wedge dz\wedge dt \qquad (477)$$

Notice, we have four differential 3-forms, namely dx^dy^dz, dx^dy^dt, dx^dz^dt and dy^dz^dt each having one gonometric x-component, one y- and one z-component. This represents a non-equilibrium thermodynamic system with a disconnected surrounding and very special behaviour with respect to fluctuations. We consider j, k as constant and vary θ and ψ. So we produce for each of the four tri-differential vectors a phase plot with its x-, y-, and z-component.

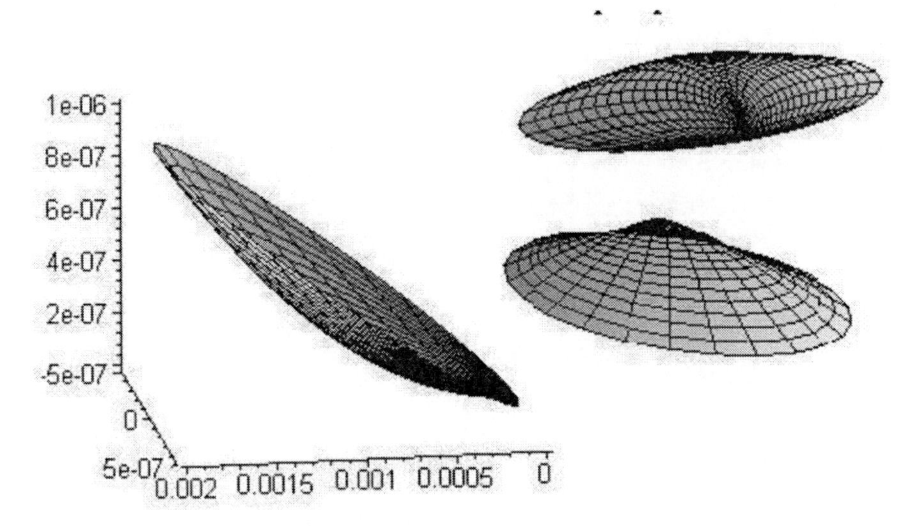

Figure 63a: Torsion phase plot (x,y,z) for tri-differential dx^dy^dt – the conch of u-quark θ =0…0,001; 1/2ψ = 0…π, looked at from three different perspectives.

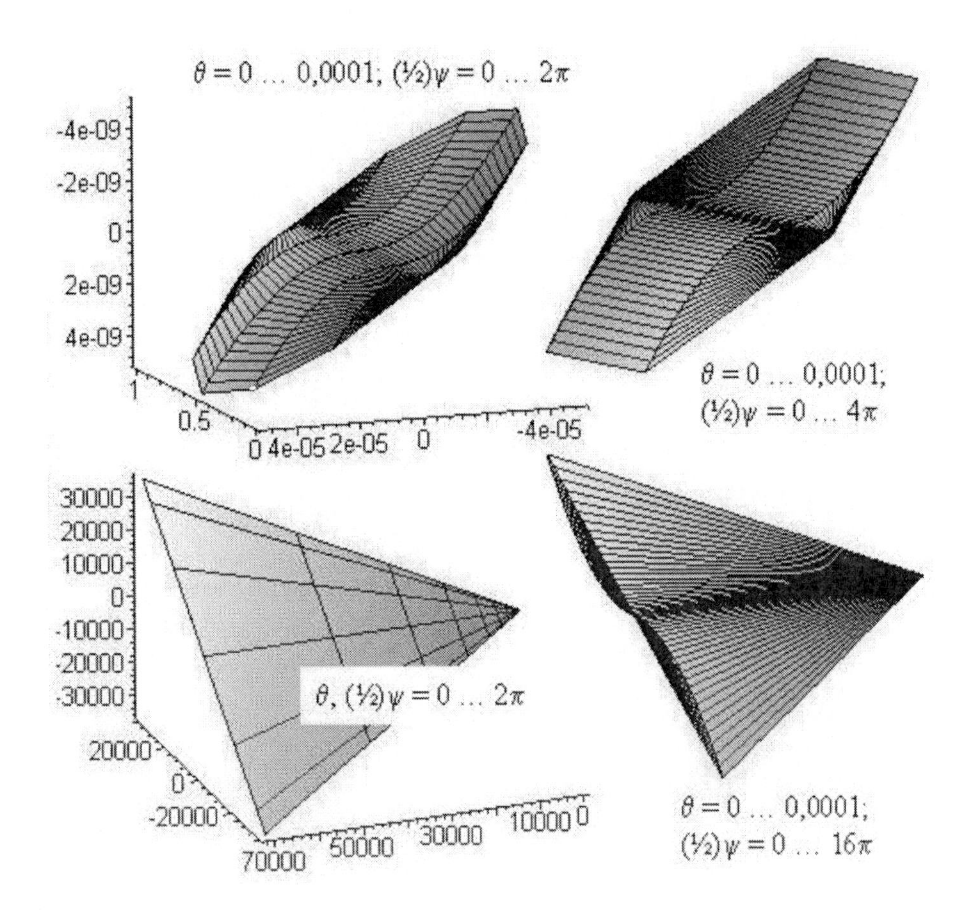

Figure 63b: Torsion phase plot (x,y,z) for tri-differential *dy^dz^dt –the child's seat* of u-quark θ = 0…0,001 and 0…2*π*; 1/2ψ = 0…2π, 4π, 16*π*, looked at from three different perspectives.

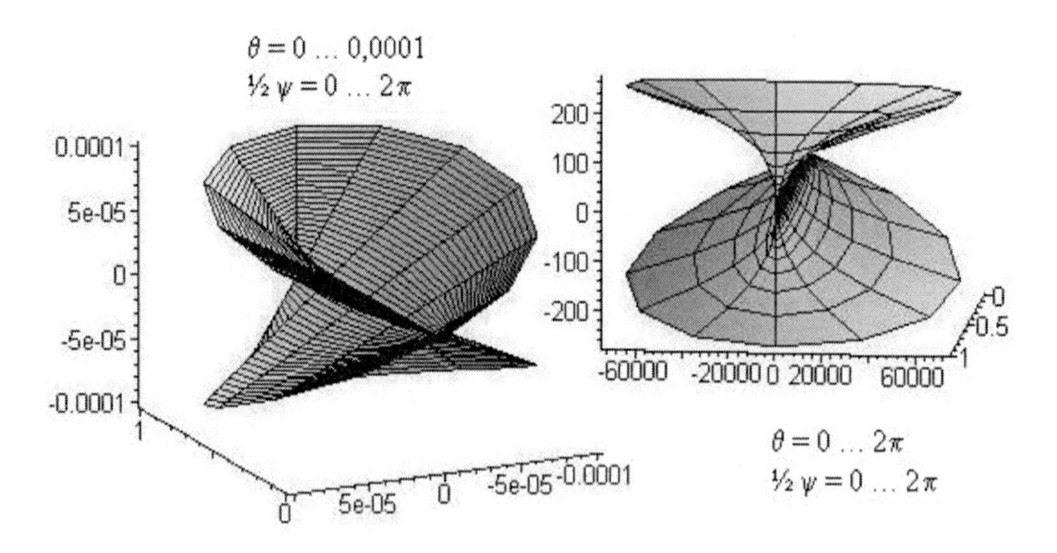

Figure 63c: Torsion phase plot (x,y,z) for tri-differential *dy^dz^dt –the basket chair singularity* of u-quark for low and high energy, θ = 0…0,001 and 0…2π', 1/2ψ = 0…2π.

There is a rich predicate logic of topological torsion that arises on the basis of such natural representation of matter in the algebra *sl*(4, ℂ). First we see that the domain of zero torsion in the {x, y, z} phase plot is essentially given by a single point. This is the 3D origin which belongs to the torsion surface for all four components of the differential trivector. The triangle is a typical high energy plot with a velocity range of $0 \leq v \leq .9999930253\ c$ (when θ = 2π). θ = arctanh(v/c) at 2π is correlated with 99,999% velocity of light.

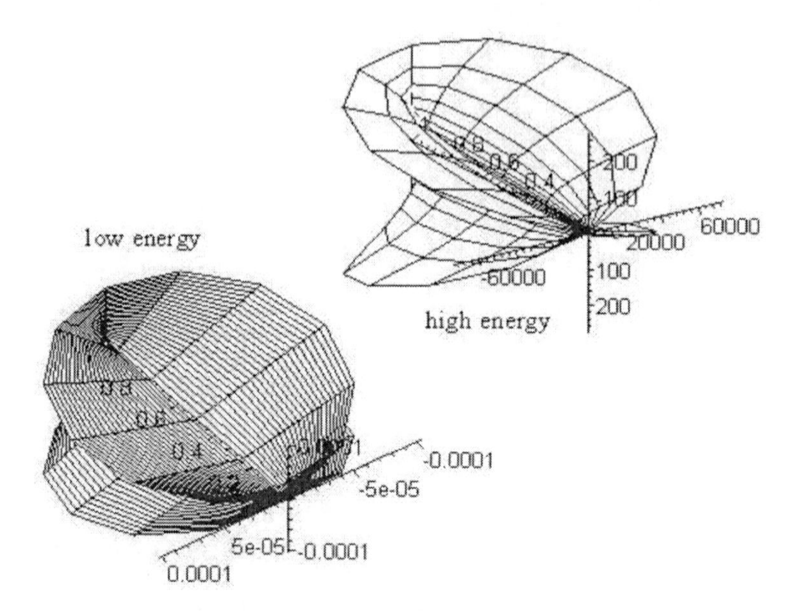

Figure 63d: Torsion phase plot (x,y,z) for *dy^dz^dt* – the point of zero torsion –basket chair singularity of u-quark for low and high energy, θ = 0…0,001 and 0…2π', 1/2ψ = 0…2π.

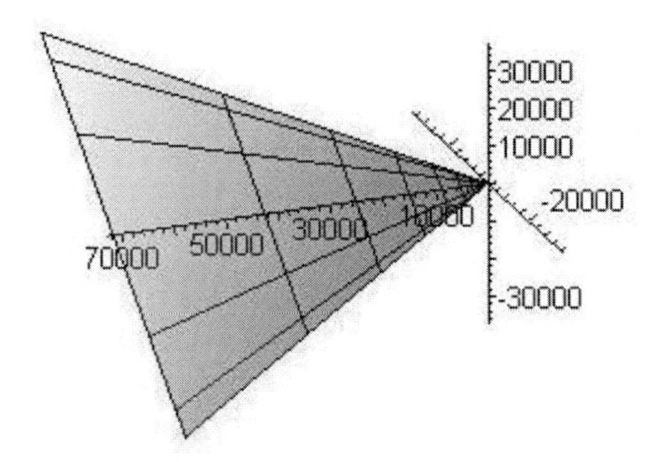

Figure 63e: Torsion phase plot (x,y,z) for *dx^dz^dt* – the point of zero torsion – triangle of u-quark for high energy, $\theta = 0\ldots 2\pi$, $1/2\psi = 0\ldots 2\pi$.

In systems beyond equilibrium, circumstances are reversed. Zero torsion is not the rule, and does not characterize the whole space, but it becomes the singular point. We can literally speak of a zero torsion singularity. The next important thing to realize is that the torsion surface as represented in the torsion phase plot does not necessarily depend on the low- high-energy dichotomy. Especially in figure 63 c – the $\{x, y, z\}$ phase plot of the *dy^dz^dt* component for the red u-quark - we realize that the figure is essentially independent of the range of arctanh(v/c). The zero point can be seen in the following

Surprisingly, the emergence of the conch surface in the *dx^dy^dt* torsion component is bound to low energy. Higher energies lead to a flattening of the seashell deformation as in figure 63f. The torsion increases linearly as we increase the distance from the origin.

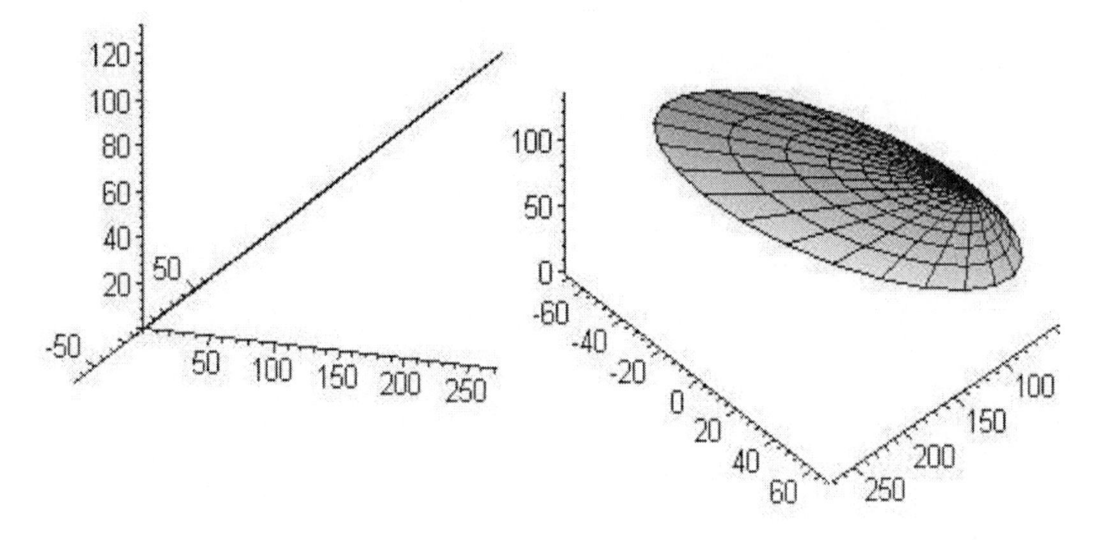

Figure 63f: Torsion phase plot (x,y,z) for *dx^dy^dt* –torsion hyperplane for u-quark high energy range ($v \leq 99{,}627c$) two perspectives $\theta = 0\ldots\pi$, $1/2\psi = 0\ldots\pi$.

Chapter 50

DARK VAN DER WAALS GAS

We saw the WIMP as a formidable enrichment of the neutrino families. But we also left it open if this was at all the most important contribution to the dark matter component of the universe. According to a famous theorem by Arthur Cayley and William Rowan Hamilton every $n \times n$ matrix with complex entries, $X \in \mathrm{Mat}(n, \mathbb{C})$, must provide the zero set of its characteristic polynomial. Kiehn has correlated this theorem with the polynomial of the classic Van der Waals equation and investigated the logical and topological consequences of such a connection. In this chapter we shall repeat some important calculations and results he presented in his volumes on »*Non-equilibrium Systems and Irreversible Processes*« as they seem to endow us with the most relevant explanation for the appearance of so called dark matter.

One of the first thermodynamic equations we have to learn in school is the idealistic equation for a perfect or ideal gas. It appears in two forms

$$P\,V = n\,R\,T \quad \text{or} \qquad P\,V = N\,k_b\,T \tag{478}$$

where P means Pressure, V volume, T temperature, R (= 8,314472 Joule /mol×Kelvin) is the universal gas constant and n the molar amount of substance, N the number of particles and k_b Boltzmann's constant. As for n and N we have the international unit of 1 Mol = $6{,}022\times10^{22}$ particles.

In a real, less ideal gas, it had been argued, there are small components or molar parts that interact with one another. Two further constants were needed, the covolume b in [dm^3/mol] and the cohesive pressure a in [bar × dm^6 /mol^2] to develop a more realistic gas equation. With the molar density $\rho = n / V$ van der Waal's equation can be written in the form

$$P = \frac{\rho RT}{1 - b\rho} - a\rho^2 \tag{479}$$

This formula was extremely successful for explaining the thermodynamic features of real gases. Kiehn put it in polynomial form for molar density

$$ab\rho^3 - a\rho^2 + (RT + bP)\rho - P = 0 \tag{480}$$

The formula displays an implicit surface in the variable space $\{P, T, \rho\}$. It is important that the cohesive forces and critical transitions of molar density are not dependent on a microscopic or quantum scale. We use moles rather than molecules as phenomenological units. This expresses the scale independence of phenomena. The fundamental topological events that emerge in a real gas are independent of size and shape. This should hold true not only for the atmosphere, but for the whole universe. Differentiating (480) with respect to molar density we determine upon the existence of a critical point on the hypersurface where we observe that

$$\frac{P_c}{T_c\,\rho_c} = const \qquad (481)$$

We can rescale thermodynamic variables to obtain dimensionless variables by using

$$T_c = \frac{8a}{27bR} \text{ critical temperature} \qquad P_c = \frac{a}{27b^2} \text{ critical pressure} \qquad (482)$$

and critical mol volume $3b$. The constant in (481) is therefore determined as equal to

$$const = \frac{3b \times P_c}{RT_c} = \frac{3}{8} = 0{,}375 \qquad (483)$$

The van der Waals equation in reduced dimensionless form reads

$$\tilde{\rho}^3 - 3\tilde{\rho}^2 + \tfrac{1}{3}(8\tilde{T} + \tilde{P})\tilde{\rho} - \tilde{P} = 0 \qquad (484)$$

We are using the notation chosen by Kiehn (2008a, p. 77).

It is this polynomial that is compared with the Cayley-Hamilton equation of a non-degenerate real 4×4-matrix. The zero set Cayley-Hamilton polynomial reads

$$\Theta(x, y, z, t) \overset{def}{=} \xi^4 - X_M\xi^3 + Y_G\xi^2 - Z_A\xi + T_K = 0 \qquad (485)$$

Here Kiehn identifies the Cayley-Hamilton polynomial with a (x, y, z, t)-local, *Universal Thermodynamic Phase Function* $\Theta(x, y, z, t)$. Coefficients $\{X_M, Y_G, Z_A, T_K\}$ are invariant with respect to similarity transformations of the Jacobian matrix and in this (restricted) sense, as he says, the method is universal. The Jacobian contains both symmetric and antisymmetric entries. The 2-form dA emphasizes the antisymmetric features of the partial derivatives of the 1-form coefficients. The eigenvalues and eigenvectors of a real matrix can be complex. As the similarity coefficients $\{X_M, Y_G, Z_A, T_K\}$ are all real, the eigenvalues are collected into three equivalence classes

4 real eigenvalues
2 real, 1 complex and 1 complex conjugate

2 complex and 2 complex conjugates.

In case that $T_K = 0$ the *UTPF* becomes

$$(\xi^3 - X_M\xi^2 + Y_G\xi - Z_A)\xi = 0 \tag{486}$$

The similarity coefficients are related to the "curvatures" of the implicite surface displayed by the molar density polynomial. The cubic factor of the Cayley-Hamiltonian can be put into direct correspondence with the van der Waals equation. We realize a correspondence having form

Van der Waals	$\xi = \widetilde{\rho}$
Linear	$\xi_1 + \xi_2 + \xi_3 + \xi_4 = X_M = 3$
Quadratic	$\xi_1\xi_2 + \xi_1\xi_3 + \xi_1\xi_4 + \xi_2\xi_3 + \xi_2\xi_4 + \xi_3\xi_4 = Y_G = \frac{1}{3}(8\widetilde{T} + \widetilde{P})$
Cubic	$\xi_1\xi_2\xi_3 + \xi_1\xi_2\xi_4 + \xi_1\xi_3\xi_4 + \xi_2\xi_3\xi_4 = Z_A = \widetilde{P}$
Quartic	$\xi_1\xi_2\xi_3\xi_4 = T_K$

The variables ξ_1, ξ_2, ξ_3, ξ_4 are identified with the local eigenvalues of the Jacobian [☺]. Forces and excitations associated with the linear curvature indicate surface tension effects. The Gauss quadratic similarity invariant is dominated by temperature, but has a pressure contribution. The cubic similarity invariant represents the pressure of molar interactions. Magnitudes *M*, *G*, *K* are 3D averages of the similarity invariants X_M, Y_G, Z_A of a 3×3 Jacobian matrix. By comparing formulas, a correspondence can be constructed between the arbitrary dynamic system and a van der Waals gas. So we obtain a close correspondence between the Cayley Hamilton invariant theory and thermodynamics.

Kiehn had suggested in the late 1970s that there is a connection between the invariants of the shape matrix of differential geometry, phase transitions in dynamic systems and thermodynamics of the Gibbs equilibrium system. In particular it was assumed that the Gauss curvature of the Gibbs function as an implicit surface in the van der Waals function must vanish. If the analogy is working, on the one hand any dynamical system in three dimensions could be used to define a basis frame, on the other hand, as every 3×3 matrix satisfies its Cayley-Hamilton polynomial, all such 3-systems could be seen as universal representations of some van der Waals gas.

Frames, Cartan matrices of connection 1-forms and material fields would, in a thermodynamic context, represent one and the same thing.

Solving (484) for renormalized pressure we obtain

$$\widetilde{P} = \frac{3\widetilde{\rho}^3 - 9\widetilde{\rho}^2 + 8\widetilde{T}\widetilde{\rho}}{3 - \widetilde{\rho}} \tag{487}$$

The surface $\widetilde{P}(\widetilde{T},\widetilde{\rho})$ contains a peculiar fold and cusp, the shape of the classic swallowtail singularity. It discloses negative pressure on several leafs close to the critical point.

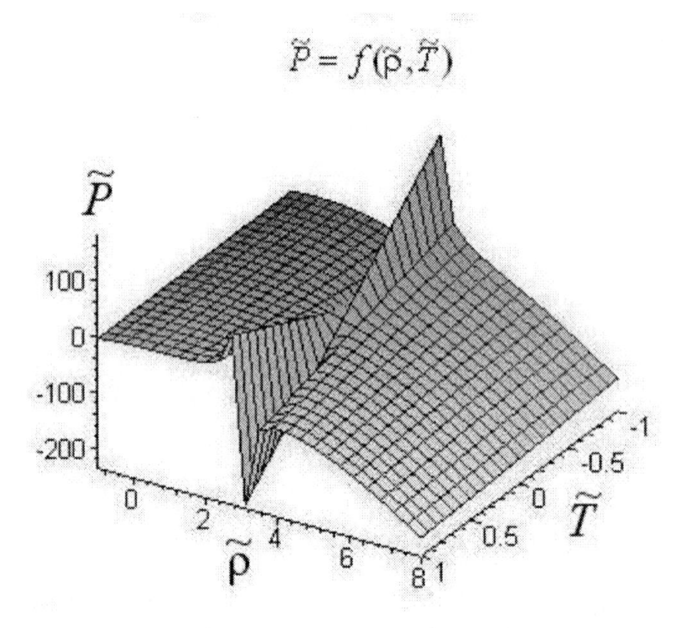

Figure 64. Negative pressure.

It is interesting to play around with such a MAPLE 3D smartplot by rescaling the reduced variables in the plot. Then one gets acquainted with the meaning of scale invariance in a sensual way. The following figure discloses two leafs of this peculiar fold with equal scales but looked at from different perspectives.

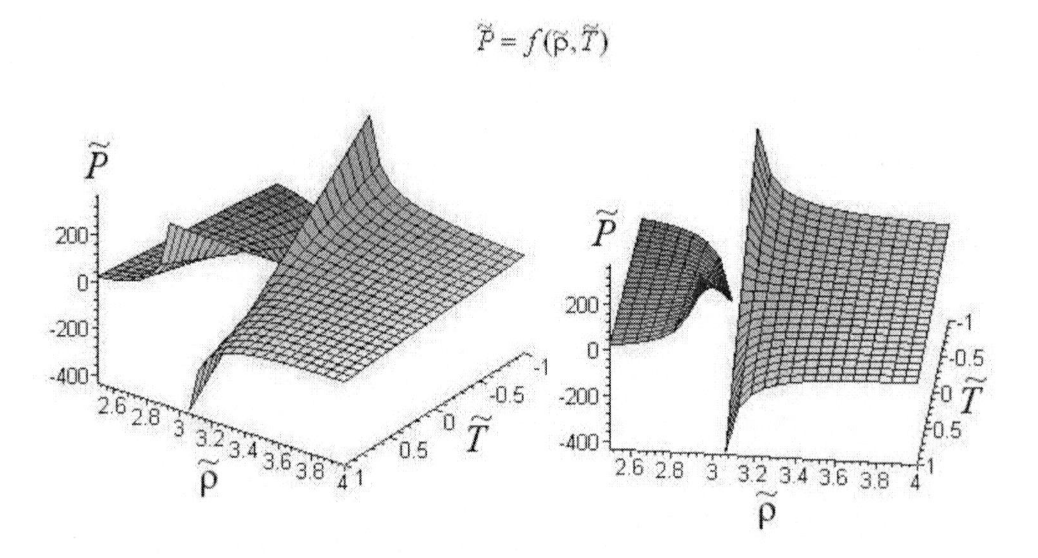

Figure 65. Two leafs with negative pressure.

Those figures are obtained when we choose the range of reduced temperature between -1 and 1 and let the reduced molar density vary between 2.5 and 4. Kiehn has transformed the characteristic polynomial (485) by substituting $\xi = s + X_M/4$ which brings forth a typical cusp catastrophe polynomial

$$\begin{aligned}
&\Phi(x, y, z, t; s) = s^4 + gs^2 - as + k = 0 \text{ with} \\
&g = -\tfrac{3}{8}X_M^2 + Y_G \\
&a = \tfrac{1}{8}X_M^3 - \tfrac{1}{2}Y_G X_M + Z_A \\
&k = T_K - Z_A(X_M/4) + Y_G(X_M/4)^2 - 3(X_M/4)^4 \\
&s = \xi - X_M/4
\end{aligned} \tag{488}$$

For a van der Waals gas we have $X_M = 3$ and $T_K = 0$ and the reduced coefficients become

$$\begin{aligned}
&g = -\tfrac{27}{8} + Y_G = -\tfrac{27}{8} + \tfrac{1}{3}(8\widetilde{T} + \widetilde{P}) \\
&a = -\tfrac{27}{8} + \tfrac{1}{2}(8\widetilde{T} - \widetilde{P}) \\
&k = -\tfrac{243}{256} - \tfrac{9}{16}\widetilde{P} + \tfrac{3}{2}\widetilde{T} \\
&s = \xi - \tfrac{3}{4}
\end{aligned} \tag{489}$$

As one of the eigenvalues is presumed to be zero the critical point has been moved to s = ¼. The transformed formula has eliminated the cubic term by transposing the critical point to the origin in terms of s. Consider the first and second derivative of the transformed phase polynomial (488) with respect to the family parameter s

(i) $\Phi = s^4 + gs^2 - as + k = 0 \quad \Rightarrow \quad k = -(s^4 + gs^2 - as)$

(ii) $\dfrac{\partial \Phi}{\partial s} = 4s^3 + 2gs - a = 0$ envelope condition

$\Rightarrow \quad a = 4s^3 + 2gs$ binoidal line $g = -2s^2$

(iv) $\dfrac{\partial^2 \Phi}{\partial s^2} = 12s^2 + 2g = 0$

$\Rightarrow \quad g = -6s^2$ spinoidal line (490)

Substituting the parameter a from the envelope condition into the equation for k yields a forth order polynomial. This gives rise to the denotation of a *thermodynamic Higgs potential.*

$$k = s^2(3s^2 + g) \qquad \text{expansion function} \tag{491}$$

The origin of this denotation is as follows. In models describing the Higgs-mechanism, the complex Higgs-field ϕ gives rise to a Higgs-potential $-m\,\phi^+\phi + \lambda\,(\phi^+\phi)^2$ with two positive real numbers m, λ. If ϕ is real, this potential displays a parabola of forth order. With an additional imaginary part, we obtain a three-dimensional rotational parabola that looks like the bottom of a champagne bottle. The expansion function k(s, g) can be called in differential geometry a reduced envelope function. If one plots this function in the appropriate domain of s, g, one obtains a pitchfork bifurcation with the typical parabola cross section. The magnitude g can be interpreted as a reduced temperature. For values below the critical point, the function k is a polynomial of fourth degree, but above the critical temperature k is quadratic. Below the critical isotherm, the expansion term k can have both negative and positive values. In the Higgs potential the set of minima forms a degenerate ground state manifold. Minima do not occur at the origin. Any deviation from the ground state manifold can contribute to the mass of those gauge fields we want to be equipped with a mass. The bottom line in the analogy is that any deviation from the manifold of minima generates massive energy. Exactly the same is holding for the expansion function of the thermodynamic Higgs field.

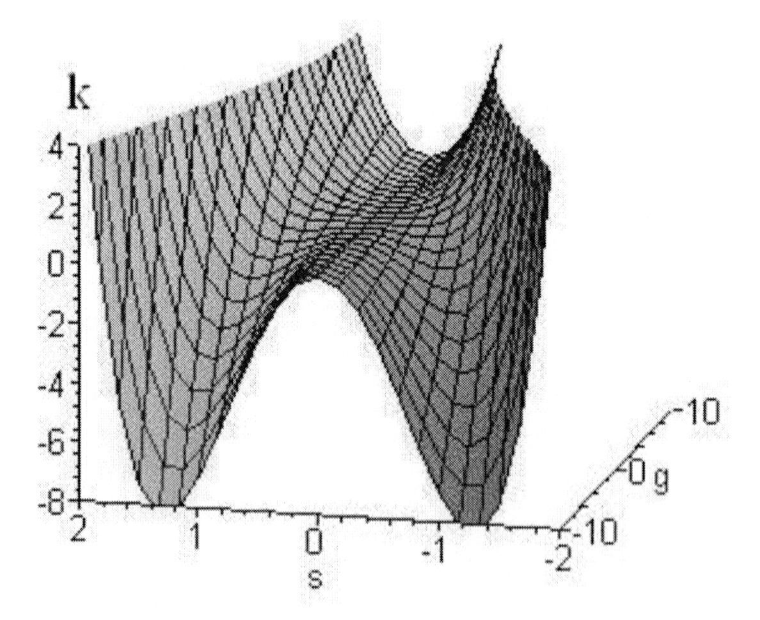

Figure 66. Thermodynamic Higgs potential of a van der Waals gas.

It has become apparent that the pinoidal line, equation (490)(ii), represents the pitchfork bifurcation in the Higgs phase function which stands for the thermodynamic Gibbs function of the van der Waals gas.

The essential difference to the original Higgs mechanism is in the fact that the thermodynamic Higgs field gives mass to universal fields independent of scale, not by that mechanism, but by negative pressure stemming from expansion/contraction-components of macroscopic spinors. How strange the emergence of the topological defect structure close to the critical point of extreme negative pressure is looking can be seen in

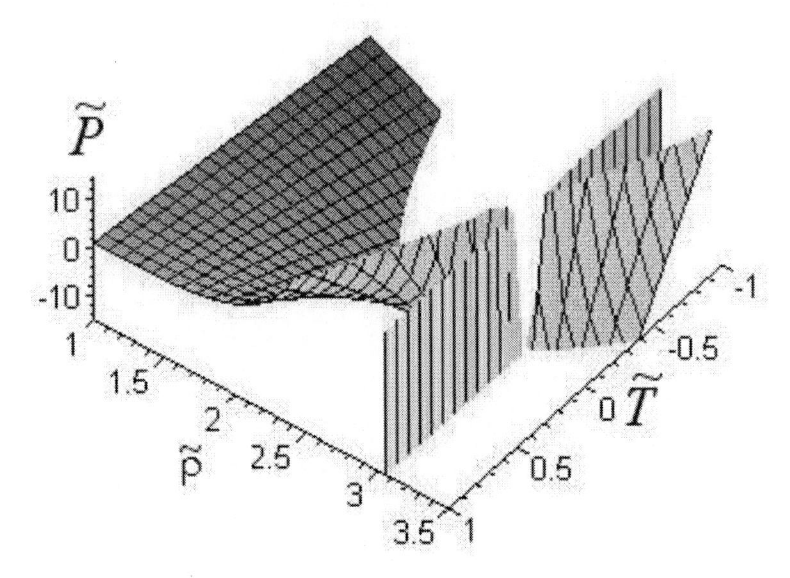

Figure 67. Negative pressure critical region in the $\{p,T,P\}$ – diagram.

For a deeper understanding of this theory, the reader is advised to read chapter 3 in Kiehn's first volume on topological thermodynamics.

Chapter 51

TOPOLOGICAL TIME

To describe in terms of mathematics the topological evolution of living forms such as the shape of an organ, of a wing, the face of a fish, is a most fascinating attempt. For a physicist, to describe wakes and solitons, the evolution of turbulent motion is a challenging task. But to understand the abstract parameter of time itself as subjected to topological catastrophes or to describe the history of its segregated domains like a biological thing would have appeared risky and strange, until recently. But it seems now that it is a most natural mathematical affair. We have to come out with the whole truth: time is a thermodynamic quantity with a topological life that follows very special rules. If there were not those rules, topological evolution of events would not have been able to create in our minds the illusion of geometry and dimensionality of space time. The stable forms of geometry are the outcome of material topological evolution.

In MAPLE Clifford the independent variables in the base variety of the Minkowski Clifford algebra are written as

$$\{x_1, x_2, \ldots, x_{16}\}. \quad (492)$$

Sauter (1930) has already used that denotation for the base variety of the $Cl_{3,1}$. Therefore the maximum magnitude of the Pfaff topological dimension of the space-time algebra is 16 and not 4. However, in some well hidden way, the specific structure of this algebra can itself be recognized as the outcome of some process of topological evolution. Since the $Cl_{3,1}$ is most naturally decomposed into a colour space, a su(2) bivector component and a quaternion timespace. Each of these three components is internally connected by the standard model Lie group which we have denoted as the '*constitutive Lie group*' of $Cl_{3,1}$. In this approach the gluon field excitations are constituted by six bivectors. Therefore practically all known subnuclear particles stem from a 1-form of action with maximum Pfaff topological dimension six. But there is a further reduction of the Pfaff dimension in a specific domain or subspace of the algebra where it becomes apparent that time, just like volume, is very essentially a thermodynamic quantity and therefore subjected to topological evolution. We consider a base unit for time, e_4, volume, e_{123} and director $j = e_{1234}$ in $Cl_{3,1}$. With *Id* these four span the timespace quaternion algebra having a base variety given by, say, $\{u, v, w, t\}$ as in the Euclidean 3-space. Again we have action 1-forms $A = A_k(u, v, w, t)\, dx^k$.

The Cayley-Hamilton polynomial equation defines a family of implicit functions in the 4-space $\{X_M(x, y, z, t), Y_G(x, y, z, t), Z_A(x, y, z, t), T_K(x, y, z, t)\}$, of real valued similarity invariants while the eigenvalues ξ_j may be complex. As indicated in connection with equation (486), we keep track of four similarity invariants

$$\begin{aligned} X_M &= \xi_1 + \xi_2 + \xi_3 + \xi_4 = Trace\,[☺_{ik}] \\ Y_G &= \xi_1\xi_2 + \xi_1\xi_3 + \xi_1\xi_4 + \xi_2\xi_3 + \xi_2\xi_4 + \xi_3\xi_4 \\ Z_A &= \xi_1\xi_2\xi_3 + \xi_1\xi_2\xi_4 + \xi_1\xi_3\xi_4 + \xi_2\xi_3\xi_4 \\ T_K &= \xi_1\xi_2\xi_3\xi_4 \end{aligned} \tag{493}$$

The four-fold degeneracy with roots $\xi = [1, 1, 1, 1]$ defines the critical point with similarity invariants $X_M = 4$, $Y_G = 6$, $Z_A = 4$ and $T_K = 1$. Notice, the critical point is not a fixed point with zero eigenvalues.

In accordance with the theory of string- and fluid surface tension, the term X_M can be interpreted as a linear deformation contributing to the energy of the dynamic system. The quantity Y_G can be interpreted as a quadratic Gauss curvature and resembles an area deformation. The coefficient Z_A in the phase function $\Theta(x, y, z, t; \xi)$ is related to a cubic interaction curvature – a triple interaction connected with a volume deformation (pressure) and the quartic contribution is related to a deformation of the space-time 4-volume. So we have symbolically

$X_M \approx$ string or surface tension
$Y_G \approx$ temperature – entropy
$Z_A \approx$ pressure – interaction
$T_K \approx$ space-time-volume Higgs expansion

We can multiply the phase function by u/ξ^4 and regard u/ξ as a length deformation, δ_{Length}, u/ξ^2 as a space area deformation δ_{Area}, u/ξ^3 as a volume deformation δ_{Vol} and consider u / ξ^4 as a space-time expansion deformation $\delta_{Exp\text{-}xyzt}$. The suggestive formula proposed by Kiehn as an improved illustrative material for the topological evolution of a universal van der Waals gas is

$$\Theta = u - X_M\,\delta_{Length} + Y_G\,\delta_{Space\text{-}area} - Z_A\,\delta_{Vol} + T_K\,\delta_{Exp-xyzt} \tag{494}$$

By the universal strong force trigonal rotation this is mapped onto the Clifford algebra such that the expanding space-time constituent *xyzt* is mapped onto a time deformation and volume deformation is mapped onto space-time volume deformation. Length tension is transformed into space-time-area tension.

$$\Theta = u - X_M\,\delta_{Spacetime\text{-}area} + Y_G\,\delta_{Space\text{-}area} - Z_A\delta_{Exp-xyzt} + T_K\,\delta_{Time} \tag{495}$$

Thus the pressure dominated interaction during strong interaction is deflected into the 4-volume and the Higgs expansion is redirected into the time dimension. The only invariant with a preserved deformation target is Y_G which remains bound to space-areas. Temperature is a quadratic phenomenon during all evolutionary ages of the universe. The important conclusion is that gravity effects related to the 4D Gauss curvature, $G = Y_G /6$ stay area related. Completing the cycle of a strong force trigonal rotation turns the $\delta_{\text{Space-time-Volume}}$ into a δ_{Time} with a coefficient Z_A of pressure-interaction.

$$\Theta = u - X_M \, \delta_{Spacetime\text{-}3\text{-}volume} + Y_G \, \delta_{Space\text{-}area} - Z_A \delta_{time} + T_K \, \delta_{Vol} \tag{496}$$

Formulas (494), (495), (496) indeed stand for the three vertices of a flavour- and colour rotation. The string tension guides the transformation of a length deformation into a space-time area-deformation or space-time-3-volume deformation (symbolized by some e_{124} or similar base unit of colour space *Ch*). Thus the surface- or string tension X_M can operate in a non-relativistic way. Surprisingly it dominates the metamorphoses within the inner spaces of subnuclear families. Only temperature is invariantly fed by quadratic interaction. In a way, the entropy is thus the only true physical invariant among the four similarity invariants. Phenomenological, during all interactions and topological changes, the entropy is bound to the pure spatial bivectors. The interesting thing is that both phenomena, pressure-interaction and Higgs expansion can feed all three of them during strong interaction: volume, space-time volume and time. In the proximity of the critical point, time can collapse or it can blow up, depending on the categorical distribution of fermions. How many universe-bubbles there appear to some meta-mathematical observer seems to be undecided. In his first volume on "Non-equilibrium Thermodynamics" Kiehn has made a remark about gravity as an entropic concept, Gauss curvature and the universal phase function. We shall contribute to this remark by a further statement about the topological evolution of the universe. This statement is the direct consequence of the marriage between topology and geometry as was proposed here.

Remark: We assume, the critical event of cosmic creation takes place within some finite space-time area covering $1{,}61624 \cdot 10^{-35}$ meters and $5{,}39121 \cdot 10^{-44}$ seconds at a density of about 10^{94} g/cm^3 and a temperature of 10^{32} K. During the supposed first moment the fundamental forces of nature are all one. After the Planck time the first symmetry breakage leads to a spinoff of the gravitational forces. This is tantamount with the creation of the invariant forms of elementary Gauss surfaces and an unbroken long range connection between gravitation, surface and entropy. This relation outlasts all evolutionary episodes of the universe. The next split-off from the GUT force delivers the strong force. The vacuum during the GUT phase had much more energy than the universe today. The Higgs field energy density splitting off gravitation is still about 10^7 times less than during the Planck period and represents a repulsive force or negative pressure. The inflationary universe expands between 10^{-35} and 10^{-33} seconds by a factor 10^{50}. After 10^{-33} seconds at a temperature of 10^{25} K strong force fermions and anti-fermions are formed. At that time the Higgs energy is rotated from the space into the time component. This makes the essential difference to the old inflationary model: we had a time dilatation caused by the emergence of the strong force-symmetry by some orders of magnitudes. This dilatation is comparable with the nowadays measured age of the universe. As soon as gravitation segregates from the initial scalar field, it is said, the Higgs field energy blows up space. But while the strong force is delivered from gravitation, the Higgs field blows up time and rotates time into 3-volume and 3-volume into space-time 4-volume and that back into time. We have dimensional unrest until the delivery of the electroweak forces. At the inset of free electromagnetic fields, the time jump has ended. Universal expansion now follows the observed Hubble course. Time is stabilized with the hadrons. But before that event we cannot say how many times and with what scales the cosmic history branches out.

Chapter 52

COMING DEVELOPMENTS

ON IDEMPOTENTS

Working with the Lie differential it becomes apparent that most of us most of the time restrict the rigor to some definite grade. That is, we calculate with 1-forms, 2-forms a.s.o. with constant grade. But to elaborate a valid theory of particle physics we sure need graded Lie differentials. This became utterly clear to me when I realized that the standard model of the forces is a natural property of the space-time algebra. The constituitive Lie algebra of $Cl_{3,1}$ is graded by necessity, and this is not by fortune, but is related to a special conformal property of motion or – if we may say – systems dynamics of space creating fields. If such is the case it is perhaps more important to study spaces of idempotents instead of vector spaces. Primitive idempotents can be generated multiplicatively in some hierarchic way by monotonously increasing their grade and complexity.

In January 2006 I defined three new products in the Clifford algebra:

$$f \oslash g = u \overset{def}{=} g - 2\{f, g\} + 4fgf \qquad \text{idempotent product}$$

$$f \boxbar g = u \overset{def}{=} g + 2\{f, g\} + 4fgf \qquad \text{anti-idempotent product}$$

$$f \boxminus g = w \overset{def}{=} f + [f, g] - fgf \qquad \text{nilpotent product}$$

The formation of idempotents by idempotents is a nonlinear matter and requires considerable understanding of combinatorial phenomena. To show us just how we may begin, consider a basis

$\mathbb{R}\text{bas}_{p,1} = \{e_1, ..., e_p, e_{p+1}\}$ with $e_1^2, ..., e_p^2 = +1$ and $e_{p+1}^2 = -1$ where 1 stands for *Id* together with the following mutually annihilating idempotents f, g involving the »*smallest dimension possible*«

$$f_1 = \tfrac{1}{2}(1 - e_1),\ f_2 = \tfrac{1}{2}(1 - e_2),\ ...,\ f_p = \tfrac{1}{2}(1 - e_p)$$

$$g_1 = \tfrac{1}{2}(1 - e_{1n}),\ g_2 = \tfrac{1}{2}(1 - e_{2n}),\ ..,\ g_p = \tfrac{1}{2}(1 - e_{pn}) \quad \text{with n = p+1}$$

Let us coordinate those objects with the *level 1 of a hierarchy* and list properties for levels

Level 1: grade 0, dimension 1
Idempotent $1 = Id$

Level 2: grade 1, dimension 2
Idempotents $f_1, \ldots, f_p$

Level 3: grade 2, dimension 2
Idempotents $g_1, \ldots, g_p$

We can easily proceed to a forth level by rescaling Clifford products as follows

Level 4: grade 2, dimension 4 calculate further
Idempotents $f_{ij} = 2f_i f_j$ example $f_{12} = 2f_1 f_2 = \frac{1}{2}(1 + e_1 + e_2 + e_{12})$

This is equal to *EC*: $f_{ij} = f_i + f_j + [f_i, f_j] - \frac{1}{2}$ where the ½ is the idempotent from first level. Take also

$g_{ij} = 2g_i g_j$ example: $g_{12} = 2g_1 g_2 = \frac{1}{2}(1 + e_{12} + e_{14} + e_{24})$
equal to *EC*: $g_{ij} = g_i + g_j + [g_i, g_j] - \frac{1}{2}$ *equation of commutators*

$fg_{ii} = 2f_i g_i$ examples $fg_{11} = \frac{1}{2}(1 + e_1 + e_4 + e_{14})$
$fg_{22} = \frac{1}{2}(1 + e_2 + e_4 + e_{24})$

Construct further the element $fg_{ii} = f_i + g_i + [f_i, g_i] - \frac{1}{2}$ for which there exists no equivalent equation EC involving the commutator [,]. There exists a fifth level in the multiplicative hierarchy of idempotents

Level 5: grade 3, dimension 4
Idempotents $fg_{ij} = f_i g_j$ example $fg_{12} = f_1 g_2 = \frac{1}{4}(1 + e_1 + e_{24} + e_{124})$
No commutator equation *EC*

Level 6: grade 3, dimension 8
Idempotents $h_{ijk} = 2fg_{ij} fg_{ik} = 2f_i g_j f_i g_k = f_i g_j + f_i g_k + [\ f_i g_j, f_i g_k\ ] - \frac{1}{2} f_i$
Example

$$h_{123} = 2fg_{12} fg_{13} = 2f_1 g_2 f_1 g_3 = f_1 g_2 + f_1 g_3 + [f_1 g_2, f_1 g_3] - \frac{1}{2} f_1 =$$
$$= \frac{1}{4}(1 + e_1 + e_{23} + e_{24} + e_{34} + e_{123} + e_{124} + e_{134})$$

Level 7: grade 4 dimension 16

Idempotents $fg_{ijk} = 4f_i g_j f_j g_k = f_i g_j + f_j g_k + 2[f_i g_j, f_j g_k] - f_k g_i$

Example:

$$fg_{123} = 4f_1 g_2 f_2 g_3 = \tfrac{1}{4}(e_1 + e_2 - e_3 - e_4 + e_{12} - e_{13} - e_{14} + e_{23} + e_{24} + e_{34} + \\ + e_{123} + e_{124} + e_{134} + e_{234} + e_{1234})$$

Next we may form conjugate products ghg^{-1} with any $g \in \exp(\mathsf{L}^{(2)})$ and unfold arbitrary grades and dimensions. The algebra where geometric products of idempotents generate new idempotents has not yet been investigated. There are many interesting questions concerning idempotents and nilpotents some of which I have initiated in the unpublished papers which I gave the nostalgic name *Al-muqabala des Idempotentes*[1]. Understanding pure states in terms of primitive idempotents is an important step in pregeometric approaches. I have just begun to set it.

When my theory had acquired a certain amount of complexity, I felt a necessity to go back to the question of observation and to take into account what we may call the category of consciousness. There came into being what I denoted as '*awareness extension field icons*' and an isomorphic concept in the graph category, namely the '*observer field graphs*'. This concept has important topological consequences. But until today I have not applied it to primordial space. Nevertheless I have decided to conclude this first book on primordial space by an explanation of this concept. It is a concept in the spirit of radical constructivism and in a way demonstrates that the development of our theories about primordial space and its connection to the material world has just begun.

IDEMPOTENT MULTIPLICATIONS

Multiplication of idempotents is defined base free and its algebraic properties can be adjusted to the current theory of Clifford algebra. There are idempotent- and anti-idempotent manifolds which can be unfolded by Coxeter reflections such that the main algebraic features of primitive and non-primitive idempotents are preserved.

As we saw, a representation of fermions in real idempotent manifolds of the space-time Clifford algebra $Cl_{3,1}$ allowed for a derivation of an improved form of the standard model symmetry in high energy physics. The rigor was based on the introduction of some further automorphism of the Clifford algebra and on a generalization of the Lipschitz method for this special graded space. Yet some of us were dissatisfied with this approach because of two reasons. First, it seems, that we need a better Ansatz in measurement theory and Frobenius observation algebra where the velocity morphism is not necessarily an isometry and where the space-time is not given in advance by a fixed geometric algebra providing us with, or equipped with, the full Lorentz group. But we wish to unfold the space and time from the

1 The Arabic textbook by Al Chwarismi, 780-846, »Alkitāb al-muktaşar fi hisāb ălğabr wa al-muqàbala« = »short book about the calculation of the supplement and adjustment« was abbreviated to „Algebra et Al-muqabala“ and latinized in the 12th century. In our title we do not use the word “algebra” but the second word “al-muqabala” in the sense of “adjustment”.

material measurement idempotent algebra. Secondly, much of the former rigor used representations with a basis, while we prefer that the laws of physics are formulated base-free. This naturally led to the question if there are minimal sets of idempotents in graded non-commutative algebras and if there exist minimal polynomials from which we can derive the manifold. Going into this, we find out about the algebra of idempotent manifolds which has surprisingly little to do with our starting platform, namely Clifford algebra of the Minkowski space-time. The algebraic manifold we reflect on can be connected with the Clifford algebra or not. It seems to exist independent of it. But I connect it.

Lipschitz in 1886 related the orthogonal transformations to automorphisms of Clifford algebras. We are always using this knowledge when we study the algebra of spinors. Today it is well known that the subspace of bivectors $\wedge^2(M)$ generated by vector space M is a Lie algebra on $\wedge(M, \mathrm{B})$ where B is a nondegenerate symmetric bilinear form. We obtain an isomorphism $\wedge^2(M) \rightarrow so(M, \mathrm{B})$. Lipschitz considered the exterior exponentials $exp^\wedge(u)$ with $u \in \wedge^2(M)$ which turn out to be finite sums (Helmstetter 2004, p. 326). Asking the question, if and how a base unit in Clifford algebra such as, say, e_2 can be transposed onto a bi- or trivector such as, say, e_{14} or e_{124} by some generalized orthogonal transformation with a similar method, a graded version of Lipschitz's method was constructed. This resulted in the constitutive Lie algebra, and the elements of it turned out graded. That is, not only the generating space M, but the whole Clifford algebra is actually perceived and used as a vector space. This latter finding would have hardly been possible without studying the transposition automorphisms τ which must be added to the main involutions of Cl. But these morphisms are essentially given by idempotents. Because every such transposition is brought forth by some reflections of the form $1 - 2f$ where f is a primitive idempotent. Going further, we would like to find out more about a possible algebra of idempotents, if there exist some useful types of multiplication and if there are discrete sets or continuous manifolds of them especially when they appear as elements of a Clifford algebra Cl. There have been made valuable attempts to classify the idempotents in special Clifford algebras (Kopczynski 1989), but a general concept of idempotent al-muqabala has not yet been given. I cut some capers, as Thirring would say, but did not come to a reasonable end. Yet, I do not want to hide this excursion I made it about five years ago.

AL-MUQABALA - ADJUSTMENT OF IDEMPOTENT ALGEBRA

Suppose $\mathrm{I} \subset Cl$ be a set with multiplication $(\mathrm{I}^2 = \mathrm{I} \times \mathrm{I}) \rightarrow \mathrm{I}$ and f, g, h elements having the property of being idempotents

S 1 I is a set yet unspecified and the map $\mathrm{I} \times \mathrm{I} \rightarrow \mathrm{I}$ is not yet known,

D 2 the map $\mathrm{I} \times \mathrm{I} \rightarrow \mathrm{I}$ is a multiplication ※ such that $\forall f, g \in \mathrm{I}\ \exists h \in \mathrm{I} : f ※ g = h$, h is definite. There is only one h to any pair f, g in I.

D 3 f, g, h are idempotents $f ※ f = f^2 = f$, $g^2 = g$, $h^2 = h$

D 4 1 is the unit scalar in both I and *Cl*

L 1 $\forall f \in \mathrm{I} \Rightarrow R = 1 - 2f$ is a Coxeter reflection as we have
$R^2 = (1 - 2f)(1 - 2f) = 1 - 4f + 4f^2 = 1 - 4f + 4f = 1$

Be aware that *f, g, h* represent any idempotents in any algebra *Cl* with multiplication, but unspecified otherwise, independent of basis, location, topology and orientation. The fact that a reflection squared gives +1 suggests that there may exist an analogue to an orthogonal group of transformations which moves the reflection, a 1-form of *f*, on a unit sphere in *Cl*. Clearly, in this case, the idempotent moves on a normed sphere too. In case that any such rotation is a product of two reflections, that insight provides us a first tool for turning I into an algebra on a manifold. As soon as we have understood this intuition, we may forget about rotations, space and all the rest of it. We just keep the statement about the form of the polynomials which bring forth the idempotents, namely, if *f, g, h* are idempotents (D 3) we must conclude that the following six expressions are idempotents too:

$$l = (1-2f)(1-2g)h(1-2g)(1-2f), \quad m = (1-2g)(1-2f)h(1-2f)(1-2g)$$
$$n = (1-2g)(1-2h)f(1-2h)(1-2g), \quad o = (1-2h)(1-2g)f(1-2g)(1-2h)$$
$$r = (1-2h)(1-2f)g(1-2f)(1-2h), \quad s = (1-2f)(1-2h)g(1-2h)(1-2f) \qquad (497)$$

Here we used all three idempotents in order to get six new ones, just to make us aware of the combinatorial structure brought in by such consideration. Now, if you have any two idempotents *f, g*, then setting *g* equal to *h* in the first formula, we obtain a new idempotent as a polynomialof degree 3 (see D 5 below). This is what we need first of all, the general idempotent:

if $f, g \in \mathrm{I} \subset Cl$ then there follows that $u \in \mathrm{I} \subset Cl$ provided we define as follows

Multiplication in I × I: $\mathrm{I} \times \mathrm{I} \to \mathrm{I}$

D 5: $f ※ g = u$ and $u \overset{def}{=} g - 2\{f, g\} + 4fgf$

where $\{f, g\}$ is the *Cl–anticommutator* and $f\,g\,f$ is a trinomial product. Obviously, we have $f※f = ff = f$. The *u* is nothing else than the *g* reflected by $1 - 2f$.

Assertion A 1: $u = f ※ g \in \mathrm{I}$

Proof:

$$u^2 \overset{def}{=} (g - 2\{f, g\} + 4fgf)(g - 2\{f, g\} + 4fgf) = gg - 2gfg - 2ggf + 4gfgf -$$

$$-2fgg-2gfg+4fgfg+4fggf+4gffg+4gfgf-8fgfgf-8gffgf+4fgfg-8fgffg-$$
$$-8fgfgf+16fgfgf=g-2gf-2fg+4fgf=g-2\{f,g\}+4fgf=u$$ q.e.d.

Notice that the definition D 5 of idempotent multiplication does not rely on Clifford algebra.

L 2: $f※g \neq g※f$ but $\langle f,g\rangle \overset{def}{=} g-f+4(fgf-gfg)$ thus we have defined

D 6: $\langle f,g\rangle = f※g - g※f$ idempotent commutator

The commutator of the *Cl*-product shall be denoted by $[f, g] = (fg - gf)$ as always. Next the question arises if the product is associative. It turns out that it is not.

L 3: $f※(g※h) \neq (f※g)※h$ non-associative multiplication

Proof: the terms $f※(g※h) = (1-2f)(1-2h)g(1-2h)(1-2f)$ and $(f※g)※h$ $=(1-2(1-2f)h(1-2f))g(1-2(1-2f)h(1-2f))$ are different in general. Thus, if we can define an algebra of idempotents $\mathrm{I} \subset Cl$, this will be non-associative by necessity. We should point out here that a general algebra of measurement such as a Frobenius algebra for the measurement of binary relative velocity turns out to be a non-associative algebra.

L 4: in case that $[f, g] = 0$ the terms $fg \in \mathrm{I}$ and $f+g-gf$ is an idempotent too. The proof is straight forward. Lemma 4 is related to Boolean algebra (Oziewicz 2005).

L 5: in case that $\{f,g\}=0$ the sum $f+g \in \mathrm{I}$

The proof is an easy exercise and takes just one line. Lemma 5 of the 'vanishing anticommutator' is related to the notion of a primitive idempotent which was introduced by Charles Sanders Peirce in his 1881 writing »*Linear associative algebra*«. Due to this definition

D 7: An idempotent $h^2 = h$ is *primitive* if it cannot be a sum of two mutually annihilating idempotents, that is

$$primitive\,(h) \Leftrightarrow h \neq f+g \text{ with (i) } f^2=f,\, g^2=g \text{ and (ii) } fg=gf=0$$

From D 7 (ii) there follows that the anticommutator $\{f,g\}$ is zero and Lemma 5 is fulfilled. Therefore the sum $h = f+g$ *is not* a *primitive idempotent*, but it is an *idempotent*; $f+g \in \mathrm{I}$. On the other hand, we must presume Lemma 6.

L 6: Let $f, g \in \mathrm{I}$. From the fact that their sum $f+g \in \mathrm{I}$ we cannot conclude that $h=f+g$ is not a primitive idempotent. The proof is in the fact that mutual annihilation is a sufficient but not a necessary condition for the anticommutator to vanish.

Now it is clear that we cannot take any number α from a field, say, $\mathbb{R}$, $\mathbb{C}$, $\mathbb{H}$, and $\mathbf{O}$ to form an idempotent αf. But we may only consider idempotents $\alpha \in Cl$ as factors which commute with the idempotents f, g.

L7: Distributivity Iff $[\alpha, f] = [\alpha, g] = 0$ and $\{f,g\} = 0 \Rightarrow \alpha(f+g) \in I$ and $\alpha(f+g) = \alpha f + \alpha g \in \mathrm{I}$

Notice, that the α is not an element in a field, and we are far from making I into a vector space. We just learn from this Lemma how the algebra is being closed by the specific properties of vanishing commutators and anticommutators.

D 8: Without at first specifying the relation with I we define the anti idempotent set C as a set of anti-idempotents having property

$$g \in \mathrm{C} \Leftrightarrow g^2 = -g$$

Multiplication in $\mathrm{I} \times \mathrm{C}$: $\mathrm{I} \times \mathrm{C} \to \mathrm{C}$ $f \in \mathrm{I}, g \in \mathrm{C}$ (498)

D 9: $f \divideontimes g = u$ and $u \overset{def}{=} g - 2\{f,g\} + 4fgf$

Assertion A 2: $u = f \divideontimes g \in \mathrm{C}$

Proof: Use the polynomial from the proof in assertion 1 and replace every ff by f and $g\,g$ by $-g$. This gives $-u$.

D 10: Let $f \in \mathrm{C}$, a Coxeter reflection of an idempotent $f \in \mathrm{C}$ is defined as the term $1 + 2f$. We verify that $(1 + 2f)(1 + 2f) = 1$.

Multiplication in $\mathrm{C} \times \mathrm{C}$: $\mathrm{C} \times \mathrm{C} \to \mathrm{C}$ $f \in \mathrm{C}, g \in \mathrm{C}$

D 11: $f \boxtimes g = u$ and $u \overset{def}{=} g + 2\{f,g\} + 4fgf$ being $\equiv (1 + 2f)\, g\, (1 + 2f)$

Assertion A 3: $u = f \boxtimes g \in \mathrm{C}$

Proof: Calculate the polynomial as in assertion 1 and replace every ff by $-f$ and $g\,g$ by $-g$. This gives $-u$. That is, product $f \boxtimes g$ is an anti-idempotent, indeed.

Example: In the Clifford algebra $Cl_{3,1}$ the following quantities are anti-idempotents

$$f = \tfrac{1}{4}(-1 + e_1 - e_{24} + e_{124}) \qquad g = \tfrac{1}{4}(-1 + e_2 - e_{34} + e_{234}) \qquad f^2 = -f$$

and $g^2 = -g$

$$h = 8p_1p_2p_3q_1q_2q_3 \qquad\qquad h^2 = -h$$

with $p_i = \frac{1}{2}(1+e_i)$, $q_i = \frac{1}{2}(1+e_{i4})$ $i = 1, 2, 3$ (minimal set in $Cl_{3,1}$)

we multiply $f \boxtimes g = \frac{1}{4}(-1+e_3-e_{14}-e_{134})$ which is an anti-idempotent $(f \boxtimes g)^2 = -f \boxtimes g$. The set $C_\lambda \overset{def}{=} \{\exp(-k\lambda) f \exp(k\lambda)\}$ is a differentiable manifold of anti-idempotents. The parameter k is a real number and $\lambda = \frac{1}{4}(-e_{23}+e_{123})$.

Continuous Idempotent Manifolds

Let $Cl_{p,q}$ a Clifford algebra with a non degenerate quadratic form with indefinite signature, indices i are from an index set P of positive signature, and j from an index set belonging to negative signature. $Cl_{p,q}$ is generated by a space $\mathbb{R}^{p,q}$

$$P \overset{def}{=} \{i / 1 \le i \le p \;\&\; e_i^2 = +1\} \qquad N \overset{def}{=} \{m / p < m \le p+q \;\&\; e_m^2 = -1\}$$

Theorem 1: The idempotent set $I \subset Cl_{p,q}$ is a manifold

Proof: Consider any triple of elements with indices $i, k, l \in P$ but $m \in N$

D 12: $\lambda_1 = \frac{1}{4}(-e_{kl}+e_{ikl})$ $\quad \lambda_2 = -\frac{1}{4}(e_m + e_{ik})$ $\quad \lambda_3 = -\frac{1}{4}(e_{il}+e_{iklm})$

L 8: $[\lambda_\eta, \lambda_\mu] = \frac{1}{2}\varepsilon_{\eta\mu\nu}\lambda_\nu$ with $\varepsilon_{\eta\mu\nu}$ antisymmetric in the indices $\eta, \mu, \nu = 1, 2, 3$

Proof of L 8 by hand or with a Clifford algebra calculator. L 8 states, the elements $\lambda_1, \lambda_2, \lambda_3$ fulfil the commutation relations of the Lie algebra $so_{Cl}(3)$ a Clifform of an algebra $so(3)$. Therefore, with the Lipschitz method of 1886 we can produce orthogonal transformations by an exterior exponential map. Let f-be an idempotent in $Cl_{p,q}$. The set

$$I = \{g(k) / g = R^{-1} f R\} \text{ with } R = \exp(\textstyle\sum_{\eta=1}^{3} k_\eta \lambda_\eta) \text{ and } k_\eta \in \mathbb{R} \tag{499}$$

is a manifold. Chosing f to be a binomial of a minimal set of six non-primitive idempotents

$$p_i = \tfrac{1}{2}(1+e_i),\; p_k = \tfrac{1}{2}(1+e_k),\; p_l = \tfrac{1}{2}(1+e_l)$$
$$q_i = \tfrac{1}{2}(1+e_{im}),\; q_k = \tfrac{1}{2}(1+e_{km}),\; q_l = \tfrac{1}{2}(1+e_{lm}) \tag{500}$$

a maximal manifold of orthogonal primitive idempotents can be derived. Namely we define

D 13: Two primitive idempotents of a Clifford algebra *Cl* are orthogonal if they mutually annihilate each other: $f \perp g \overset{def}{=} f\,g = 0$.

The action of the group $SO_{Cl}(3)$ preserves the orthogonality relation, as it preserves all the relevant properties of idempotents. If the product of two non-primitive factor idempotents such as $p_2\, q_3$ is a primitive idempotent, then after the orthogonality transformation the factors are still non-primitive and their product will be primitive. Suppose that an idempotent is the sum of two mutually annihilating primitive idempotents. Then after the orthogonality transformation the two primitives will still annihilate each other and their sum will be equal to the transformed non-primitive idempotent. A further investigation should prove that there exists a generalized graded method for every Clifford algebra with a bilinear form φ which results in a morphism $Cl \rightarrow \mathcal{L}(Cl, \varphi)$, where $\mathcal{L}$ is a constitutive Clifford Lie algebra. The definition of multiplication in an idempotent manifold by Coxeter reflections as in D 5 should turn out to be a general and useful operation.

Chapter 53

PRE-GEOMETRIC OBSERVER FIELDS

NEW CONSTRUCTION OF PHYSICS

Investigating the origin of the standard model and compiling results, it became evident that geometric Clifford algebra represented a powerful means to understand the interconnection between symmetries in matter and space-time groups. Quarks can be understood as eigenforms arising in the Clifford algebra of the Minkowski space-time using the Lorentz metric. Then it also became possible to reflect upon the relation between observer fields and observed fields. At first much of this relationship remained unclear. There prevailed a typical master-slave like interval between observer and observed in the sense of control systems (governor, cybernetes) and controlled (object). Both space and forces of nature disclosed objective behaviour in the guise of invariant qualities absolutely independent of the cybernetics of circular observer systems (SOC).

Space and time were allowed, so to say, to behave like objects decoupled from the entailment network[1] of the interactive observer systems. This fact strongly contradicts findings and demands of second order cybernetics, that is, cybernetics of observing systems (von Foerster 1979). Briefly put, according to system theory any objective quality or object comes upon as an eigenform (or eigenobject) within a circular – and thus recursively – self-organising, interacting system of observers and devices as a particular view determined by limited and dynamic reference frames.

Heinz von Foerster began with physics but ended in social science. Still he was ready to answer the question what the observed is or can be apart from the self organising objective of the observer mesh entailed by instruments, physicists, perceptions, societies and so on. His answer resembled that of the radical constructivist Ernst von Glasersfeld: "*The environment does not contain any information. The environment is as it is.*" (Foerster 2003, p. 189). Independent of observations no object can have any quality, because it is the interaction in which such property is acquired. Without the observers the observed is just as it is. It resembles no-thingness, since beyond observing interaction no quality is constituted and no pre-given predicate can be decided on. Therefore we can ask questions as the following:

---

[1] Denotation used by Gordon Pask (1961): »*entailment mesh*«

i) Beyond observation by physicists, do elementary particles, space and time have no measurable qualities at all? Or
ii) Is there any observing assembly – different from the human mind and instruments – providing qualities of matter and space as eigenforms within such particular observer systems?
iii) Do fermions have charge, bosons have strangeness and weight, space have extension a. s. o. independent of the human science culture? If so, we must confess, charge, colour, flavour, gravity, metric a. s. o. seem to appear, nevertheless, as objective features of matter.
iv) Would that, in any case, contradict second order cybernetics?

Apart from the physics peculiarities, these system theoretic statements of measurement foundations can be traced back to Gordon Pasks "Approach to Cybernetics". Independent of our final decision on this matter, we can give a list of initial demands for constructing physics.

i) Considering our world as a process of nature in "natural history", in some way compatible with an epistemological demand to be able to "think back" to the big bang, we must ask if there were elementary particles – photons, fermions, bosons a.s.o. in the begin or briefly after. If there were, we must confess: there must have been an observer system – an "entailment mesh" – capable to provide fermions as eigenforms of participatory observation.
ii) Observers are man, animals, machines or physical fields learning about their environment and impelled to stabilize some events which occur in it. [2]
iii) There are no observer relations without energy transfer.
iv) Both energy and information are important in observation, and both can be measured.

Therefore, if energy, matter and space-time can be traced back until to the earliest of times and independent of the human activity, there must exist an observer mesh. Since, beyond observer interactions there could not be a measure for any of those qualities. Energy and extension would have no measure, information, if only as a reduced abstract quality, could have no bit and byte. Temporal motion could have no order and space-time neither seconds nor light-years. Consequently, observation – just as consciousness, reflection, learning, recursive unfolding of eigenforms – eventually beginning at some instability of no-thing – must arise with the physical world as natural endowment of the existing. This solves the old conflict brought into science by Norbert Wiener.

"Wiener could never quite stop himself from thinking of cybernetics as somehow a part of physics" (Wiener, N. 1948) said Ranulph Glanville (2008, 19). Cybernetics is not a part of physics, but physics is compatible with cybernetics in the above sense. It is exactly in that way that the apparent conflict between radical constructivism and physics is settled and main statements of second order cybernetics are qualified. Intelligent awareness comes upon as one of the first qualities of the cosmic observer fields. It is therefore that we say mind begins even before the stars are born.

---

[2] a slightly changed version of Pask's list given on page 18, also quoted in Glanville (2007) on page 20.

Scientific consequences of Wiener's feeling are not unsubstantiated. There is a domain where circular self organisation and space-time are one. Here cognition meets extension and apparently the map becomes the territory. After all, the map may be brought forth by self referencing the territory. When Ashby (1956) defined cybernetics as the study of systems open to energy but closed to information and control, physicists should immediately have asked what could oppose the realization that small fermion systems such as hydrogen clusters, too, are forming such interactively – in terms of energy transfer – open but organisationally closed systems. One single hydrogen atom consists essentially of four fermions, three quarks and one electron. If we localize the electron, we destroy the atom, as has already pointed out Friedrich von Weizsäcker.

According to Chomsky and other genetic structuralists universal grammar requires logical syntax with negation. As Bernard Scott (2007) explained, in Piagets genetic epistemology such circular logic is imminent in the logic of human action and the concept of reversibility. In agreement with Piaget, Scott says that actions may be undone. Interestingly nothing hinders us to provide the same entailment mesh for physical systems. Especially fermion systems in geometric Clifford algebra show the typical circular logic – but embedded into the geometry – and thus allow for universal grammar in terms of syntax and negation (Schmeikal 2002). Rather we may say, Piagets theory of knowledge is closer to physics than to biology when it comes to test for reversibility of action. Translations and rotations of rigid bodies done by children resemble physical motion, whereas biological system dynamics brings on motion that leads beyond reversibility. Biological development cannot be undone.

It is prevailed not only throughout developmental psychology, structuralism and interactionism, but also in radical constructivism that in living beings, integration of sensori-motor schemata into coordinated wholes both generates *object permanence* in the environment and *object-subject differentiation* in the observer. But observing a cluster of water molecules, hydrogen atoms and the like, who says that the quark is not sensing the proton, the electron is not sensing the nuclei or that the fields do not move according to circularly processed information fed back from the strong and electroweak interaction to the fermions? There is a tacit logic of interaction which provides a semantic base (Scott 2007, 198) that, "*when digitalised as units of meaning, gives rise to a syntax*". There is every reason for us to sense the same universal logic also in systems of elementary quantum fields.

Consequently we shall give an introduction into what the observer calls awareness extension field icons and observer field graphs. We shall make some genetic statements about the organisation of observer field graphs not only in social, but also in pure physical systems. We shall understand that matter is intelligent and that if life should be regarded as holy or god-given, quantum-fields should no less.

Chapter 54

REFERENCE FRAMES OF AWARENESS

DISAPPEARANCE OF I

Where are the material things we perceive? Our immediate answer is: "out there in space and time". But out there, there are no space and time. So what? Where and what is the I that perceives the matter out there? Can we locate it in our body? Or is it in sense perception? In sensation, emotion or feeling? Is it hidden in thought? A ghost in cognition? A muddle? Is it in mental formations or is it pure consciousness?

The investigation of that cherished notion of a minds separate I is at least as old as civilization. When the Buddha searched for the reasons of suffering he found five »adhesion groups« (upādāna-khandha): 1.) material body, 2.) emotion and feeling, 3.) perception, 4.) mental formations and 5.) consciousness. In meditation practice all those turn out as transient, frail and void of ego. In whatever we perceive as material, feeling, sensual, thought out or imagined there is not any invariant that we can validly denote as personal and I. What we seem to be turns out as impersonal. In modern times the turning away from a central processing unit of cognition and departure from the "*users illusion of the self*" (Norretranders 1998) has led to Second-Order Cybernetics (SOC) and Trans-Disciplinary Science (Müller and Müller 2007).

THINKER AND FEELER

The observer doing science is a human being. Once the observer has managed to observe thought as a transient process, he is aware of the thinker being thought. Consciousness is its own content. In silent cognition that consciousness fades. Next the observer might say, if the I is transient in thought, it might be invariant in emotion. So with thought resting in tranquillity there might arise clear perception of emotion and feeling. [1] Who is the feeler? Which I has the emotion? There is no thought. So we can only ask that question briefly after the thinking has set in again. Back we go to observe sensation, emotion and feeling. What is there to arise in

[1] There are modern books on such meditation practice (Thanissaro Bhikkhu 1996, 75 f.); one basic exercise: "Remain focused on the body in & of itself, but do not think any thoughts connected with the body. [...] In a "scanning" or "body sweep" practice, mindfulness means remembering to stick with the process of scanning the body, while alertness would mean seeing the subtle sensations of the body being scanned."

body sensation, -consciousness? Which feelings? Sensations, scorchers, emotions, sweet feelings, clouds of those, mixed feelings, pent up emotions, callous, insensate, numb, unfeeling, sultrily, pressing, dragging, neutral, inert, intense, intensive, pleasant, tingling, delightful, deranging, incommode, disturbing, painful, unpleasing, sucking, wrenching. We are giving vent to our feelings. We may have several experiences of evidence, such as: the pleasant does not depend on cognition. But what is most obvious, most lucidly clear: no feeling is in space, and no feeling is in time. Emotions and feelings are just being there as pure perceptions. They move, but their motion is beyond time. It has no temporal order, just as they have no spatial order. In case the observer experiences that this or that feeling is, say, located in one of his or her extremities, thought has set in. And the observer has located his sensation in his or her thought image. The feeling becomes located by a control act of thought. This shows very clearly – in meditation – how space and time are cognitive concepts constructed by human observers. Matter is not in space-time, but human observers have constructed material reality such that it appears to be in space and time. Emotion and feeling are pre-geometric events.

The Process and the Illusion of I

If such is the case and space-time is a concept of cognition rather than a vessel of matter, then we must find out about the connection between matter, perception, thought, feeling and consciousness. If the personal I of the observer had to be an illusion there could be a »relational invariant« responsible for the appearance of the illusion. In this way the I is not just a falsely cherished notion of a blended human mind. But it is, in a way, part of the cosmic bionic process. Last not least the interval between science as based on the concept of ego and another reality without such concept represents no less than the tension between secular and sacral perception of being. This interval between secular and sacred, then, is an illusion too. But as an illusion it is part of the real. It acts, though as an illusion. What is the cause for the emergence of such illusion? What is the process like in which it is embedded? Again this is known since long. [2] In terms of second-order cybernetics: "Our life is a never ending process imbedded in a variety of other never ending processes. This process performs sequences of establishing relations or references which inevitably call on presuppositions. Most of these presuppositions are beyond our reach or command". (Schmidt 2007, 325 f.)

Sebastian Jünger and others have dwelt on relationality as the basic principle of consciousness. Where clear thought and mathematics are concerned that has a consequence, namely mathematics of bionic field theory to some considerable extent has to rely on

[2] »DIE ANATTA-LEHRE. Unser sogenanntes individuelles Dasein ist, ebenso wie die gesamte Welt, nichts weiter als ein bloßer Prozeß dieser in den fünf Daseinsgruppen zusammengefaßten, sich unaufhörlich verändernden geistigen und körperlichen Vorgänge. Dieser Prozeß war bereits für unermeßliche Zeiten vor dieser unserer augenfälligen Geburt im Gange und wird sich auch nach dem Tode für unermeßliche Zeiten fortsetzen. Die vorhergehenden Texte haben gezeigt, daß diese fünf Daseinsgruppen weder einzeln noch in ihrer Gesamtheit eine wirkliche, in sich bestehende Ich-Einheit (attā) bilden und daß auch außerhalb dieser Gruppen keinerlei Ichheit oder Wesenheit als deren Besitzer zu finden ist. Der Glaube an eine wirkliche und beharrende Ichheit, Persönlichkeit oder an eine „ewige Seele“ muß daher angesichts der ausnahmslosen Veränderlichkeit und Bedingtheit alles Geschehens als eine bloße Illusion gelten « . (Nyanatiloka Mahatero 1989, p. 26)

categories rather than set theory and algebra alone. But we have to be aware of some considerable epistemic difference between categories and SOC. We shall have to think about relational invariance in mathematics too. It seems that consciousness is well hidden in every domain of life, and it is not a property in the usual sense.

The Invariant Beauty of Meaning In Sense Qualities

The red of the rose extends in and beyond space-time. While you look at the red rose with a tranquil mind the red spreads out, and someone said in his ecstasy (Krishnamurti 1976) that colour is god. This experience may signify an experience of the observer in deep meditation where thought – the intrinsic observer – fades or calms down. Though the writer is familiar with the event and deeply trusts in its appearance, he hesitates to tag it "god". Rather we should call it beauty or meaning or the »*deep quality*« in deep vision. The holy Hildegard von Bingen denoted her perception as the »*cloud of the living light*«. Anyway, in each sense perception there is a sense quality which must not be confused by the physical, chemical a.s.o. quality which thought assigns to it.

The red of the rose	is not a wave with length of 630 nm
the blue of the sky	is not a wave having wavelength of 470 nm
perceived colour	is not an electromagnetic wave
the standard pitch a'	is not an acoustic wave withfrequency 440 Hz
the Danube waltz	is not a sequence of tones
the audible	is not a set of longitudinal air oscillations
the odour of Vanillin	is not the same as its chemistry
the quality of smell	is not a chemical formula
the taste of a dish	is not a set of ingredients
flavour	is not chemical composition
the touch of cheek	is not force per face (dyn /cm^2)
sense quality of touch	is neither pressure nor vibration nor temperature
feeling	is not a physical quality
emotions	have no measure

sensations	have sense quality, but are beyond topology and metric
the meaning of thought	is not in its logic structure nor in its syntactic or semantic rules
the meaning of mathematical truth	is not a formula is not in the priory decidability of a statement is not the computability of the terms and equations

But the meaning of a mathematical statement is in its derivation and the fact that the rigor is carried out by a living mind. Also the meaning of thought is in the experiencing of the procedure of thinking. The thinker is the thinking. Feelings, emotions and body sensations are disclosed in living awareness. They are not facts of objective science. We repeat here what R.D. Laing has already said: experiences are not facts of science, and scientific facts need not be experienced. (Laing 1983) The essential in sense quality is not the physical or chemical structure but the fact that it has no measure. Take for example the Danube waltz. The melody is a Gestalt brought on by the tone intervals rather than by the tones themselves. In a way, in awareness we are listening to the void wrapped into the audible structure of chimes.

With the appearance of second order cybernetics (SOC) we have turned from the platonic idea to action. Even a seemingly simple mathematical truth such as $a^2+b^2=c^2$ is not a priory written into the cosmic landscape of truth, but the meaning of it is bound to the life action of some thinkers derivation. It is not a Platonic idea that $1+1=2$. But it needs a living mind in a living world whose devices disclose the assumption.

A Brief Episode in Science History

The second order cybernetics and progress in systemic dialectic logic has come upon, briefly after the level of cognitive completeness [3] has passed through its minimum (Müller 2007). That was about after 1960. The shift from little science to big science (De Solla Price 1974) that was formerly guided by a search for unifying models in single fields now turned over into a complete change of research and development programs. We began to invent trans-disciplinary programs and paradigms. We looked for a common language in the many fields of big science (Jantsch 1972). This shift from, say, unity of physics, to unifying language for all science was first a shift from little science to trans-disciplinary understanding and second a shift from consciousness as a mechanically functioning information system to awareness plus algebra as a human living system in the sense of Wolfgang Köhler who conceived "Gestalt" as a biological dynamic systems feature as far from equilibrium as possible (Köhler 1969, p. 62).

In the course of the observed paradigm shift various authors from Hegel to Spencer-Brown, Niklas Luhman to Sebastian Jünger have pointed at the dynamic relational features of consciousness. The process of nature in consciousness requires a double differentiation – also called trifferation – in the relational system of observation (Schmidt 2007, p. 326). As such we are still dealing with an object of mathematics. Namely consciousness – though it can no

[3] as given by the level of perceived knowledge in relation to the level of perceived ignorance

longer be determined by a set of elements – is still conceivable in terms of algebra and co-algebra. The author (Schmeikal 2006) has shown that observation in quantum field theory requires that the observer perceives himself in the observed and vice verse, and that consciousness as a life process of nature contains a factor which is not mathematical, namely awareness. Awareness and Gestalt are energetic realities mediated by emptiness, and as science concepts are absolutely necessary for the understanding of the conscious mind (again a second order cybernetic event: »*understanding understanding*«). We have to link the operational structures of the conscious mind to the fundamental algebraic systems of physics such as geometric space-time algebra. We also have to understand the proximity of certain symmetries in high energy physics, biology and sociology. This requires unifying approaches and languages in mathematics which are essentially given by category theory (Mac Lane 1997) and Clifford algebra (Ablamowicz 2004). In the following section we shall lay the foundations for a categorical theory of consciousness.

THE MEANING OF AWARENESS IN CONSCIOUSNESS

Without consciousness there is no thing. Speaking about itself it already presupposes itself. Observing a relation between two it assumes itself as having seen A and having become aware of B und thus further as being aware of itself as realizing the relation between A and B. This has been called presupposition in trifferation. We may denote anything that we may observe and become conscious of by a symbol, say ○. This may signify an event, a cognitive pattern, a mathematical formula, a social subsystem, a general system, chaotic or not, a biological state. Briefly, we speak of an observer field ○ or of an observed field ○. That makes no difference, especially in pre-geometric physics. We are aware of one of the main unifying functions of consciousness, namely, the awareness. Awareness is energy, is biological, is neurologically active, is a psychological reality, shortly: it extends beyond the limitations of disciplines, and is a really trans-disciplinary phenomenon. Awareness has the capability to unite those two ○ ○ that were formerly separate, in order to bring forth

 a trifferation – a hethitan symbol of "god".

It is not by fortune that the awareness-function connecting two formerly separate fields is represented by this old symbol. But this is not the only aspect of consciousness, and it is not the first one. The first might rather be seen in our ability to make a difference between two observer fields. This is the ability of thought to distinguish between patterns in realization and to make them different both cognitively and energetically. An energetic difference always points beyond cognition and thus involves emotion. So we begin by a state that we wish to denote as

Differentiation or separation or disjunction

1.) 1st state ○ ○ »separation«

Next we have identification

2.) 2nd state (○) »integration«, »inclusion«, »reflection« or »identification«

also called (self-)awareness of a field which is an inclusive (self)observation or presupposition. The third state involves awareness of distinction

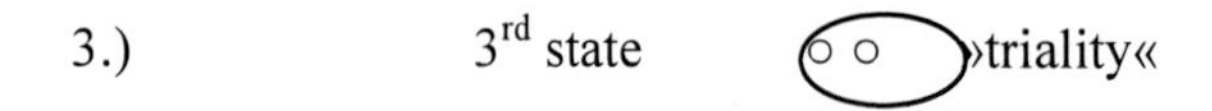

which is the same as trifferation (5). The forth and last action of awareness in consciousness allows for a distinction of different observer fields of the same observed field. In that way it brings forth the possibility to unify two trifferation fields, namely it turns

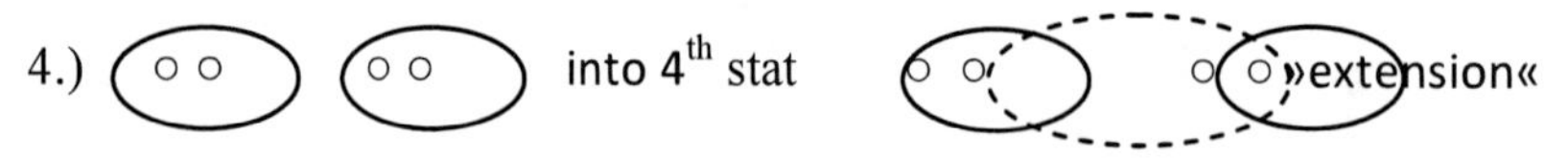

Clearly, the »extension of consciousness« implies that two observers are aware of the same observed. This can simply be put into form

Note, 4* provides an axiom equivalent with the 4th state because 4* together with 3 imply 4.

A QUESTION OF RADICAL CONSTRUCTIVISM

Consider the prime statement or axiom

meaning that two different observers observe the same observed field. It is, at first, not at all evident that this is a valid axiom. According to second order cybernetics (SOC) the postulate of objectivity – "The properties of the observer shall not enter the description of his observations" – cannot be maintained. But the observation of the observer is more important than the object. This shift from the »observed« to the »observer« signifies the difference between first and second order cybernetics.

In SOC, strictly, it cannot easily be said that two observers observe the same object. Each perceives the observed through his own system of perception. As is stated in Humberto Maturana's Theorem Number One "*Anything said is said by an Observer*". As Karl Müller has made us aware, Heinz von Foerster added to this Corollary Number One "*Anything said is said to an observer*" (von Foerster 2003, p. 283). Foerster originally made use of the distinction between primary and secondary qualities invented by John Locke. Müller quotes the example of propositions given by Foerster (2003, p. 201). "*In the proposition 'spinach is green', 'green' is a predicate; in 'spinach is good', good is a relation between the chemistry of spinach and the observer who tastes it.*" Pondering over such differences in terms of second order cybernetics, Foerster ended up with the radical statement: "*The environment does not contain any information. The environment is as it is*" (von Foerster 2003, p. 189). Karl Müller has compiled this in a figure (Müller 2007, p. 419)

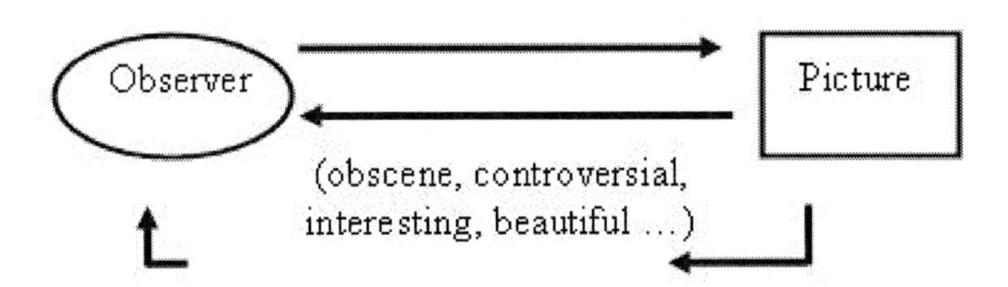

Müller explains that the focus lies on the relation between an observer, a picture and some apparent property provoked by that picture. Going into the aphorism, Müller concludes "*Thus, it can be assumed that SOC was based on the more radical view where primary qualities, too, do not occupy a privileged epistemic position and, thus, become observer dependent as much as 'obscene' or 'beautiful'.*" Therefore, identity too becomes observer dependent. So it must be clarified within SOC how identity comes upon.

Seen from the most radical standpoint it can neither be assumed as objective nor as evident that two observers observe the same. This does not only concern the special properties which are regarded in different ways – and clearly as deviating from each other – but it also holds for the *identity* of the observed. There may not be any marker – given independently of the dyadic relation between observer O_i and observer O_j – which allows for a definite identification of the observed. So we may end up with a rebuttal of axioms 4 and 4*. But that could somehow disprove science as a whole. Because it is quite obvious that we are able to somehow identify the actors, events, relations, systems and subsystems we are speaking about even in SOC. For example we mean a certain picture when we draw the above figure, just as we speak about a definite observer who denotes that picture as obscene, interesting or beautiful. Therefore, we have to answer how the identity comes upon. The answer can indeed be given in terms of SOC.

The identity of an observed comes upon as an eigenform in a minimal cognitive system. Such minimal systems are essentially given by triadic relations between at least two observers and one observed context (Müller 2007, p. 434). We shall not go deeper into the instrumentation of a state 4* design. But we shall keep in mind that this is really an important problem which is posed by the most radical versions of constructivism, but answered by second order cybernetics. By the way, we are quite familiar with it.

Suppose the three of us are convinced, today we have spoken with »Paul«. What is it that makes us so sure it was Paul, the look of his face, his voice, the fact he repeated a few words we have expected? But how much more than that is Paul! Are we ever aware of the whole of Paul? You see, we just build up a little information on Paul that we take as the real Paul. But none of us can ever be ready in constructing Paul as he actually is, not even Paul himself. This is what Heinz von Foerster is saying when he points out "The environment does not contain any information. The environment is as it is." Paul too, and both observers of Paul are just as they are. It is just that by observing we turn Paul into what makes us identify him as what he may be. Having this in mind we claim it is possible both A and B are observing the same.

Material observer fields

Today we can be quite sure that self organizing matter – biological matter – behaves intelligently. Though we may say that is not human intelligence, it is an even more general form of intelligence which gives birth to the animal- and thereon to human cognition. Deeply convinced of the meaning of this observation I had to answer a question by Anatol Rapoport replying to my paper on nonlinar self-referencing social systems. He asked "do you believe that molecules, atoms and electrons do have intelligence?" I said "yes, they move in the field of cosmic consiousness". The answer had fallen out radical. But Rapoport did not refuse it. As for myself, I felt the quantum field was intelligent, but I could not yet design the cybernetic system responsible for that premature intelligence of physical phenomena. As a matter of fact, electrons and protons sense their environments. They are vigilantly aware of the surrounding nuclei. How this happens? Suppose there are two protons sensing the oxygen atom in a water fragment H_2O – usually these fragments form tetrahedrons, icosahedrons and water polymer structures through which the proton can tunnel – or take two electrons surrounding some proton; we draw a field icon involving

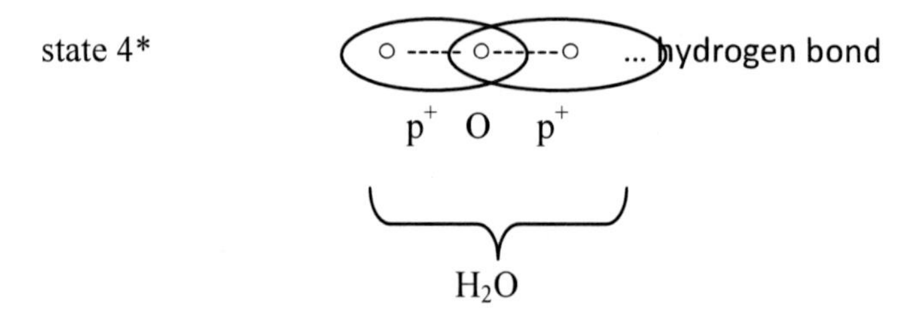

The protons "identify" the oxygen nucleus by sensing each other in interaction with the nucleonic field. We have a rather complex fermion interaction, both strong and electroweak, and an electromagnetic communication channel in the far infrared (FIR). Once any one of the two protons leaves that metastable triadic observer field it emits an FIR quantum of electromagnetic energy. Thus it contributes to the self organizing process of the water polymerase. In any case, we regard an icon of the type

as possible and generative for the construction of some algebraic or non-algebraic observer-field-category (OFC): Two observer fields give identity to the observed field through their interaction. On that ground we can develop the most general category of observer field graphs in physics. From this we can proceed further to space and time and geometric algebra. But those field graphs make sense in any real life situation as well. Consider a parent aware of herself as observing her child watching a red rose. This could be signified as

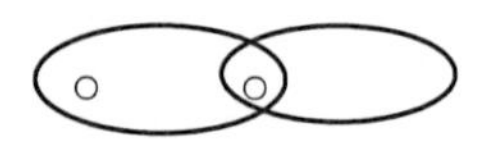

Note, an observer field has the same symbol as an observed field, namely ○.

Chapter 56

THE CATEGORY OF CONSCIOUSNESS

States one to four allow for the generation of general field categories. Going on slowly, we first construct main levels of complexity. The first level consists of a single observer field, the 2nd of two arrangements of 2 fields according to the 4 axioms, the 3rd level of all arrangements of 3 fields generated, and so on. We only have to bear in mind that two unequal fields such as

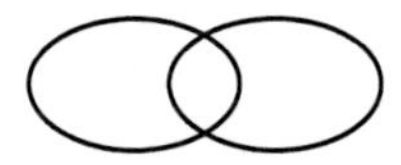

are indeed separate and therefore are not different from a state

So our symbols must not be interpreted as strictly set-theoretic, though there is a similarity.

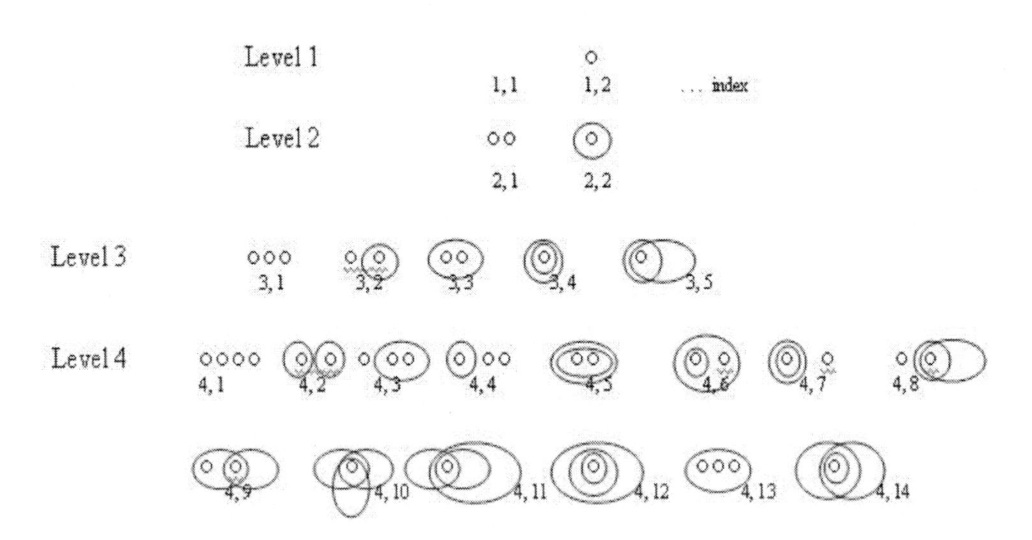

Figure 68. "awareness extension field icons" for categories n=1,…, 4.

SEMIOTIC FUNCTION IN AWARENESS FIELD ICONS

With the semiotic function a system may represent its own action. Maturana (quoted after Scott 2007, p. 198) said the system may interact with its own interactions. This refers to an icon having form

Scott (2007, 198) observed there is an accompanying „awareness of awareness".

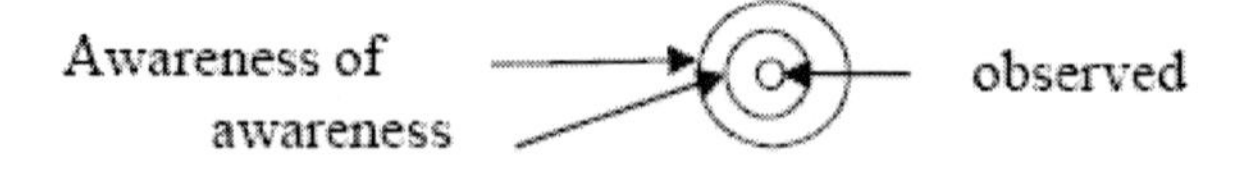

The icon M1 represents the same as "*presupposition*" of consciousness in Schmidt (2007, 326). It also corresponds with what Mead says about social interaction, namely that its logic "*arises as an abstraction from the experience of interaction*". An organism thus may represent its own interaction. The same holds for systems of quantum fields such as water polymers involving dynamic hydrogen bonds.

The icons of figure 1 can be represented by very simple directed graphs.

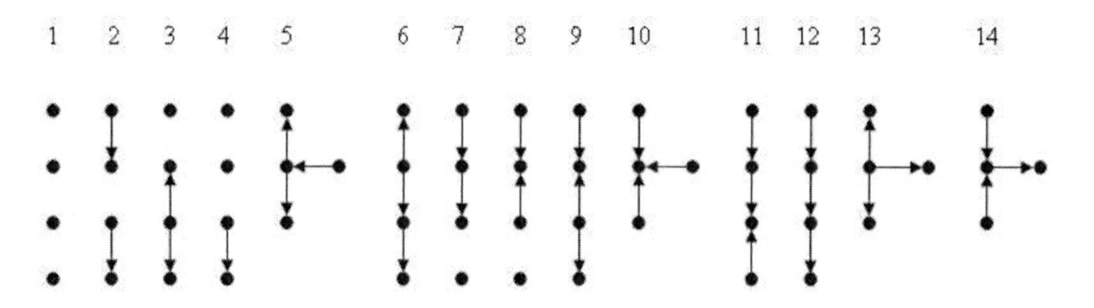

Figure 69. "observer field graphs" for 4 observer fields (Level 4).

Note, each pair of vertices is connected by at most one directed arrow. We can omit the enumeration of vertices since the graphs are the same if knots are permuted. A graph to order n has at least none and at most n-1 arrows.

GENERATING FIELD GRAPHS

The »awareness extension field icons« are utterly equivalent to the »field graphs«. These graphs are directed graphs (also called diagram schemes) and can be represented by sets of observer fields O , arrows A and two functions

$$A \underset{codom}{\overset{dom}{\rightrightarrows}} O \qquad (501)$$

The diagram schemes are generated by two rules of recursion:

1) at any step in the iteration add none or one vertex
2) at any step add none or at most one arrow with a definite direction
3) avoid cycles involving 3 or more vertices

In this way most complex couplings of energy, that is, extension fields of awareness can be represented graphically. For example

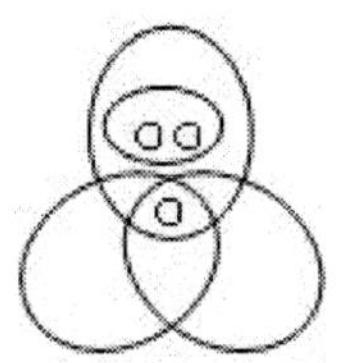

There are three fields of awareness each of which couples to one observer. This single observer is separate from a pair of observer fields linked in one awareness which together with the single observer is part of one of the three fields of awareness. This is definitely represented by the directed graph

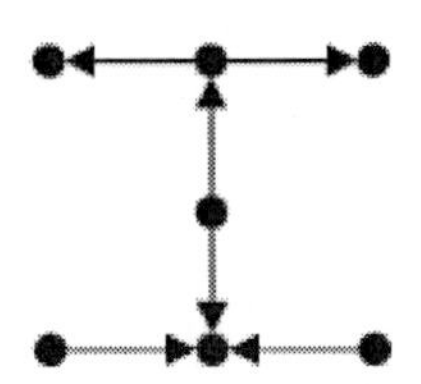

having seven vertices. This might represent a social situation just as well as a high energy physics event: for example, three fields isolating a meson in a nucleon. We see the meson built up by two quarks is nested in two observer fields. The most exterior of those superimposes with two further separate fields in the centre of which there is nested a third fermion. In such graphs, the set of composable pairs of arrows is

$$A \times_O A = \{\langle g, f \rangle | g, f \in A \,\&\, \mathrm{dom}\, g = \mathrm{cod}\, f\} \qquad (502)$$

called the "product over A". Categories offer the advantage to represent different complex systems by two functions or respectively morphisms among arrows.

A nesting figure as above can also be pinned down by arrangements of brackets

{.{{ }{ }}{..{...{ }.}..}...}

which would be definite but require one more symbol (dot). The dot allows for a colouring of brackets which is necessary. For example, consider the figure

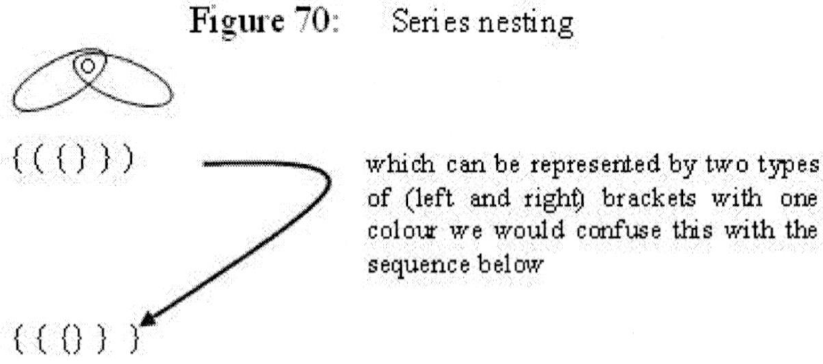

Figure 70: Series nesting

{ ({ }) }

which can be represented by two types of (left and right) brackets with one colour we would confuse this with the sequence below

{ { () } }

Figure 70. Series nesting.

which is a series nesting:

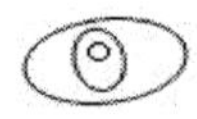

Category of Observer Field Graphs

A category is a graph with two additional functions

$$O \xrightarrow{id} A,\; A \times_O A \longrightarrow A,$$
$$c \mapsto id_c, \qquad \langle g, f \rangle \mapsto g \circ f, \tag{503}$$

called identity and composition $g \circ f$ such that

$$dom(id\,a) = a = cod(id\,a),\; dom(g \circ f) = dom\, f,\; cod(g \circ f) = cod\, g \tag{504}$$

Given two objects p and q in a category of observer field graphs there is at most one arrow $p \longrightarrow q$. Therefore the category **Cofg** is a *preorder in Grph*. Observe that a speciality of *Cofg* is that it contains at each level n the series of nested observer fields

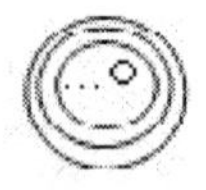

equivalent to Kauffman's nested

parenthesis {{{{{...}}}}}

The infinite nest of brackets $W = \{\{\{\{\{\{\{\ldots\}\}\}\}\}\}\}$ is one of the many special objects that are eigenforms for the operation of framing. We have the eigenform equation

$$W = \{W\} \tag{505}$$

Heinz von Foerster used an infinite concatenation of boxes to create a simple model for his concept of an "*object as token for eigenbehaviour*"

$$F(X) = \quad X\boxed{} \tag{506}$$

$$F(F(F(\ldots))) = \quad \boxed{\boxed{\boxed{\ldots}}} \tag{507}$$

If X denotes the infinite nest of boxes we obtain X = F(X). Such an infinite nest does not exist in the physical world, we believe, but it exists in our imagination as a cognitive image which somehow completes the incomplete of the material word. It is known that the infinite concatenation can be circumvented by lambda calculus invented by Church and Curry in the 1930s. Their argument can be applied quite generally to the »awareness extension field icons« and respectively the observer field graphs of *Cofg*.

THE GREMLIN OF INFINITE SELF REFLECTION

Kauffman (2005) has demonstrated how the eigenform of infinite self-reflexivity can be constructed by the aid of the »lambda calculus« (Church 1941, Curry 1958, Barendregt 1981) without apparent excursion to infinity: We want to find J = F(J). Church and Curry propose to construct the operator G having property G(X) = F(X(X)). That is, G operating on X makes a duplicate of X and lets X act on its duplicate. It is clear that X represents well formed terms in the linguistic logic calculus by Church. Letting G act on herself, we obtain G(G) = F(G(G)) and thus the finite eigenform G(G). Now this procedure can be applied to quite general problems of the observer field graph category **Cofg**.

The category *Grph* contains *connected and disconnected* graphs. This partition is of considerable importance in modelling the process of observation. Namely, we can be sure there are historic intervals between distinct observer systems. Such separations may be differences of instrumentation or cultural distinctions. The constructed differences can be a mercy or a malediction. The construction of a difference between electrons and protons was a boon, the difference between witches, bitches and women was a bane, social nemesis, to blur witchcraft with aircraft a global mistake with lethal social, political and cultural consequences. It is therefore that any fastidious theory of observer fields should keep track of partitions.

For now, we consider the comma », « as a separator in the preorder of directed graphs X, Y, a.s.o. that is (X, Y) denotes a disconnected pair of graphs. Let G represent a recursion form acting on the separator such that

$$G(X, Y) \equiv G(X), G(Y) \tag{508}$$

Let X, Y = F(X, Y) and define (509)

$$G(X, Y) \stackrel{\text{def}}{=} F(X(X), X(Y)) \quad (510)$$

Substititute G for X and Y for Y. This gives us

$$G(G, Y) = GG, GY = F(GG, GY) \quad (511)$$

So the GG, GY represent separated eigenforms of F. We can say that G keeps track of the separation which is established by the arrangement of observer fields X, Y. Clearly, in a similar way we can build a recursion that keeps track of a connection between graphs. Thus »*separation*« and »*connection*« between observer and observed in observer field graphs become recursive eigenforms themselves. They characterize a history of division and of connection for every real system.

EIGENFORMS OF MATTER

The standard model of physics describes conservation laws of subnuclear matter and thus the symmetry properties of interactions of strong and electroweak forces. In many writings it has been shown that the standard model can best be represented by manifolds of Lie algebras within quadratic Clifford algebras $Cl_{3,1}$ and $Cl_{4,1}$. There are six coloured quarks, the up, down, charm, strange, bottom and top quarks acted on by rank 2 Lie algebras having form $\mathcal{L}^{(2)}$ = $sl_{Cl}(2, R) \times so_{Cl}(3, R)$ and dimension 8. These Lie algebras provide the colour and flavour symmetries together with the isotropy of the electroweak forces. Last but not least the $\mathcal{L}$ fix the corresponding leptons. That is, their exponential maps are stabilizer groups for the electron e^-, the electronneutrino ν_e, the myon μ^-, the myon- and tauonneutrinos ν_μ, ν_τ and the tauon τ^-. Each lepton is represented by a corresponding primitive idempotent of the Clifford algebra. Moreover, each of those lepton idempotents f absorbs the corresponding Lie group exp($\mathcal{L}$) that stabilizes the fermion family such that we have for example $f g = f = g f$ for every group element $g \in$ exp($\mathcal{L}$). It is quite realistic to think about discontinuous nestings

$$a(b(c(d\ldots(f)\ldots))) = f$$

where the a, b, c . . . are discontinuous arrangements of elements of the Lie groups exp($\mathcal{L}$) generated by the algebras $\mathcal{L}$. In this way the representative primitive idempotents of the six fundamental leptons turn out as separate eigenforms.

DOES MATTER HAVE HISTORY?

It does not seem that the extension of matter in space and time happens independent of cognition. Rather the categories, in the mathematical sense, of space and time, are subject to cognitive construction. It is not impossible that there is extension of matter beyond space and time. In this case there would be energy fields which have no measure and orientation. Those fields nevertheless feed both matter and consciousness. Therefore orientation and measure of space-time would turn out as somehow located at the interface between emotion and thought. Emotion means to move things. Emotions are deeper than space and time and time is thought. Time is bound to thought, and history a selection of reality events in space. As Flusser said, without history there would also be no natural history. Without cognition no big bang. The problem of parallel universes would appear in an entirely new garment.

The Real

Awareness is a vital component of material life having both emotion and consciousness. It is a pregeometric form of energy. Yet it is capable to bring on geometric relations in matter. It is probably worth considering energetic awareness as not separate from life itself. It may not be so important whether we sense it or think it, namely: that *living systems are based on living awareness*. Awareness in action may in some sense be part of the dark in as much as it acts in fine feeling and emotion without even involving any conscious thought or image. Awareness acts even in the subconscious – no less materially than the conscious system.

We cannot limit or define the range and domain of the observer just as we cannot limit and define the observed. Consider for a moment observation as originating in some neuroscientists brain. He may observe something like the continuation of geometric relations from the outer material world into the inner organization of the human (his own) neocortical system. In this case the virtual limitation between the observer and the observed appears to be in the actor. Part of the outer is in the inner world. But *the real* in the sense of an *efficacy* or *sphere of action* is not solely given by the difference between observers and observed, but by the vivid energy of awareness which constitutes the system. The energy of awareness can shift the limits and interfaces between observers "outside" to anywhere. The real observer field is neither a single actor nor a group of actors, but rather an indeterminate and fluctuating subfield or –system. Any reality which "surrounds" some definite observer, as a living reality extends over, pulsates, shrinks and expands over other such fields, unfolds, annihilates and eventually decays. There are many subjects claiming they are observers. But besides those who know that and who say so, there are other real systems of matter unable to communicate that to us. We may believe they are without awareness and entirely unconscious. But they are not.

There are infinitely many and unimaginable parallel observer fields (worlds) dynamically integrated in awareness. There is the *existing* becoming apparent in our awareness of the *insisting* void – what some denoted as the Nibbhana (or Nirwana). Where the existing contrasts the insisting, the interval between observer and observed becomes apparent. Here the awareness contemplates the interval between inner and outer between that which ex-ists from that which in-sists (and thus negates existence). Where is that first location where the existing separates from Nirwana? That place of cosmic awareness is within the observer, is in the very centre of the subject. In "reality" there is no stable difference between subjects and objects. There is not even a stable difference between different observers – no stable time-evolution of a system of observer fields. When the observer becomes aware that the cosmic about which he is speaking is in his innermost, the interval between observer and observed fades. Object and subject are the same. In that way the teachings of radical constructivism shall have to be unfolded. The real as the world existing is unstable, and such is the insisting. Both energy and the void decay and thereby create space and time as properties in the field of awareness. This event of creation may take place anywhere and at any time in any domain of the observer fields. This denotes the creation of physical fields in space and time.

TOOLBOX

The layout of quantum mechanic systems in graded Lie algebra can be based on the combinatorial structure of geometric Clifford algebra. This has the advantage that the group structures of the standard model can easily be identified and connected with differential geometry. For the more advanced mathematicians among us it is clear that the multivector grade is not a generic Clifford algebra concept, and there exist grade free products more general than the Clifford product. They will perhaps prefer to follow the line of Rota-Stein 'cliffordization' and combinatorial analysis. The study of more general product behaviour is connected with the historic fibre of invariant theory which has been for a while somewhat out of time, just like combinatorics or topology. False colours have been put upon many of us, but always only from the limited point of view of some other special groups or schools of science. There are many lines of separation even within the field of exact science. One such line concerns the interval between geometry and topology.

One of the main aims of this book was to open up the demarcation line between geometry and topology. Another to bring logic, geometry, algebra and physics closer together. This is done because I am convinced the origin of these bifurcations in scientific knowledge has constructivist roots in the inner structure of the mind. Our mathematics is so to say the mirror of a morphogenetic root structure of human thought. It is therefore that, at first, I have chosen the arduous and, for some of us seemingly primitive concept of the geometric Clifford product as a starting point. This is recursively applied on homogeneous multivectors, - i.e. sums of decomposable multivectors of definite grade -, and extended by bilinearity. But as we make good progress in the understanding of the constitutive Lie group of the Minkowski algebra we shall also find out how physical motion transgresses different grades. Where the mathematics is concerned this means that we begin to comprehend why systems dynamics in a context of algebra-cogebra is entirely grade free. However, the present first fruit begins with a most basic presentation of primordial systems within Clifford algebra. After all the fundamental equations of motion, the great discoveries of Schrödinger, Pauli, Dirac, de Broglie, Bohm and others of the bygone century are utterly connected with the name of and the mathematics we have ascribed to Clifford. As we use a specific matrix representation of the complex Majorana algebra and of the special linear group $Sl(4, \mathbb{C})$, it becomes apparent that quantum mechanics is neither depending on grade, nor on scale. But its fascinating dynamic behaviour is emerging from a quite universal phenomenology.

Clifford algebra is an associative algebra which is generated by a vector space V over some field F in a recursive way and it contains V. The field may be given by real numbers, complex numbers, quaternions, octonions, non standard reals or other number fields. As we are dealing with associative algebras in the first line, we concentrate on reals, complex and quaternion numbers. Therefore we begin the toolbox with the introduction of quaternion numbers and proceed with Clifford algebras.

Toolbox 1

QUATERNIONS

A quaternion number is written as a 4-dimensional vector of the form

$$x + yi + zj + uk \qquad \text{with} \qquad x, y, z, u \in \mathbb{R} \tag{1}$$

and i, j, k are unit vectors which squared give -1. They are generalized imaginary units. Two quaternion numbers

$$q_1 = x_1 + y_1 i + z_1 j + u_1 k$$
$$q_2 = x_2 + y_2 i + z_2 j + u_2 k$$

can be added

$$q_1 + q_2 = (x_1 + x_2) + (y_1 + y_2)i + (z_1 + z_2)j + (u_1 + u_2)k \tag{2}$$

Quaternion base units follow the multiplication table

	i	j	k
i	– 1	k	– j
j	– k	–1	i
k	j	– i	–1

Therefore there is a quaternion product of two numbers

$$\begin{aligned} q_1 q_2 = {} & (x_1x_2 - y_1y_2 - z_1z_2 - u_1u_2) + (x_1y_2 + y_1x_2 + z_1u_2 - u_1z_2)i + \\ & + (x_1z_2 - y_1u_2 + z_1x_2 + u_1y_2)j + (x_1u_2 + y_1z_2 - z_1y_2 + u_1x_2)k \end{aligned} \tag{3}$$

We can represent quaternions q_1, q_2 and by pairs of complex numbers

$$\begin{aligned} & q_1 = (a, b), \qquad a = x_1 + y_1 i \qquad \text{and} \qquad b = z_1 + u_1 i \\ & q_2 = (c, d), \quad c = x_2 + y_2 i \qquad \text{and} \qquad d = z_2 + u_2 i \end{aligned} \tag{4}$$

Denoting complex multiplication by the dot, we can represent the quaternion multiplication in terms of the numbers *a, b, c, d*.

$$q_1\,q_2 = (a\cdot c - b\cdot d^*, b\cdot c^* + a\cdot d) \tag{5}$$

where * means complex conjugation. Using definition (4) of bicomplex numbers we obtain for the quaternion product a pair of complex numbers as below

$$\begin{aligned} q_1\,q_2 &= (x_1x_2 - y_1y_2 - z_1z_2 - u_1u_2 + (x_1y_2 + y_1x_2 + z_1u_2 - u_1z_2)i, \\ &+x_1z_2 - y_1u_2 + z_1x_2 + u_1y_2 + (x_1u_2 + y_1z_2 - z_1y_2 + u_1x_2)i) \end{aligned} \tag{6}$$

This is the bicomplex version of product equation (3). Within the context of quaternion algebra the 4-dimensional real space $\mathbb{R}^4$ has traditionally been called the quaternionic space or Hamiltonian space $\mathbb{H}$. It is a representation space for the Clifford algebra $Cl_{0,2}$, that is, a Clifford algebra generated by a space with no spatial base units and two time-like directions. It can also be comprehended as the Cartesian product of two complex planes, or $\mathbb{C}^2$. These different perceptions of the real space $\mathbb{R}^4$ enable us to classify the representations that we can introduce for a quaternion number $q = (a, b) = (x + y\,i, z + u\,i)$. This classification makes clear why there has prevailed some considerable amount of logic inexactness and ambiguity where the meaning of the 4-fold ring of reals ${}^4\mathbb{R}$ is concerned:

$\mathbb{H}$ - representation q
$\mathbb{C}^2$- representation by complex pair (a, b)
$\mathbb{R}^4$ – representation of complex pair $(x + i\,y, z + i\,u)$
$\mathbb{R}^4$ – representation for vector form $(x + y\,i + z\,j + u\,k)$
$\mathbb{R}^4$ – base free representation as quadruple (x, y, z, u)
The 3-dimensional anti-Euclidean space $\mathbb{R}^{0,3}$ consists of vectors $p = x\,e_1 + y\,e_2 + z\,e_3$ with a negative quadratic form

$$p\,p = -x^2 - y^2 - z^2 \tag{7}$$

The orthonormal basis of $\mathbb{R}^{0,3}$ can be represented by the unit quaternions $\{i, j, k\}$ as they provide the appropriate multiplications rules. The 8-dimensional Clifford algebra $Cl_{0,3}$ of the anti-euclidean space is isomorphic as an associative algebra to the direct sum $\mathbb{H} \oplus \mathbb{H}$. This is conceived as a 2-fold ring of quaternion numbers ${}^2\mathbb{H}$. Thus a pair of quaternions generalizes the pair of complex numbers as we proceed from anti-euclidean space to its Clifford algebra. As a rule, it often leads to shortest calculations if we calculate with bicomplex numbers.

Some use the cross × for quaternion multiplication. We avoid this in order not to confuse it with the cross product. When quaternions are used, it follows from the context that, say, in the Fueter equation, the quaternion product between a standard unit quaternion and a partial derivative of a quaternion function is meant. We may, in a text like this, simply write $j\,\partial_z\psi$. Sometimes, the quaternion j might be confused with the unit director in a Clifford algebra. Clarity should be derived from the context. Using the bicomplex representation of the product, in accordance with (5) we state that

$$(a,b)+(c,d)=(a+c,b+d) \tag{8}$$

$$(a,b)(c,d)=(ac-bd^*,bc^*+ad) \tag{9}$$

The quaternion product is

$$[(a,b)(c,d)](e,f)=(a,b)[(c,d)(e,f)] \quad \text{associative} \tag{10}$$

$$(a,b)[(c,d)+(e,f)]=(a,b)(c,d)+(a,b)(e,f) \quad \text{left distributive} \tag{11}$$

$$[(c,d)+(e,f)](a,b)=(c,d)(a,b)+(e,f)(a,b) \quad \text{right distributive} \tag{12}$$

has identity elements for addition and multiplication

$$(a,b)+(0,0)=(a,b) \tag{13}$$

$$(a,b)(1,0)=(a,b) \quad \text{and zero for multiplication} \tag{14}$$

$$(a,b)(0,0)=(0,0) \tag{15}$$

Numbers $(\lambda, 0)$ with $\lambda\in\mathbb{R}$ are scalar quaternions. Multiplication of a quaternion number with a scalar quaternion is a distributive operation as we have

$$\forall_{q=(a,b)} (\lambda,0)(a,b)=(\lambda a,\ \lambda b) \tag{16}$$

Multiplication with scalars is commutative

$$(\lambda,0)(a,b)=(a,b)(\lambda,0) \tag{17}$$

The quaternion number field forms a non-commutative ring. In analogy with the complex number field there exists the main involution of quaternion conjugation. The conjugation involution * satisfies the following rules

$$(a,b)^*=(a^*,-b) \tag{18}$$

formulated in the $\mathbb{R}^4$-algebras

$$(x+yi, z+ui)^* = (x-yi, -z-ui) \tag{19}$$

$$(x+yi+zj+uk)^* = x-yi-zj-uk \tag{20}$$

$$(x, y, z, u)^* = (x, -y, -z, -u) \tag{21}$$

Conjugation distributes over addition

$$(q_1+q_2)^* = q_1^* + q_2^* \tag{22}$$

$$(q^*)^* = q \tag{23}$$

The square norm is defined by the quaternion product

$$Q(q) = q\,q^* \qquad \text{norm in } \mathbb{H} - \text{representation} \tag{24}$$

$$Q((a,b)) = a\,a^* + b\,b^* \qquad \text{in } {}^2\mathbb{C} - \text{representation} \tag{25}$$

$$Q(x+yi,\, z+ui) = x^2+y^2+z^2+u^2 \;\text{ in } \mathbb{R}^4 - \text{representation} \tag{26}$$

$$Q(q_1\, q_2) = Q(q_1)\, Q(q_2) \tag{27}$$

We introduce the inverse of a nonzero quaternion by defining, in non-standard notation ($\neq q^{-1}$)

$$q_{inv} = \frac{1}{Q(q)} q^* \qquad \text{for} \quad q \neq 0 \tag{28}$$

With the inverse we can define a division of two quaternion numbers. But since the multiplication is non-commutative, we have to differ between left and right-inverses.

In analogy to complex functions and analysis we can introduce quaternion functions and quaternion differential geometry. The development of quaternionic analysis was initiated by Rudolf Fueter in 1935. In analogy to the Cauchy-Riemann equations he denoted the objects of analysis the *regular functions*.

Let $q = x + y\,i + zj + u\,k$ be a quaternion in standard notation and

$$\psi: \quad q \to \psi(q) \qquad \text{a quaternion function} \tag{29}$$

ψ is called regular if it satisfies the partial differential equation

$$\frac{\partial\psi}{\partial x}+i\frac{\partial\psi}{\partial y}+j\frac{\partial\psi}{\partial z}+k\frac{\partial\psi}{\partial u}=0 \tag{30}$$

where the omitted multiplication sign requires the quaternion product rule. In the complex plane the equation (30) has the analogue of the Cauchy-Riemann equations condensed into a single equation

$$\frac{\partial f}{\partial x}+i\frac{\partial f}{\partial y}=0 \qquad \text{with} \tag{31}$$

$$f:\ a\to f(a) \qquad \text{a complex holomorphic function.} \tag{32}$$

Equation (30) is called the Fueter-equation. It is represented in the Majorana algebra Mat(4, $\mathbb{R}$) by the four PDEs

$$\begin{aligned}
&\partial_x\psi_1-\partial_y\psi_2-\partial_z\psi_3-\partial_u\psi_4=0\\
&\partial_x\psi_2+\partial_y\psi_1+\partial_z\psi_4-\partial_u\psi_3=0\\
&\partial_x\psi_3-\partial_y\psi_4+\partial_z\psi_1+\partial_u\psi_2=0\\
&\partial_x\psi_4+\partial_y\psi_3-\partial_z\psi_2+\partial_u\psi_1=0
\end{aligned} \tag{33}$$

The conjugate Fueter equations were derived by Lanczos in 1929.

$$\begin{aligned}
&\partial_x\psi_1+\partial_y\psi_2+\partial_z\psi_3+\partial_u\psi_4=0\\
&\partial_x\psi_2-\partial_y\psi_1-\partial_z\psi_4+\partial_u\psi_3=0\\
&\partial_x\psi_3+\partial_y\psi_4-\partial_z\psi_1-\partial_u\psi_2=0\\
&\partial_x\psi_4-\partial_y\psi_3+\partial_z\psi_2-\partial_u\psi_1=0
\end{aligned} \tag{34}$$

The matrix representation of quaternions in Mat(2, $\mathbb{C}$) is:

$$\mathbb{H}\to \text{Mat}(2,\mathbb{C}): \qquad q=x+y\,i+zj+u\,k\mapsto\begin{bmatrix} x+yi & z+ui\\ -z+ui & x-yi\end{bmatrix} \tag{35}$$

The matrix representation of quaternions in Mat(4, $\mathbb{R}$) is:

$$\mathbb{H}\to \text{Mat}(4,\mathbb{R}): \qquad q\mapsto\begin{bmatrix} x & -y & u & -z\\ y & x & -z & -u\\ -u & z & x & -y\\ z & u & y & x\end{bmatrix} \tag{36}$$

Those are embeddings which preserve addition and multiplication. From these representations we can conclude that the quaternion algebra is a subalgebra of both Mat(2, $\mathbb{C}$) and the Majorana algebra Mat(4, $\mathbb{R}$). The matrix algebra Mat(2, $\mathbb{C}$) is isomorphic with the Clifford algebras of the Euclidean space and of the 3-dimensional Minkowski space $\mathbb{R}^{1,2}$ with signature $\{-++\}$.

Recently the concept of regularity has been enlarged and differentiated. It has become necessary to evaluate solutions to the Fueter equation according to better concepts of holomorphicity and regularity.

Attention To New Insight

We began with two complex structures J_1, J_2 on $\mathbb{H}$. Identifying $\mathbb{C}^2$ with $\mathbb{H}$ by means of a mapping which associates with the pair (4) the quaternion $a + b\,j$ we considered quarternions q in a bounded domain $\Omega \subset \mathbb{H}$. A quaternion function $f = f_1 + f_2\,j \in C^1(\Omega)$ was called regular on Ω in the sense of Fueter if

$$Df = \frac{\partial f}{\partial x} + i\frac{\partial f}{\partial y} + j\frac{\partial f}{\partial z} + k\frac{\partial f}{\partial u} = 0 \qquad \text{(equ. (30))}$$

Given in addition the *structure vector* $\psi = (1, i, j, -k)$, f is called (left) ψ-regular on Ω if

$$D'f = \frac{\partial f}{\partial x} + i\frac{\partial f}{\partial y} + j\frac{\partial f}{\partial z} - k\frac{\partial f}{\partial u} = 0 \qquad (37)$$

Sudbery (1979), Shapiro and Vasilevski (1995) and Nono (1985) have worked on properties of regular functions that Perotti (2005) used for his fundamental rigor. To put it brief, every holomorphic map (f_1, f_2) on Ω defines a ψ-regular function $f = f_1 + f_2\,j$. Now, the author showed that on every domain Ω there exist ψ -regular functions that are not J_p-holomorphic for any quaternion p. Holomorphic maps w.r.t. a complex structure J_p can be constructed as follows. Let $J_p = p_1 J_1 + p_2 J_2 + p_3 J_3$ be the complex structure on $\mathbb{H}$ defined by a unit imaginary 'pure quaternion' $p = p_1\, i + p_2\, j + p_3\, k$ in the sphere $S^2 = \{p \in \mathbb{H} \mid p^2 = -1\}$.

Perotti consideres the space of holomorphic maps from (Ω, J_p) to $(\mathbb{H}, L_p)$ where L_p is the complex structure defined by left multiplication by p on a copy $\mathbb{C}_p = \langle 1, p\rangle$ (recall equ. (5), (16), (17)) of $\mathbb{C}$ in $\mathbb{H}$.

$$Hol_p(\Omega, \mathbb{H}) = \{ f : \Omega \to \mathbb{H} \,\big|\, \bar{\partial}_p f = 0 \text{ on } \Omega \} = Ker\,\bar{\partial}_p \qquad (38)$$

with the Cauchy-Riemann operator w.r.t. the structure J_p.

$$\bar{\partial}_p = \tfrac{1}{2}(d + p\, J_p^* \circ d). \tag{39}$$

Lichnerowicz had proved that holomorphic maps between Kähler manifolds minimize the energy functional in their homotopy classes. After defining the energy of a map $f : \Omega \to \mathbb{C}^2$

as $$\mathrm{E(f)} = \tfrac{1}{2}\int_\Omega \|df\|^2 dV = \tfrac{1}{2}\int_\Omega \langle g, f^* g\rangle dV \tag{40}$$

it could be concluded that "*If f is ψ–regular on Ω, then it minimizes energy in its homotopy class relative to $\partial\Omega$.*" It turns out that "*almost all*" *linear ψ–regular functions are non-holomorphic*. After developing a criterion for holomorphicity, repeating arguments of Chen and Li, the author gives the explicit expression of the 6th-degree real homogeneous equation satisfied by the complex coefficients of a linear J_p-holomorphic ψ–regular function.

From such a fine proof we can learn that fastness of calculation, - due to the use of two complex structures, - may not be a proper criterion for the validity of an equation of motion w.r.t. physics events. Also we can see how important it is to understand the combinatorial structure of monogenic functions in general Clifford analysis. Quaternion analysis, sure, is only a special case in general analysis which is altogether geometric, multivectorial, combinatorial, and topological.

HISTORIC REMARK

William Rowan Hamilton wanted to construct a product for vectors in Euclidean 3-space analogous to the product of complex numbers in the complex plane. He searched for a multiplication rule for two vectors **a**, **b** such that the quadratic norm of their product $|\mathbf{a}\ \mathbf{b}|$ would be equal to the product of their norms $|\mathbf{a}||\mathbf{b}|$. However, such a product rule does not exist. He also tried to generalize the complex number systems for three dimensions which led to contradictions. His great success was achieved when he turned over to 4 dimensions and included the scalar by forming the element $q = x + y\,i + z\,j + u\,k$. The differential operator

$$\nabla = i\frac{\partial}{\partial x} + j\frac{\partial}{\partial y} + k\frac{\partial}{\partial z}$$

is due to Hamilton. Rudolf Fueter was the first who studied solutions of equation (30). He specified the combinatorial structure of rational polynomials which constituted the series expansion of left- and right regular quaternion functions. The bicomplex approach was discovered by Corrado Segre in 1982. The bicomplex algebra is fast, but contains an anomaly which has consequences for quaternion analysis: singular numbers unequal zero that lack an inverse. Among the most elementary solutions of the Fueter equation there is the dynamic element of a soliton. The soliton replaces the classical plane wave.

Toolbox 2

RINGS

The terms "algebra" and "ring" as we use them, are synonymous. Therefore, as far as this book is concerned, we could in principle just as well speak of Clifford rings instead of Clifford algebras. But we want to address directly the structures of scalars and n-tuples of scalars as rings and n-fold rings and the additive and multiplicative structures on modules as algebras.

A non-empty set $(R, +, .)$ with two morphisms
$+$ and $. : R \times R \to R$ satisfying (41)

(i)	$(R, +)$	a commutative group with neutral element 0
(ii)	$(R, .)$	a semigroup
(iii)	$(a + b). c = a. c + b. c$	. distributive morphism
(iv)	$a. (b + c) = a. b + a. c$	

A ring R is *commutative* if $(R, .)$ is commutative (42)

A ring R is *associative* if $(R, .)$ is associative (43)

(i) an element $e \in R$ is denoted as left unit if $e\,a = a$
(ii) element $e \in R$ is denoted as right unit if $a\,e = a \quad \forall a \in R$. (44)

a ring with unit is called a *unital ring* (45)

The *opposite ring* R^{op} is the additive group with the
opposite multiplication $a \circ^{op} b = b \,.\, a$ (46)

(i) A subgroup I of $(R, +)$ is denoted left ideal if $R \,.\, I \subset I$
(ii) is denoted right ideal if $I \,.\, R \subset I$ (47)

a *bilateral ideal* is at the same time left and right ideal (48)

In a commutation ring every ideal is bilateral (49)

A ring morphism is a mapping $f: (R, +, .) \rightarrow (S,+, \circ)$
which satisfies (50)

(i) $f(a+b)=f(a)+f(b)$

(ii) $f(a.b)=f(a)\circ f(b)$

(iii) $f(e_R)=f(e_S)$ if R, S are unital rings

The kernel of a ring homomorphism $f: R \rightarrow S$ is an
ideal $I_f = \ker f = \{a \in R / f(a) = 0\}$ (51)

Every ideal is the kernel of some appropriate homomorphism. Consider the projection

$$\pi_I: \quad R \rightarrow R/I$$

R/I is called the residue class ring or a factor ring. Let a, $b \in R$, the ring structure is given by

(i) $(a+I)+(b+I)=(a+b+I)$

(ii) $(a+I)(b+I)=(ab+I)$ (52)

If A is a subset of R, an ideal (left or right) I_A is said to be generated by A if it is the smallest ideal I_A with $A \subset I_A$. If A has finite cardinality we call I_A *finitely generated.* I_A is the intersection of all ideals which contain A. The direct sum $A \oplus B$ of two ideals A, B is their Cartesian product under the condition $A \cap B = 0$.

$R = A \oplus B \oplus \ldots \quad A \cap B = 0, \ldots$ the ring R is decomposable (53)

In decomposable rings every element $r \in R$ can be decomposed in a definite way as $r = a + b + \ldots$ with $a \in A$, $b \in B$ ... Rings are indecomposable if they cannot be written as a direct sum of ideals. Next we define predicates of ring elements.

(i) $ldif(a,0) \overset{def}{\Leftrightarrow} a \in R \,\&\, \exists_{b \neq 0}\, ab = 0$; a is left divisor of zero

(ii) $rdif(a,0) \overset{def}{\Leftrightarrow} a \in R \,\&\, \exists_{b \neq 0}\, ba = 0$; a is right divisor of zero

(iii) $dif(a,0) \overset{def}{\Leftrightarrow} ldif(a,0) \vee rdif(a,0)$; a is divisor of zero (54)

$idem(a) \overset{def}{\Leftrightarrow} a^2 = a$; a is idempotent (55)

$nil_k(a) \overset{def}{\Leftrightarrow} a^k = 0$; a is nilpotent of order k (56)

$$uni(a) \overset{def}{\Leftrightarrow} R\ is\ unital\ \&\ a^2 = e\ ;\ a \text{ is unipotent} \tag{57}$$

$$reg(a) \overset{def}{\Leftrightarrow} \exists_{b \in R} aba = a\ ; \qquad a \text{ is regular} \tag{58}$$

(i) $linv(a) \overset{def}{\Leftrightarrow} \exists_{b \in R} ab = e\ ;$ a is left invertible

(ii) $rinv(a) \overset{def}{\Leftrightarrow} \exists_{b \in R} ba = e\ ;$ a is right invertible

(iii) $inv(a) \overset{def}{\Leftrightarrow} linv(a) \vee rinv(a)\ ;$ a is invertible (59)

$$cent(a) \overset{def}{\Leftrightarrow} \forall_{b \in R}\ ab - ba = 0\ ; a \text{ is central} \tag{60}$$

$$prim(h) \overset{def}{\Leftrightarrow} h \neq f + g \text{ with } idem(f)\ \&\ idem(g)\ \& \\ \& f g = g f = 0 \tag{61}$$

An idempotent $h^2 = h$ is *primitive* if it cannot be a sum of two mutually annihilating idempotents. From the above primary statements there follow several Lemmas and theorems relevant for geometric algebra such as for example the following lemmas

[L1] $idem(f)\ \&\ idem(g)\ \&\ [f, g] = 0 \Rightarrow idem(fg)\ \&\ idem(f + g - gf)$

[L 2] $idem(f)\ \&\ idem(g)\ \&\ \{f, g\} = 0 \Rightarrow idem(f + g)$

From (61) there follows that the anticommutator $\{f,g\}$ is zero and Lemma 2 is fulfilled. Therefore the sum $h = f + g$ is not a *primitive idempotent*, but it is an *idempotent*; *idem*($f + g$). On the other hand, we must presume Lemma 3.

[L 3] $\neg[(idem(f)\ \&\ idem(g)\ \&\ idem(f + g)) \Rightarrow \neg prim(f + g)]$

That means, let *idem*(f) *and idem*(g) . From the fact that the sum $f + g$ is an idempotent we cannot conclude that $f + g$ *is not* a primitive idempotent. The proof is in the fact that mutual annihilation is a sufficient but not a necessary condition for the anticommutator to vanish. You see, there is a certain predicate logic that becomes relevant as soon as we investigate the interrelation between algebraic properties and predicates of ring elements.

There is a whole set of relevant definitions to orthogonality. One of the most simple is in terms of mutual annihilation of idempotents

$$\perp(f, g) \overset{def}{\Leftrightarrow} fg = 0 = gf \qquad \text{orthogonality} \tag{62}$$

(i) $A_L(\phi) \overset{def}{\Leftrightarrow} \phi \subset R \,\&\, A_L(\phi) = \{b \in R \,/\, ba = 0, \forall a \in R\}$ defines the left annulator

(ii) $A_R(\phi) \overset{def}{\Leftrightarrow} \phi \subset R \,\&\, A_R(\phi) = \{b \in R \,/\, ab = 0, \forall a \in R\}$ defines the right annulator

and

(iii) $A(\phi) = A_L(\phi) \cap A_R(\phi)$ defines the annulator (63)

[Th 1][i] Decomposition theorem: If the left ideal $I \subset R$ is generated by an idempotent $f \in \mathrm{R}$ and $I = Rf$, then R is decomposable into left ideals $R = A \oplus \mathrm{A_L}(f)$.

[Th 1][ii] If the ideal J is generated by a central idempotent f then R is decomposable into $R = J + \mathrm{A}(f)$

[Th 1][iii] Every left or right ideal in a unital ring which is a direct summand in a decomposition, is generated by an idempotent f. If I is an ideal f is central. Decompositions have the form $R = Rf + R(1 - f)$ where $R(1 - f) = \mathrm{A_L(f)}$ is a left or right annulator.

RING EXAMPLES

are given by the real numbers $\mathbb{R}$, the ring of complex $\mathbb{C}$, the ring of quaternion numbers $\mathbb{H}$ and the ring of octonions O. On the linear space $\mathbb{R}^2$ there are various interesting bilinear products. One of those is component-wise multiplication

$$(x_1, y_1)(x_2, y_2) = (x_1 x_2, y_1 y_2) \tag{64}$$

which brings forth the double-ring $^2\mathbb{R}$. The real double ring contains two automorphisms, the identity and the swap

$$swap\text{: } {}^2\mathbb{R} \to {}^2\mathbb{R} \; ; \; (a,b) \xrightarrow{swap} (b,a) \tag{65}$$

The swap acts like the complex conjugation in $\mathbb{C}$ as we have

$$\underset{(a,b)\in {}^2\mathrm{R}}{\forall} \; a(1,1) + b(1,-1) = a(1,1) - b(1,-1) \tag{66}$$

The element $j = -1$ stands for the unit imaginary, since $j^2 = -1$. The real double unit is Id = (1, 1). Equivalently pairs of real numbers $(a, b) \in \mathbb{R}^2$ are called *Study numbers*. They have the Study-conjugate $(a + j\,b)^{-} = a - j\,b$ and hyperbolic norm $a^2 - b^2$. The elements of $^2\mathbb{R}$ can be written in the hyperbolic polar form

$\rho(\cosh\chi + j\sinh\chi)$ for $a^2 - b^2 \geq 0$. They can be represented in symmetric Mat(2, $\mathbb{R}$):

$$a + j\,b = \begin{bmatrix} a & b \\ b & a \end{bmatrix} \tag{67}$$

In graded Lie algebras such as $\mathcal{L}^{(2)} \subset Cl_{3,1}$ the 4 – fold ring of real numbers ${}^4\mathbb{R}$ plays an important role in connection with the maximal Cartan subalgebras and colour spaces ch_χ. For consider the element $\Lambda_3 = 2\lambda_3 = ½\,(e_{24} - e_{124})$ and the idempotent $f = ½\,(1 - e_1)$. Both are Clifford numbers in colour space ch_1. Let $n \in \mathbb{N}$ and verify the identities

$$\Lambda_3^{2n} = f \qquad \text{and} \qquad \Lambda_3^{2n-1} = \Lambda_3 \tag{68}$$

That is, exponentiation acts on Λ_3 like a swap. The colour space can be decomposed into a sum of two ideals

$$ch_1 = ch_1\, f + ch_1\, \hat{f} = I_1 \oplus I_2 \quad \text{with main involuted } \hat{f} \tag{69}$$

where $I_1 = \{f, \Lambda_3\}$ and $I_2 = \hat{I}_1 = \{\hat{f}, \hat{\Lambda}_3\}$. From the equations (68) there follows that spaces I_1, I_2 are isomorphic with $Cl_{1,0} = \{1, e_1\} \simeq {}^2\mathbb{R}$ – the real double-ring. Thus we obtain a typical colour space decomposition

$$ch_1 = \mathbb{R} \oplus \mathbb{R} \oplus \mathbb{R} \oplus \mathbb{R} = {}^4\mathbb{R} \quad \text{the fourfold real ring} \tag{70}$$

In ${}^4\mathbb{R}$ the basis of ch_1 can be represented by the quadruples Id =(1,1,1,1); e_1 = (1,1,–1,–1); e_{24}=(1,–1, 1,–1); e_{124} =(1,–1,–1,1). The last example to be mentioned concerns the representation of Clifford algebra by generalized "*Dirac-Clifford matrices*". From the algebra point of view, these objects together with the pure scalar, the multiplicative unit and Clifford multiplication constitute a ring. Especially for dimension 4 this is called the *Dirac ring* (Mignaco and Linhares 1993).

Toolbox 3

TENSOR- AND GRAßMANN ALGEBRA

Graßmann algebra is a prerequisite for Clifford algebra. Let F be an unital commutative ring and V an F-linear space. The tensor algebra $T(V)$ is formed by the direct sum of tensor products of V:

$$T(V) = F \oplus V \oplus (V \otimes V) \oplus \ldots = \bigoplus_k T^k(V) = \bigoplus_k \bigotimes^k V \qquad (71)$$

We identify F with V^0. The unit of F in $T(V)$ is denoted as Id. The injection

$Id_V : F \to T(V)$ is the "*unit map*" (72)

It is assumed that V can be spanned by a set $\{e_i\}$ of linearly independent elements, the generators. The number of factors in a "word" $a_1 \otimes \ldots \otimes a_k \in T^k(V) = \otimes^k V$ is called the rank of the tensor. Products of tensors are formed by concatenation of words. Concatenation is associative by definition. We define the wedge-ideal which identifies all but antisymmetric tensors

$$I_\wedge = \{a \otimes x \otimes x \otimes b \mid a, b \in T(V), x \in V\} \qquad (73)$$

The Graßmann algebra is the factor algebra of $T(V)$ where the elements of the Ideal $I_\wedge$ are identified with zero.

$$\wedge V = \frac{T(V)}{I_\wedge} = \pi_\wedge(T(V)) \qquad (74)$$

with the projection $\pi_\wedge(T(V))$ from $T(V)$ onto $\wedge V$. By the aid of categories it can be shown that the Graßmann algebra over a linear space V is a universal algebraic object and is therefore defined uniquely up to isomorphy. The detailed relations equivalent to the factorization are

$$e_i \otimes e_i = 0 \mod I_\wedge \qquad (75)$$
$$\pi_\wedge(e_i \otimes e_i) = e_i \wedge e_i = 0$$

$$\pi_{\wedge}(e_i \otimes e_j) = e_i \wedge e_j = -e_j \wedge e_i$$

As a consequence of factorization, words of generators can be ordered according to ascending or descending indices. A basis can now be given by Graßmann monomials from length zero to length n.

$$G_{bas} = \{Id; e_1,\dots,e_n; e_1 \wedge e_2,\dots,e_{n-1} \wedge e_n;\dots; e_1 \wedge \dots \wedge e_n\} \quad (76)$$

G_{bas} consists of $\binom{n}{k}$ words of length k and $\dim(G_{bas}) = 2^n$.

Toolbox 4

CLIFFORD ALGEBRA

We introduce the basis free definition of a quadratic form

(i) $Q(\alpha x) = \alpha^2 Q(x)$

and

(ii) $2B_p(x, y) = Q(x - y) - Q(x) - Q(y)$ polar bilinear form. (77)

The presentation of a Clifford algebra is often introduced by generators and relations

$$Cl(V,Q) \overset{def}{\Leftrightarrow} \langle X, R \rangle = \langle \{e_i\}, e_i e_j + e_j e_i = 2g_{ij} \rangle \tag{78}$$

with base unit $e_i \in X$ span V and g_{ij} is the symmetric bilinear form which represents Q in the basis. The anticommutation relations in (78) express the orthogonality of base unit vectors. The notation $Cl_{p,q}$ may be preferred to $Cl(V)$. In that case V is an $\mathbb{R}$−linear space of dimension $p + q$ where the quadratic form has p positive and q negative eigenvalues. V is then the linear space with indefinite metric $\mathbb{R}^{p,q}$. We can introduce different types of Clifford algebras in various ways. Clifford himself proceeded as follows.

Graßmann's exterior algebra $\Lambda\mathbb{R}^n$ is an associative algebra with dimension 2^n. It is based on the antisymmetric wedge product

(i) $e_i \wedge e_j = -e_j \wedge e_i$ for $i \neq j$

(ii) $e_i \wedge e_i = 0$ (79)

Clifford kept the first rule, but altered the second

(ii') $e_i e_i = 1$

In this way we obtain an orthonormal basis for the Euclidean n-space with positive definite metric. A Clifford algebra of this type is written Cl_n. Four years earlier (1878) Clifford had considered the multiplication rules (i) and

(ii'') $e_i e_i = -1$

So we obtain the Clifford algebra $Cl_{0,n}$ of the 'negative definite' space $\mathbb{R}^{0,n}$. Lounesto (2001, p. 190) had introduced the Clifford algebra by generators and relations in the following manner.

An associative algebra $Cl(Q)$ over a unital ring F with a non-degenerate quadratic form Q on V is called Clifford algebra if it contains V and $F\,Id$ as distinct subspaces, so that

(i) $\forall_{x \in V} \quad \boldsymbol{x}^2 = Q(\boldsymbol{x})$

(ii) V generates $Cl(Q)$ as an algebra over F

(iii) $Cl(Q)$ is not generated by any proper subspace of V (80)

We can also begin with the demand that the quadratic form of a vector should be equal to the associative (Clifford) product

$$\boldsymbol{x}^2 = |\boldsymbol{x}|^2 \tag{81}$$

from which there would follow the orthogonality relations for base unit vectors (Lounesto 2001, p. 8). Many authors introduce the Clifford algebra by factoring the tensor algebra. Chevalley (1954) constructs

$$Cl(Q) = \frac{\otimes V}{I(Q)} \qquad \text{where } I(Q) = \{x \otimes x - Q(x)Id \,/\, x \in V\} \tag{82}$$

The same procedure is suggested by many others (Degirmenci and Karapazar 2005). Fauser (2002) goes a little bit deeper. After demonstrating the phenomenological problems that are brought in by generators and relations, he substantiates the functor *Cl* by factorization. He uses this categorical denotation to justify the importance of the *gebra/cogebra*-duality. There are various arguments and discoveries which naturally lead to this duality. One of those is *construction by deformation* as was carried out by Woronowicz (1989), the other was the study of non-grade preserving isomorphisms j of the tensor algebra $T(V)$ by Zbigniew Oziewicz and the Clifford algebra of multivectors developed by him.

Keeping track of the consequences of equation (78) we may end up with the observation of Marcel Riesz (1957) that a wedge product can always be constructed consistently in a Clifford algebra by antisymmetrization of the base elements:

$$e_i \wedge e_j = \tfrac{1}{2}(e_i e_j - e_j e_i) \tag{83}$$

But there remains an important difference. The exterior algebra $\wedge V$ is an indecomposable algebra. The only idempotents in a Graßmann algebra are 0 and *Id*. So it is only in the Clifford algebras where primitive idempotents generate various spinor representations. This fact follows from the quadratic form introduced into the Clifford algebra. We could believe, as the Graßmann basis spans a $\mathbb{Z}$-graded linear space, that this is inherited by the Clifford algebra. But, on the contrary, the Clifford algebra does not respect this grading, but only a much weaker filtration. We know from the double covers of rotation groups by spin groups that the Clifford algebra is only $\mathbb{Z}_2$-graded. As Fauser emphasized, the usually defined grade projection operators $\langle\ldots\rangle_k : \wedge V \rightarrow \wedge^k V$ are foreign to the concept of a Clifford algebra and belongs to the underlying Graßmann algebra. It is even possible to employ different $\mathbb{Z}_n$-gradings at one time. He gave several examples, using groups by generators and relations, to show that isomorphy can be established by non-grade preserving transformations of generators. This discovery has ultimately led me to the construction of the constitutive Lie group of a Clifford algebra. *This observation*, as Fauser legitimately states, *is crucial for any attempt to identify algebraic expressions with geometric objects*.

In order to factor out the Clifford algebra from the tensor algebra $T(V)$ we have to introduce an ideal I_{Cl} as we have done before to obtain the Graßmann algebra. This ideal is now

$$I_{Cl} = \{a \otimes (x \otimes y + y \otimes x) \otimes b - 2g(x,y)\, a \otimes b \mid a,b \in T(V); x,y \in V\} \tag{84}$$

and the Clifford algebra is

$$Cl(V,Q) = \frac{T(V)}{I_{Cl}} \tag{85}$$

Woronowicz studied the theory of deformed Graßmann algebras. The latter were obtained by projecting out all symmetric tensors from the tensor algebra. This projection maps the tensor product onto the exterior: $\pi_\wedge(\otimes) \rightarrow \wedge$. If one proceeds to deformed symmetries, one obtains deformed Graßmann algebras $\wedge_q$ with some deformation magnitude q. In my answer to letter five of Zbigniew Oziewicz I asked the question what would be the minimal set of idempotents to generate all idempotents and how do the polynomials in two or more variables look? I offered a guess based on transposition involutions. Oziewicz also returned a guess: "*One can only guess: maybe Hecke-potents,* $\rho^2 = a\rho + b$, *for scalars a and b, would unify and generalize some of your Theorems and formulas?*" What is important in this answer is the direction indicated. Namely, Fauser pointed out, the presentation of the symmetric Graßmann algebra in terms of generators and relations reads:

$$S_n = \langle X, \{\mathbf{R}_1, \mathbf{R}_2, \mathbf{R}_3 \rangle$$
$$\mathbf{R}_1 : s_1^2 = 1$$
$$\mathbf{R}_2 : s_i s_j s_i = s_j s_i s_j$$

$$\mathbf{R}_3 : s_i s_j = s_j s_i \qquad \text{if } |s_i - s_j| \geq 2 . \tag{86}$$

X contains n–1 generators s_i. The projection operator onto the alternating part of the algebra is

$$\pi_\wedge = \frac{1}{n!} \sum_{red.words} (-1)^{length(w)} w \tag{87}$$

where w runs in the set of reduced words. For S_3 we obtain

$$\pi_\wedge = \frac{1}{3!}(1 - s_1 - s_2 + s_1 s_2 + s_2 s_1 - s_1 s_2 s_1) . \tag{88}$$

A slight generalization of relation $\mathbf{R}_1$ which allows for a quadratic relation for the transposition introduces the Hecke algebra

$H_n = \langle X, \{\mathbf{R}_1, \mathbf{R}_2, \mathbf{R}_3\} \rangle$

with

$$\mathbf{R}_1 : \tau_1^2 = a\tau + b \qquad \text{(Fauser 2002, p. 26)} \tag{89}$$

where $\mathbf{R}_2$, $\mathbf{R}_3$ are still the braid relations of the “Artin braid group”. Taking care of the additional parameters, one gets

$$\pi_{\wedge_q} = \frac{1 - \tau_1 - \tau_2 + \tau_1 \tau_2 + \tau_2 \tau_1 - \tau_1 \tau_2 \tau_1}{(1 + q + q^2)(1 + q)} \tag{90}$$

with $a = 1 - q$ and $b = q$. It is remarkable that the so produced algebra is the same like an undeformed Clifford algebra with a properly chosen non-symmetric bilinear form (Fauser and Ablamowicz 2000, Fauser 1999).

Oziewicz (1997) had the idea to investigate *non-grade preserving* isomorphisms j of $T(V)$ and to study their projection under an ungraded switch generator onto the exterior algebra. This is displayed by the following commuting diagram

$$\begin{array}{ccc} T(V) & \xrightarrow{\quad j \quad} & T_j(V) \\ \downarrow \pi_\wedge & & \downarrow \pi_\wedge \\ \Lambda V & \xrightarrow{\quad \gamma \quad} & \Lambda_j V \end{array}$$

He proved that $\wedge_j V$ is a Clifford algebra with respect to an arbitrary bilinear form induced by $j^2 : T(V) \to T_j(V)$. This work of Oziewicz culminated in the discovery of the Clifford co-algebra as the natural companion of the Clifford algebra (Oziewicz 1997 and 2001). He put it into the following form

$$Cl(\cap) \equiv Cl(g^{-1}) \xrightarrow{Model} Cl(M, g^{-1} \in M^{\otimes 2}), \qquad (91)$$
$$Cl(\cup) \equiv Cl(g) \xrightarrow{Model} Cl(M, g \in M^{*\otimes 2})$$

"These both structures together we call the Clifford convolution $Cl(g, g^{-1})$" (Oziewicz 2005, 2001, 2003, Fauser and Oziewicz 2001). He derived this result independent of Helmstetter who in 1987 proved that the Clifford coalgebra linear map $M^{\wedge} \to M^{\wedge} \otimes M^{\wedge}$ is an algebra morphism from Clifford algebra $Cl(\cup)$ to an algebra that is the $\mathbb{Z}_2$-graded tensor product of the Graßmann algebra with the Clifford algebra (Helmstetter 1987 §5, Proposition 35, p. 38, quoted after Oziewicz).

$$Cl(\cap): \quad Cl(\cup) \longrightarrow Gra\beta \otimes Cl(\cup) \qquad (92)$$

In an enormous effort and repeated investment into foundations of both physics and mathematics Oziewicz had shifted the value of a basic explanatory dichotomy from inside algebra to the outside »*gebra-cogebra*« differentiation.

The reader is invited to recall the part on unity of motion, the Dirac equation and equation (324) where a four-component spinor is decomposed into left and right handed chiral neutrino waves. This is exactly the spot where Oziewicz hooks in, and he transposes the left-right dichotomy onto the interval between gebra and cogebra. He asks: *can we understand the Dirac operator without Clifford algebra, $Cl \otimes Cl \longrightarrow Cl$? Who needs Clifford coalgebra, $Cl \longrightarrow Cl \otimes Cl$?* And he answers that already *Bargman and Wigner (1948) described composed systems in terms of the tensor product* ½ ⊗ ½ = 0 ⊕ 1. *Does this tensor product need the Clifford coalgebra*? Indeed it does!

Oziewicz proposed to describe higher spin states as the tensor product of *Cl* –modules and to consider the addition of linear energy-momenta in terms of the tensor product of *Cl* –modules. This cannot be done without Clifford cogebra. In a short paragraph titled *Quark-antiquark system* he locates the dichotomy in a *theory of the quark-antiquark bound system, neutrino-antineutrino composed system, boson, photon, and of the electromagnetic field as composed from a pair of ½-spin states (de Broglie 1932, 1934, 1940). Louis-Victor de Broglie (1892-1987) proposed that the photon is composed of the neutrino-antineutrino pair. We refer to Bargmann and Wigner (1948), Dvoeglazov (1997-2000) for the problem of the higher spin-states. The quantum theory of positronium was developed by Fauser and Stumpf (1997), and the general theory of families of particles from Cl-spinors was advocated by Wilczek and Zee (1982), Królikowski (1993) and by Finkelstein (2001).*

A holistic view that is based on the gebra-cogebra dichotomy sure has consequences on the 'half formulations' in Clifford algebras such as $Cl_{3,1}$, $Cl_{4,1}$ and $Cl_{4,2}$. When we constructed the constitutive Lie group $\mathcal{L}^{(2)}$ for the Minkowski algebra $Cl_{3,1}$ we respected an important non-grade preserving isomorphism mapping the tensor algebra $T(V) \to T_j(V)$ such that the Pauli principle is preserved while space-like multivectors are carried from grade to grade, that is, e.g. in ch_1 the e_1 to e_{24} and the e_{24} to e_{124}. Such a flavour rotation equalizes grades 1, 2 and 3 of spatial multivectors thereby expressing their equivalence within the finite structure of the Klein 4-group. By constructing the higher spin group $\ell^{(2)}$ we replace the spin group derived from the Lipschitz group by a Lie group that resembles the standard model.

We consider the six isomorphic lattices consisting of 16 commuting idempotents generated by the four annihilating primitive idempotents of each colour space and each maximal Cartan subalgebra (Lounesto 2001, p. 227) in $Cl_{3,1}$. The mutually annihilating primitive idempotents signify the potential physical pure states. They act as cursors to spinors of any relevant construction that respects the strong force. We investigate *non-grade preserving* isomorphisms j of $T(V)$ which preserve those six lattice structures, but alter the grade and dimension of the pure states. In this way, the dimension of a primitive idempotent may change from 4 to 15, but all its orthogonality relations are preserved. I called this the *unfolding of the idempotent manifold.* It is a deconvolution of primitive idempotents together with all their nonlinear algebraic relations. This was meant when I said that nature at her deepest level does not differ between grades. But it is necessary to understand that this indifference is rooted in the algebraic structure of the maximal Cartan subalgebra. Nature "does not differ between grades" in a very specific way! Not just in any way! It is indifferent, and thereby allowing for a fractal conformal deconvolution of motion which is compatible with the standard model. At its deepest level the process of nature treats space like a set rather than like a measurable space with definite dimension. It is therefore that we can begin with the construction of space by starting off at a concept of extension with no definite dimension, but moving dimension. So we make use of the mathematical indifference of Clifford algebras against grades of vectors. This approach is most related and connatural with the theory pursued by Oziewicz.

CLIFFORD ALGEBRA $CL_{3,0}$ OF EUCLIDEAN 3-SPACE

The Clifford algebra $Cl_{3,0}$ is generated by the 3-dimensional Euclidean space having unit vectors e_1, e_2, e_3 with positive signature (1) and satisfying anti-commutation relations (2)

$$\text{(i)} \quad e_1^2 = 1, \ e_2^2 = 1, \ e_3^2 = 1$$
$$\text{(ii)} \quad \{e_1, e_2\} = \{e_1, e_3\} = \{e_2, e_2\} = 0 \tag{93}$$

We use to establish a 1-1 correspondence with the Pauli algebra of unitary 2 x 2 matrices with complex entries.

$Cl_{3,0}$						$Mat(2, \mathbb{C})$
Id				I_2		scalar
e_1	e_2	e_3	σ_1	σ_2	σ_3	vector
e_{12}	e_{13}	e_{23}	$\sigma_1 \sigma_2$	$\sigma_1\sigma_3$	$\sigma_2\sigma_3$	bivector
e_{123}				$\sigma_1\sigma_2\sigma_3$		director

As you see, any exterior product $e_j \wedge e_k$ can indeed be represented by a matrix product of Pauli spin matrices $\sigma_j \sigma_k$. The unary $e_1 \wedge e_2 \wedge e_3$ represents something like an imaginary unit. It is a unit matrix with non-vanishing diagonal entries i, the pseudo-scalar.

Another View of Angular Momentum Algebra

Consider the three angular momentum operators $J_1 = \frac{1}{2}\, e_1$, $J_2 = \frac{1}{2}\, e_2$ and $J_3 = \frac{1}{2}\, e_3$ given by the base unit vectors that generate the Clifford algebra $Cl_{3,0}$ of Euclidean 3-space. We recall the well known commutation relations with the imaginary unit i. We substitute the i by the unit director or pseudo scalar $e_{123} = e_1 \wedge e_2 \wedge e_3 \in Cl_{3,0}$. Then the shift operators $(J_x \pm i\, J_y)$ are transposed onto

$$J_+ = \tfrac{1}{\sqrt{2}}(e_1 - e_{13}) \quad \text{and} \quad J_- = \tfrac{1}{\sqrt{2}}(e_1 + e_{13}). \qquad (94)$$

Though they are graded, their commutators with J_3 are preserved

$$[J_3, J_+] = J_+ \text{ and } [J_3, J_-] = -J_- \qquad (95)$$

From there we can go on. Understanding physical motion as graded motion, the story unfolds until to the standard model. Depending on the identity Id of the Clifford algebra $Cl_{3,0}$, we obtain for the sum of squared components

$$J^2 = J_1^2 + J_2^2 + J_3^2 = \tfrac{3}{4}\, Id \qquad (96)$$

From now on everything is well known. Yet, there are a few things that remained unfinished until today. Let us go into it.

Rotation Roots
– Completing Angular Momentum Algebra –

Algebraic theory of angular motion, both relativistic and non relativistic, is incomplete. Even in the language of the Pauli algebra the Lie manifolds of spin phenomena and angular momentum have not been sufficiently exhausted. Spin in the Clifford algebra of Euclidean 3-space does not involve only the root space A_1 of the $su(2)$, but at least the space D_2 with 4 roots bound to the Cartan algebra $\{e_{12}, e_3\}$ of the Lie group $L_{3,0}(Cl) \simeq SO(4, \mathbb{C})$. This has dimension 6 and rank 2. In non relativistic quantum mechanics as is formulated in the geometric Clifford algebra $Cl_{3,0}$ – also called Pauli algebra – angular motion provides *2 rotation roots* for quantization. In this toolbox chapter the algebra $\mathcal{L}_{3,0}(Cl)$ together with its roots, its raising- and lowering operators for the spin-vector $\frac{1}{2}\, \sigma_3$ and –bivector $\frac{1}{2}\, \sigma_{12}$ are calculated. The eigenform for spinor spaces and shift operators is constructed.

Uplifting HEPhy Lie groups to Clifford geometric algebra (Schmeikal 2009) I found myself confronted with a peculiar incompleteness. It is scientific general knowledge (Lounesto 2001) that the Pauli matrices generate a complex two-dimensional representation of the Clifford algebra $Cl_{3,0}$ for the Euclidean 3-space. Therefore the denotation *Pauli-algebra* for the Clifford algebra $Cl_{3,0}$ has become a universal concept of language. It is further believed since long that the Lie algebra responsible for nonrelativistic angular motion is the $su(2)$ together with its angular momentum operators [1] J^2, J_3, and $J_\pm$. But this is false. The Lie algebra $\mathcal{L}_{3,0}(Cl)$ controlling the symmetries of nonrelativistic angular motion has dimension 6 and rank 2. It is best represented by the $so(4, \mathbb{C})$ isomorphic with $sl(2, \mathbb{C})\times sl(2, \mathbb{C})$ with rootspace D_2. The classical spin vector ½ σ_3 constitutes the spin eigenforms and spinor space just as well as the nowadays preferred rotation bivector ½ e_{12} = ½ $\sigma_1\wedge\sigma_2$. Those two belong to the same Cartan subalgebra in $Cl_{3,0}$. In the Clifford algebra of Euclidean 3-space we have 3 isomorphic Cartan subalgebras given by a triple of commuting pairs $[\sigma_1, \sigma_{23}]$, $[\sigma_2, \sigma_{13}]$, $[\sigma_3, \sigma_{12}]$. In accordance with this finding, the manifold of shift operators $J_\pm \in Cl_{3,0}$ commuting with a fixed third component – whether denoted as J_3, t_z, or τ_0 – form a graded manifold and naturally correspond with our concept of graded motion (Pezzaglia Jr., 2000). It is this concept which can be pursued into the geometric Clifford algebras generated by Minkowski spaces with indefinite metric (Schmeikal 2006). From that approach then there follows the full symmetry of the standard model. In this section we shall go into the non-relativistic part of the rigor.

It is well known that the Clifford algebra of Euclidean 3-space can be represented by matrices in many ways. Representations of its elements by matrices Mat(4, $\mathbb{R}$), Mat(3, $\mathbb{R}$) and Mat(2, $\mathbb{C}$) are only the most convenient ones.

THE LIE GROUP $L_{3,0}$

In the *Lie group guide to the universe* we specified the functor

$$\wedge\mathbf{V} \xrightarrow{ClgL} \mathcal{L}(Cl) \tag{97}$$

in order to construct a Lie group $L_{p,q}(Cl) \equiv L(Cl_{p,q})$ directly associated with the quadratic Clifford algebra $Cl_{p,q}$. The functor *ClgL* takes us from the category of general associative Graßmann algebras to the category of general non-associative linear Lie algebras over linear spaces of Clifford algebras. We have gone into that problem for the Minkowski space-time and solved it entirely for the Clifford algebra $Cl_{3,1}$ with the Lorentz metric. Surprisingly, the case of the Pauli algebra has not been completely clarified. Let us do that now.

We consider a Lie algebra $\mathcal{L}$ as a vector space over some field $\mathcal{F}$ as usual, and multiplication is given by the commutator bracket over Clifford multiplication. Thus is constituted the functor *ClgL*. If g, h are two elements of the Lie algebra $\mathcal{L} \equiv \mathcal{L}\,(Cl_{p,q})$ then $[g, h]$

---

[1] Such *operators* are not distinguished quantities. Rather they are elements of Clifford algebras like any others and should properly be called »*Clifford numbers*« or »*geometric numbers*«.

are in the algebra too. Besides this first constitutive feature we assume the usual properties of distributivity, Jacobi identity and consequently antisymmetry. Then $\mathcal{l}$ generates a Lie group $\boldsymbol{L}$ by the Clifford exponential mapping. It has a general group element

$$M = \exp c^i g_i \qquad \text{with} \qquad c^i \in F\,,\; g_i \in \mathcal{l} \tag{98}$$

where the g_i represent generators of the Lie group which span the tangent space of the group manifold at a local origin. We define the *adjoint representation* as an automorphism

$$ad\, h_i :\; h_j \longrightarrow [h_i, h_j] \qquad \text{and equivalently as} \tag{99}$$

$ad\, h_i(h_j) = [h_i, h_j] = C_{ij}^k\, h_k$ with structure constants C_{ij}^k and summing over k.

The matrix $M(h_i)$ associated with the adjoint representation is constituted by the entries

$$(M_i)_{jk} = C_{ik}^j \tag{100}$$

Identifying the matrices M_i with the Graßmann monomials forming the basis of $Cl_{3,0}$ we first realize that the scalar Id and pseudoscalar j commute with all basis vectors and therefore do not contribute to the Lie algebra. First, consider the remaining 6 normalized quantities

$$M_1 \cong \frac{e_1}{2},\; M_2 \cong \frac{e_2}{2},\; M_3 \cong \frac{e_3}{2},\; M_4 \cong \frac{e_{12}}{2},\; M_5 \cong \frac{e_{13}}{2},\; M_6 \cong \frac{e_{23}}{2} \tag{101}$$

For those we calculate the following non-vanishing matrix elements:

$(M_1)_{42} = C_{12}^4 = 1$, $\quad (M_1)_{53} = C_{13}^5 = 1$, $\quad (M_1)_{24} = C_{14}^2 = 1$ and
$(M_1)_{35} = C_{15}^3 = 1$, $\quad (M_2)_{41} = C_{21}^4 = -1$, $\quad (M_2)_{63} = C_{23}^6 = 1$ ...
until to $\quad (M_6)_{54} = C_{64}^5 = -1$, $\quad (M_6)_{45} = C_{65}^4 = 1$

The full adjoint matrix representation consists of six real matrices

$$M_1 \overset{adj}{=} \begin{pmatrix} 0 & 0 & 0 & 0 & 0 & 0 \\ 0 & 0 & 0 & 1 & 0 & 0 \\ 0 & 0 & 0 & 0 & 1 & 0 \\ 0 & 1 & 0 & 0 & 0 & 0 \\ 0 & 0 & 1 & 0 & 0 & 0 \\ 0 & 0 & 0 & 0 & 0 & 0 \end{pmatrix},\; M_2 \overset{adj}{=} \begin{pmatrix} 0 & 0 & 0 & -1 & 0 & 0 \\ 0 & 0 & 0 & 0 & 0 & 0 \\ 0 & 0 & 0 & 0 & 0 & 1 \\ -1 & 0 & 0 & 0 & 0 & 0 \\ 0 & 0 & 0 & 0 & 0 & 0 \\ 0 & 0 & 1 & 0 & 0 & 0 \end{pmatrix} \tag{102}$$

$$M_3 \overset{adj}{=} \begin{pmatrix} 0 & 0 & 0 & 0 & -1 & 0 \\ 0 & 0 & 0 & 0 & 0 & -1 \\ 0 & 0 & 0 & 0 & 0 & 0 \\ 0 & 0 & 0 & 0 & 0 & 0 \\ -1 & 0 & 0 & 0 & 0 & 0 \\ 0 & -1 & 0 & 0 & 0 & 0 \end{pmatrix}, \quad M_4 \overset{adj}{=} \begin{pmatrix} 0 & 1 & 0 & 0 & 0 & 0 \\ -1 & 0 & 0 & 0 & 0 & 0 \\ 0 & 0 & 0 & 0 & 0 & 0 \\ 0 & 0 & 0 & 0 & 0 & 0 \\ 0 & 0 & 0 & 0 & 0 & 1 \\ 0 & 0 & 0 & 0 & -1 & 0 \end{pmatrix}$$

$$M_5 \overset{adj}{=} \begin{pmatrix} 0 & 0 & 1 & 0 & 0 & 0 \\ 0 & 0 & 0 & 0 & 0 & 0 \\ -1 & 0 & 0 & 0 & 0 & 0 \\ 0 & 0 & 0 & 0 & 0 & -1 \\ 0 & 0 & 0 & 0 & 0 & 0 \\ 0 & 0 & 0 & 1 & 0 & 0 \end{pmatrix}, \quad M_6 \overset{adj}{=} \begin{pmatrix} 0 & 0 & 0 & 0 & 0 & 0 \\ 0 & 0 & 1 & 0 & 0 & 0 \\ 0 & -1 & 0 & 0 & 0 & 0 \\ 0 & 0 & 0 & 0 & 1 & 0 \\ 0 & 0 & 0 & -1 & 0 & 0 \\ 0 & 0 & 0 & 0 & 0 & 0 \end{pmatrix}$$

There are three isomorphic maximal Abelian subalgebras. Of these we select the one which traditionally contains the spin vector $\frac{1}{2}\sigma_3$:

$$\mathfrak{h} = \{M_3, M_4\} \simeq \{\tfrac{1}{2}\sigma_3, \tfrac{1}{2}\sigma_{12}\} \tag{103}$$

Let $\hat{\mathfrak{h}}$ denote the complement of $\mathfrak{h}$ in the algebra $\mathcal{L}_{3,0} \equiv \mathcal{L}(Cl_{3,0})$. The

$$\hat{\mathfrak{h}} = \{M_1, M_2, M_5, M_6\} \tag{104}$$

has four elements. We have $\dim(\hat{\mathfrak{h}}) = \dim(\mathcal{L}_{3,0}) - \mathrm{rank}(\mathcal{L}_{3,0}) = n - r$. Since the Cartan subalgebra $\mathfrak{h}$ has been chosen, a linear basis of eigenvectors can be uniquely defined in the complementary subspace. In this way $\hat{\mathfrak{h}}$ is decomposed into a direct sum of invariant rays. The eigenvectors are habitually denoted as e_α. The greek indices run over a set of $n - r$ values. In cases like $\mathcal{L}_{3,0}$ they are paired and denoted as $\{e_\alpha, e_{-\alpha}\}$ corresponding with the fact that $n - r$ is even, namely 4. We can take the matrices $\{M_3, M_4\} = \{\frac{1}{2}(e_3), \frac{1}{2}(e_{12})\}$ as a linear basis $\{h_1, h_2\}$ in a Cartan subalgebra $\mathfrak{h}_0$. The eigenvalues $\alpha_1, -\alpha_1, \alpha_2, -\alpha_2$ are the roots corresponding with root space $D_2 = A_1 \times A_1$ and Lie algebra $so(4, \mathbb{C}) \simeq sl(2, \mathbb{C}) \times sl(2, \mathbb{C})$. Calculating the eigenvectors and eigenvalues of matrices M_3, M_4 we find the roots ± 1, $\pm i$.

Four Rotation Roots

So we have not only one root corresponding with the classical $J_\pm$ but we obtain two rotation root spaces A_1, - one for the subalgebra of vectors, and another for the bivectors. We shall denote the raising and lowering operators with the symbols $J_\pm$ and $I_\pm$. We calculate

$$J_+ = \frac{1}{\sqrt{2}}(M_1 - M_5) \qquad J_- = \frac{1}{\sqrt{2}}(M_1 + M_5) \qquad J_3 = M_3 \tag{105}$$

$$J_+ = \frac{1}{\sqrt{2}}\begin{pmatrix} 0 & 0 & -1 & 0 & 0 & 0 \\ 0 & 0 & 0 & 1 & 0 & 0 \\ 1 & 0 & 0 & 0 & 1 & 0 \\ 0 & 1 & 0 & 0 & 0 & 1 \\ 0 & 0 & 1 & 0 & 0 & 0 \\ 0 & 0 & 0 & -1 & 0 & 0 \end{pmatrix}, \quad J_- = \begin{pmatrix} 0 & 0 & 1 & 0 & 0 & 0 \\ 0 & 0 & 0 & 1 & 0 & 0 \\ -1 & 0 & 0 & 0 & 1 & 0 \\ 0 & 1 & 0 & 0 & 0 & -1 \\ 0 & 0 & 1 & 0 & 0 & 0 \\ 0 & 0 & 0 & 1 & 0 & 0 \end{pmatrix}$$

$$I_+ = \frac{1}{\sqrt{2}}(iM_6 + M_5) \qquad I_- = \frac{1}{\sqrt{2}}(iM_6 - M_5) \qquad I_0 = M_4 \tag{106}$$

$$I_+ = \frac{1}{\sqrt{2}}\begin{pmatrix} 0 & 0 & 1 & 0 & 0 & 0 \\ 0 & 0 & i & 0 & 0 & 0 \\ -1 & -i & 0 & 0 & 0 & 0 \\ 0 & 0 & 0 & 0 & i & -1 \\ 0 & 0 & 0 & -i & 0 & 0 \\ 0 & 0 & 0 & 1 & 0 & 0 \end{pmatrix}, \quad I_- = \frac{1}{\sqrt{2}}\begin{pmatrix} 0 & 0 & -1 & 0 & 0 & 0 \\ 0 & 0 & i & 0 & 0 & 0 \\ 1 & -i & 0 & 0 & 0 & 0 \\ 0 & 0 & 0 & 0 & i & 1 \\ 0 & 0 & 0 & -i & 0 & 0 \\ 0 & 0 & 0 & -1 & 0 & 0 \end{pmatrix}$$

These matrices in the adjoint representation can be reduced down to matrices of size 4×4 or 3×3 and 2×2 . Observe the difference to the classical situation where we had only one rootspace corresponding with the $su(2)$. But now we have two spin components: one corresponding with the spin vectors σ_1, σ_2, σ_3 and a second one for the bivectors σ_{12}, σ_{13}, σ_{23}. The duality between the classic spin vector σ_3 and spin bivector σ_{12} commuting with σ_3 is directly transposed onto the duality between real and imaginary roots. After all, the root space is represented by the following simple figure:

Root space $A_1 \times A_1$ for $\mathcal{C}\ell_{3,0} \cong \mathrm{sl}(2, \mathbb{C}) \times \mathrm{sl}(2, \mathbb{C})$

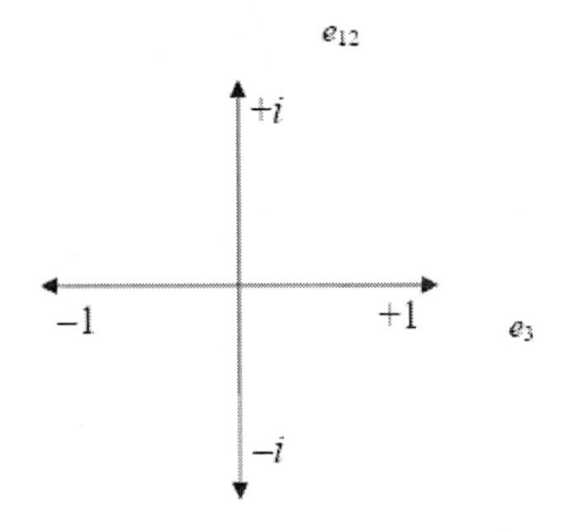

The whole structure can be best understood from the viewpoint of the Clifford algebra. consider the three angular momentum operators $J_1 = ½\, e_1$, $J_2 = ½\, e_2$ and $J_3 = ½\, e_3$ given by the base unit vectors that generate the Clifford algebra $Cl_{3,0}$. We recall the well known commutation relations with the imaginary unit *i*. We substitute the *i* by the unit director or pseudo scalar $e_{123} = e_1 \wedge e_2 \wedge e_3 \in Cl_{3,0}$. Then the classic shift operators $(J_x \pm i\, J_y)$ are translated into

$$J_+ = \frac{1}{\sqrt{2}}(e_1 - e_{13}) \quad \text{and} \quad J_- = \frac{1}{\sqrt{2}}(e_1 + e_{13}) \qquad \text{(equations 104)}$$

Though the $J_\pm$ are graded, their commutators with J_3 are preserved . We have

$$[J_3, J_\pm] = \pm J_\pm \qquad (107a)$$

and

$$[I_0, I_\pm] = \pm i\, I_\pm \qquad (107b)$$

The imaginary root in the I_0-shift operator algebra cannot be avoided. Yet, by multiplying $M_4 \simeq ½\, e_{12}$ with *i* we can give both elements J_3, $I_0 \in$ ♒ a positive definite signature and transpose (15b) to

$$[i\, I_0, I_\pm] = \mp I_\pm \qquad (107c)$$

THE SPIN MANIFOLD AND ITS EIGENFORM CLIFFORD MANIFOLD OF STEP OPERATORS

Despite the existence of Clifforms of Lie algebras $su_{Cl}(2)$ and $so_{Cl}(4)$ we regard the Clifford algebra as the proper space of the manifolds. We confirm the significance of the angular momentum $½\, \hbar\, \sigma_3$, as the quantity $½\, e_3 \in Cl_{3,0}$. Next we realize that the normalized Clifford product of J_+ with J_- is equal to that primitive idempotent

$$\tfrac{1}{2} J_+ J_- = \tfrac{1}{2}(Id + e_3) = f_3 \quad \text{with} \quad f_3 f_3 = f_3 \qquad (108)$$

which brings forth a minimal left ideal: the algebraic *Pauli spinor space* associated with the angular momentum J_3.

$$S = Cl_{3,0} f_3 \stackrel{def}{=} \left\{ \begin{pmatrix} \psi_1 & 0 \\ \psi_2 & 0 \end{pmatrix} \right\} \text{ with } \quad \psi_1, \psi_2 \in \mathbb{C} \qquad (109)$$

We shall call ½ $J_+ J_-$ an eigenform of the angular momentum algebra associated with the *spinor space S*. Now consider the general multivector element of the Clifford algebra $Cl_{3,0}$ with the eight coordinates x_j:

$$X = x_1 Id + x_2 e_1 + x_3 e_2 + x_4 e_3 + x_5 e_{12} + x_6 e_{13} + x_7 e_{23} + x_8 e_{123} \qquad (110)$$

We are searching for general elements $J_{\pm}(X)$ of angular algebra satisfying commutation relations (107). Separate independent solutions can be calculated for the { $J_+(X)$ } and { $J_-(Y)$ }. We then obtain

$$J_+ = x_2(e_1 - e_{13}) + x_3(e_2 - e_{23}), \quad J_- = y_2(e_1 + e_{13}) + y_3(e_2 + e_{23}) \qquad (111)$$

Using a Clifford algebra calculator, we can easily confirm these elements actually satisfy the right commutation relations. Their normalized product is equal to

$$\tfrac{1}{2} J_+(X)\, J_-(Y) = = (x_2 y_2 + x_3 y_3)(Id + e_3) + (x_2 y_3 - x_3 y_2)(e_{12} + e_{123}) \qquad (112)$$

Referring to Kauffman (2005) we construct eigenforms for quantum pure states and manifolds of raising- and lowering elements. Verify that the quantity $f = \tfrac{1}{2} J_+(X)\, J_-(Y)$, the Clifford product of $J_+(X)$ and $J_-(Y)$, is an eigenform in the Clifford algebra $Cl_{3,0}$. Namely, we define recursively the bilinear term

$$f_{n+1} = f_n f_n \text{ having form}$$
$$f_n = a(Id + e_3) + b(e_{12} + e_{123}) \qquad (113)$$

Clearly that form is recursively preserved, that is, we obtain after the next step of recursion the same form with other coefficients A, B:

$$f_{n+1} = A(Id + e_3) + B(e_{12} + e_{123}) \qquad \text{with} \qquad A = 2(a^2 - b^2), \quad B = 4ab$$

Therefore, the f_n moves stepwise on a plane within the same real vector space, spanned by $\{Id, e_3, e_{12}, e_{123}\}$.

$$E(f) = span_C\{Id, e_3, e_{12}, e_{123}\} \qquad \text{the space of spinor eigenforms} \qquad (114)$$

Observe that $E(f)$ is itself an eigenform within $Cl_{3,0}$. It reproduces itself by Clifford multiplication. It is a closed 4-dimensional real associative subalgebra of $Cl_{3,0}$. We have $E(f)E(f) = E(f)$ which signifies the closedness and it adopts the * (either reversion or conjugation) from the Clifford algebra. The space of *spinor-eigenforms* $E(f)$ is therefore a C*-algebra too. Those four elements represent the essential components of an angular momentum algebra associated with J_3. If you remember, the coefficient a was equal to $x_2 y_2 + x_3 y_3$ and the b was $x_2 y_3 - x_3 y_2$. So the f_n performs a nonlinear motion in the 4-dimensional subspace generated algebraically by the *core spin space*

$$\mathsf{J} = \{e_3, e_{12}\} \subset Cl_{3,0} \tag{115}$$

This is no other than the first Cartan subalgebra we have chosen to identify the Lie group $L_{3,0}(Cl)$ *of and in* $Cl_{3,0}$. Thus we can be convinced: for an angular momentum algebra of spin in Euclidean 3-space we need both, *bivector* e_{12} and *spinvector* e_3.

Generally, such a form may diverge or collapse towards zero or converge to a fixed point in the eigenform. We obtain the fixed point for f_n at first under the condition $\mathsf{J}_-(Y) \equiv \mathsf{J}_+(X)^* = \mathsf{J}_+(X)^\dagger$ (J_+ reverted)

$$\mathsf{J}_+ = x_2(e_1 - e_{13}) + x_3(e_2 - e_{23}),\quad \mathsf{J}_- = x_2(e_1 + e_{13}) + x_3(e_2 + e_{23}) \tag{116}$$

$$f = (x_2^2 + x_3^2)\, Id + (x_2^2 + x_3^2)\, e_3 \tag{117}$$

We know that at a given basis $\{e_1, e_2, e_3\}$ the $Cl_{3,0}$ contains six primitive idempotents ½ $(Id \pm e_j)$. One of those is f_3 = ½ $(Id + e_3)$. Clearly the f_3 represents the fixed point to form (114). That fixed point occurs on the Thales' circle where

$$x_2^2 + x_3^2 = \tfrac{1}{2} \tag{118}$$

and the diameter is equal to $1/\sqrt{2}$.

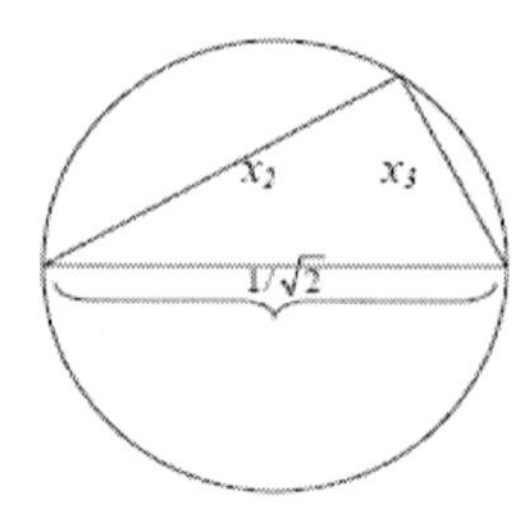

LOGIC OF LIE ALGEBRA $\mathcal{L}$ (CL) OVER CLIFFORD ALGEBRA

The logic of defining relations for Lie algebras over general spaces has not been sufficiently worked out. At least, the situation is not yet fully satisfying. The logic depends on which of the existing relevant definitions of Lie algebra we chose. There is a double characteristic which makes an axiomatic difference. Either we wish that the algebra is given by a certain vector space L over some field $\mathcal{F}$ with an additional binary operation given by the Lie commutator. Or we demand that the elements of the Lie algebra can be represented by a Lie bracket. This is not the same. That is, there exist two peculiar axioms before we go on to demand linearity, Jacobi identity and so on.

In their pioneering works Cartan (1930) and Chevalley (1954), (see Knapp 1988) have paved the way for the pair of *Lie groups* and *Lie algebras*. They reached agreement that Lie algebras of interest are finite-dimensional real vector spaces with a multiplication law having certain properties. Any vector subspace closed under the product operation is again a Lie algebra. Many authors who wrote after Chevalley (Sattinger and Weaver 1986), (Baker 2002) deal with closed linear groups and/or begin with the Cartan-Weyl theory before defining general Lie groups. They skip elementary Lie theory and questions of logic structure. Knapp (2002) has reviewed Bakers book in due consideration of these historic works. He reconsidered the Cartan-Weyl theory by enhancing the object to complex algebras. Allowing for real and complex scalars, he stated that in any case an ideal in a Lie algebra is a Lie subalgebra such that $[X, Y]$ is in the ideal if X is in the ideal and Y is in the whole algebra. A simple Lie algebra is a nonzero Lie algebra whose only ideals are 0 and the whole algebra. By mere convention, a 1-dimensional Lie algebra is not considered as simple. Extending the proposition to Lie algebras over Clifford modules to obtain the associated Lie algebra $\mathcal{L}_{p,q} \equiv \mathcal{L}(Cl_{p,q})$, some peculiar confusion arises.[2] Is the Clifford algebra $Cl_{3,0}$ simple ?

A maximal ideal $\mathcal{C}$ which satisfies $[L, \mathcal{C}] = 0$ is called the centre of L. We have

$$\mathcal{C}_{3,0} \overset{def}{=} Cen(Cl_{3,0}) = \mathbb{R} \oplus \Lambda^3\mathbb{R}^3 \simeq \mathbb{C} \text{ the centre of } Cl_{3,0} \tag{119}$$

Clearly, the centre of $Cl_{3,0}$ is a maximal ideal and is not trivial. Just its commutators with the whole vector space vanish. But $0 \times Id \in \mathcal{C}_{3,0}$ Therefore, if we define Lie algebras as follows:

Axioms

A_0: The vector space of $\mathcal{L}_{p,q}(Cl)$ is ½ΛV Graßmann's

A_1: $X, Y \in \mathcal{L}_{p,q}(Cl) \rightarrow [X, Y] \in \mathcal{L}_{p,q}(Cl)$ defined by Clifford product

⋮

the centre is taken over into the Lie algebra. Thus, in such cases where the underlying Clifford algebra has a maximal ideal, - like is the case with $Cl_{3,0}$ -, its Lie algebra $\mathcal{L}_{3,0}(Cl)$ could not be called »*simple*«. The Axiom A_1 is too week to substantiate »*simplicity*« of Lie algebras over a Graßmann "Hauptgebiet". Unfortunately, the axiomatic design of the form A_0, A_1, a. s. o. was retained even by forward-looking scientists like Magnea (2002) and myself (2005). Being aware that algebras like the Euclidean $Cl_{3,0}$ by Clifford algebra software packages are denoted as *simple*, we should secure the proposition by another axiom, say, A_1'

A_1': $$\forall_{Z \in L} \exists_{X,Y \in L}\; Z = [X, Y] \tag{120}$$

2 Some of that confusion has even been realized by technical encyclopedia: „A Lie algebra is called simple, if it has no non-trivial ideal and is not abelian. At the Lie algebras simplicity is used differently. This can lead to confusions. If one considers Lie algebra as an algebraic structure, to postulate it must not be commutative is unnatural." (german entry in Wikipedia 2007/08; translated verbatim by the author)

Clearly, within $Cl_{3,0}$ the director e_{123} cannot be represented by any commutator. Therefore the ideal $\mathcal{C}_{3,0} = \text{span}_{\mathbb{C}}\{Id, e_{123}\}$, is excluded by axiom A_1'. Now we have achieved that the Lie algebra $\mathcal{L}_{3,0}(Cl)$ is a simple Lie algebra with a Cartan subalgebra spanned by $\{e_{12}, e_3\}$ while the Clifford algebra $Cl_{3,0}$ contains the space of eigenforms $E(f)$ which is nothing else than the maximal Cartan subalgebra of $Cl_{3,0}$. Thus, by demanding simplicity and semi-simplicity of maximal Lie algebras we reduce the size of the maximal Cartan subalgebra. Therefore the largest Lie algebra $\mathcal{L}_{3,0} \subset Cl_{3,0}$ does not have 8 dimensions and rank 4, but dimension 6 and rank 2. In agreement with the classic conventions we realize $\mathcal{L}_{3,0} \stackrel{\text{def}}{=} \mathcal{L}_{3,0}(Cl) \simeq sl(2,\mathbb{C}) \times sl(2,\mathbb{C})$ can now be denoted a *simple algebra*.

Toolbox 5

HIGHER SPIN IN THE MINKOWSKI ALGEBRA

The Clifford algebra of the Minkowski space-time with Lorentz metric is sometimes briefly called Minkowski algebra. It is generated by the Minkowski space $\mathbb{R}^{3,1}$ with signature $\{+++-\}$. The general element $\mathcal{X} \in Cl_{3,1}$ is

$$\mathcal{X} = \{x_1 Id + x_2 e_1 + x_3 e_2 + x_4 e_3 + x_5 e_4 + x_6 e_{12} + x_7 e_{13} + x_8 e_{14} + \\ + x_9 e_{23} + x_{10} e_{24} + x_{11} e_{34} + x_{12} e_{123} + x_{13} e_{124} + x_{14} e_{134} + x_{15} e_{234} + x_{16} j\}$$

where j denotes the oriented space-time volume e_{1234}. It seems that climbing higher on the ladder of abstraction we lost our attention for the reality of space and time through sheer negligence. We now have categories, gebras and cogebras, calculate without basis, but still have not comprehended how the higher spin components of the so called $SU(3)$ follow naturally from the space-time geometry. Some even say that geometry is no way to understand that, others take an oath on topology, and a few are prepared to compromise between both views. Actually the problem appears to be quite simple to solve if one understands the structure of the Clifford algebra $Cl_{3,1}$. We are looking for a general multivector w fulfilling commutation relations

$$[\tau_3, w_+] = \tfrac{1}{2} w_+ \qquad [\tau_3, w_-] = -\tfrac{1}{2} w_- \tag{121}$$

Solutions are

$$w_+ = a\, e_2 - b\, e_3 - a\, e_4 + c\, e_{12} - d\, e_{13} - c\, e_{14} - d\, e_{234} - b\, e_{1234} \tag{122}$$

$$w_- = a\, e_2 - b\, e_3 + a\, e_4 + c\, e_{12} - d\, e_{13} + c\, e_{14} + d\, e_{234} + b\, e_{1234} \tag{123}$$

Following an idea of Kauffman we find an eigenform

$$f = \tfrac{1}{2} w_+ w_- \tag{124}$$

Recall what Kauffman (2005) said about eigenforms. Verify quantity $f = \frac{1}{2} w_+ w_-$, the Clifford product of w_+ and w_-, is an eigenform in the Clifford algebra $Cl_{3,1}$. We define recursively the bilinear term

$$f_{n+1} = \tfrac{1}{2} f_n f_n \qquad \text{with} \qquad (125)$$

$$f_n = A(Id + e_{24}) + B(Id - e_{124}) + C(e_{23} + e_{123}) + D(e_{34} + e_{134})$$

and

$A = a^2 - c^2$, $B = b^2 - d^2$, $C = ad - bc$, $D = cd - ab$. We find out that this form is recursively preserved, just as it was in the angular momentum algebra. Therefore, the f_n moves within the *space of isospinoreigenforms*

$$E(f) = span_R\{Id, e_1, e_{23}, e_{24}, e_{34}, e_{123}, e_{124}, e_{134}\} \qquad (126)$$

Observe $E(f)$ is itself an eigenform within $Cl_{3,1}$. It reproduces itself by Clifford multiplication. A form f_n may diverge or collapse towards zero or converge towards a fixed point. With the input of (124), (125) we obtain the general solution form

$$f_{n+1} = \alpha\, Id + \beta e_{23} + \gamma e_{123} + \delta e_{24} + \varepsilon e_{34} + \phi e_{134} \qquad (127)$$

The regained f_{n+1} is a polynomial with nonlinear coefficients having a special structure, e. g. for the first and last coefficients we have

$$\alpha = a^4 + b^4 + c^4 + d^4 + 2(a^2b^2 + c^2d^2 - a^2c^2 - a^2d^2 - b^2c^2 - b^2d^2),$$

$$\phi = abc^2 + abd^2 + a^2cd + b^2cd - a^3b - ab^3 - c^3d - cd^3$$

Anyway, as long as the w_+, w_- have the symmetric structure (122), (123) the isospinor eigenform moves in 6-dimensional space

$$E(f) = span_R\{Id, e_{23}, e_{24}, e_{34}, e_{123}, e_{134}\} \qquad (128)$$

which is a bit smaller than in (126). That larger space comes in as soon as the symmetry of equations (122), (123) is broken and coefficients for w_+ deviate from those for w_-. Only the space of (126) is closed for Clifford multiplication. Therefore the f_n are eigenforms. That linear space which generates the space of isospinor eigenforms by Clifford multiplication is called the *core isospin space*.

$$\mathrm{J} = \{e_1, e_{123}, e_{124}\} \subset Cl_{3,1} \qquad \text{core isospin space of } Cl_{3,1} \qquad (129)$$

Actually there is a stable manifold of fixed points f_n in $E(f)$. The equation (124) denotes only one of those, though apparently a rather prominent one. This situation must be investigated more closely. In the 2004 work on transposition involutions there have been used two isospin multivectors that were denoted as u and v. Those are special solutions within the manifolds of (122) and (123). Since these manifolds can be split such that the $SU(3)$ commutation relations are preserved.

$$\text{Split } u_+ = b\, e_3 - d\, e_{13} + d\, e_{234} - b\, e_{1234}$$

$$u_- = b\, e_3 + d\, e_{13} + d\, e_{234} + b\, e_{1234} \quad \text{with} \quad [\tau_3, \pm u_\pm] = \mp\tfrac{1}{2} u_\pm \tag{130}$$

$$v_+ = a\, e_2 - a\, e_4 - c\, e_{12} + c\, e_{14}$$

$$v_- = a\, e_2 + a\, e_4 + c\, e_{12} + c\, e_{14} \qquad \text{with} \quad [\tau_3, \pm v_\pm] = \pm\tfrac{1}{2} v_\pm \tag{131}$$

and take all constants $a = b = c = d = ¼$. Then we obtain the special forms of u_+, u_-, v_+, v_- in the standard representation of $\mathcal{L}_{3,1}$.

$$u_+ = \tfrac{1}{4}(e_3 + e_{234}) - \tfrac{1}{4}(e_{13} + e_{1234}) = \lambda_4 + \lambda_5 \tag{132}$$

$$u_- = \tfrac{1}{4}(e_3 + e_{234}) + \tfrac{1}{4}(e_{13} + e_{1234}) = \lambda_4 - \lambda_5 \tag{133}$$

$$v_+ = \tfrac{1}{4}(e_2 + e_{14}) - \tfrac{1}{4}(e_4 + e_{12}) = \lambda_6 + \lambda_7 \tag{134}$$

$$v_- = \tfrac{1}{4}(e_2 + e_{14}) + \tfrac{1}{4}(e_4 + e_{12}) = \lambda_6 - \lambda_7 \tag{135}$$

Please, observe quantities λ_4, λ_5, λ_6, λ_7 are generators of the constitutive Lie group $\mathcal{L}^{(2)} \simeq$ a real Clifform $sl_{Cl}(3)$ of the Lie algebra $sl(3, \mathbb{R})$. For completeness, consider the isoform $f_n = ½\, v_+ v_-$ as derived from general v-spin (68). We have

$$f_n = \tfrac{1}{2} v_+ v_- = (a^2 + c^2)(Id + e_{24}) - 2ac(e_1 + e_{124}) \tag{136}$$

which approaches a fixed point in the proximity of $a = c = \frac{1}{2\sqrt{2}}$. Only in that case is the quantity $½\, v_+ v_- = \text{const.}(Id - e_1)(Id + e_{24})$ a primitive idempotent. In the current work this has been denoted as $f_{13} \in ch_1$. It is fit to represent the pure state of a u-quark. It is therefore that we say that

A pure state gives rise to an isospin split in the Lie manifold.
The isospin manifolds form pure state equivalence classes.

We can carry out the same rigor for the general $u_\pm$ spin given by formulas (67). We obtain

$$f_n = (b^2 + d^2)(Id - e_{124}) + 2bd(e_1 - e_{24}) \tag{137}$$

which converges towards a pure state, iff $b = d = \frac{1}{2\sqrt{2}}$.

In that case quantity ½ $u_+ u_-$ = ½(Id + e_1)½(Id – e_{24}) is the Clifford conjugate of f_{13}, namely f_{12}, a fermion with strangeness.

Toolbox 6

GRADE EQUIVALENCE IN $CL_{3,1}$

LOGIC PROJECTOR FIELDS AND SPLITS

The algebraic Weyl spinor components could be separated by projector fields

$$\eta_0^{\pm} = \tfrac{1}{2}(1 \pm \gamma_0) \text{ with the Dirac matrix } \gamma_0 \tag{138}$$

Generally there are paravector spaces spanned by "*erzeugende Einheiten*". A paravector space can be conceived as a Clifford algebra $Cl_{1,0}$ spanned by one unital element and a scalar $\{1, e\}$. Projectors, as we used them until today are given by equations

$$P_{\pm} = \tfrac{1}{2}(1 \pm e); \quad P_{\pm}^2 = P_{\pm}; \qquad P_+ P_- = 0; \qquad P_+ + P_- = 1 \tag{139}$$

The important meaning of those geometric entities for quantum mechanics has first been realized by John v. Neumann (1932). He saw the correlation between the statistical interpretation of the Schrödinger equation and the *idempotent decomposition of unity*. Carl Friedrich v. *Weizsäcker* (1958a,b) also used the field projectors when he constructed his logic interpretation of quantum mechanics. His approach has not been appreciated enough. But it is very important.

Any paravector of form $x = x^0 + x^1 e$ can be represented in the space of projectors by

$$x = (x^0 + x^1)P_+ + (x^0 - x^1)P_- \tag{140}$$

The old calculations of the Hamilton operator by v. Neumann (1932) were based on the decomposition of unity $P_+ + P_- = 1$. This moved C.F. v. Weizsäcker to establish the "*quantum theory of the unit alternative*". It was essentially given by *logic complementarity of pure states*. However, this did not respect the whole ortho-modular lattice of logic binary connectives. Only recently it has been realized that the complete 16-element lattice of binary connectives has a significant meaning for physics (Schmeikal 2000). The elder complementarity relation of binary pure states was essentially determined by a $Cl_{1,0}$-multiplication table

	1	e
1	1	e
e	e	1

satisfying the discrete group relations of $\mathbb{Z}_2$.

Investigating extensions we must go further and consider unital algebras with m unital elements a, b, c, ... (with $a^2 = b^2 = c^2 = 1$). We can form logic combinations of idempotent factors

$$\tfrac{1}{2}(1 \pm a)\tfrac{1}{2}(1 \pm b)\tfrac{1}{2}(1 \pm c)... \quad (141)$$

The units constitute more complicated group tables and finer grading. In this more general situation we obtain a decomposition of unity such like

$$1 = \sum_{\beta=0}^{2^m-1} P(\beta) \quad (142)$$

In the year 2000 we defined the Clifford algebras in neutral signature $Cl_{m,m}$ as »*logic Clifford algebras*«. In those algebras we obtain logic projectors

$$P_\Lambda = \frac{1}{2^m}\prod_{i=1}^{m}(1 \pm e_{i,i+m}) \quad (143)$$

giving a decomposition of unity into 2^m terms. We can easily figure out how the design of P_Λ constitutes a truth table for m-ary logic connectives.

We shall now show that the elements of the maximal Cartan subalgebra of the constitutive Lie group $\mathcal{L}^{(2)}$ are algebraically equivalent and act independent of their grade. We have linked this with the idea of a spatial extension independent of grade. We can think of a point set increasing its fractal dimension while being mapped or rather unfolded from dimension one to dimension three. That is, we define geometric space-time extensions as graded abelian subspaces of some unital algebra. So we investigate the ternary extension of the Minkowski algebra. The ternary extension gives rise to the strong force non-relativistic inner motion. It brings forth causality violating dynamics and chromodynamic transitions satisfying the Pauli principle. We define

The ternary extension is a paravector space $\{1, a, b, c\}$
with $[a, b] = [a, b] = [a, c] = 0$ and $a^2 = b^2 = c^2 = 1$. (144)

It is interesting to see that the works of a few scientists like v. Neumann, Weyl, Weizsäcker forebode the important role of discrete structures which became important in modern Clifford algebra. From about 1930 onward, a few mathematicians and physicists began to anticipate that various algebraic units and relations such as the logic lattice, the cyclic group $\mathbb{Z}_2$, the relation between dihedral group, Klein 4 -group and quaternion group played a special role in quantum physics. Hermann Weyl established a one-one correspondence between the units of the Klein 4-group "K_4" and the Pauli matrices $\{\sigma\}$ times *i*, that is, the quaternion group. He based that correspondence on what he called "*geschickte Eichung*", skillful gauging. Historically, the difference between an abelian basis $\{a, b, c\}$ and a non commutative euclidean basis $\{\sigma_1, \sigma_2, \sigma_3\}$, where the σ_i are standard Pauli matrices, has been explained by Herman Weyl as what he denoted the difference between vector-space and ray-space (*Vektorkörper and Strahlenkörper*). The Klein-4 group provides a simple algebraic type of triality by its cyclic product rule: the product of any two base elements gives the third. As we shall see that K_4-triality gives rise to the strong force symmetries. Therefore we postulate a K_4-group table for ternary extensions.

K_4 table for extension units

Id	a	b	c
a	Id	c	b
b	c	Id	a
c	b	a	Id

The main diagonal entries report the fact that *a, b, c* are unital elements; we have $a\,a = b\,b = c\,c = Id$. The off diagonal entries implicate that step 1 quantities ½ (1 ± a) a.s.o. represent non primitive idempotents whereas step 2 polynomials are primitive idempotents. Namely, we have

$$\eta_1 = \tfrac{1}{2}(1+a) \qquad \eta_2 = \tfrac{1}{2}(1+b) \qquad \eta_3 = \tfrac{1}{2}(1+c)$$
$$\eta_1^c = \tfrac{1}{2}(1-a) \qquad \eta_2^c = \tfrac{1}{2}(1-b) \qquad \eta_3^c = \tfrac{1}{2}(1-c) \tag{145}$$

Step 1 non primitive idempotents with conjugation η^c. Quantities

$$\begin{aligned}
\psi_0 &= \tfrac{1}{2}(1+a)\tfrac{1}{2}(1+b) = \tfrac{1}{4}(1+a+b+c)\\
\psi_1 &= \tfrac{1}{2}(1+a)\tfrac{1}{2}(1-b) = \tfrac{1}{4}(1+a-b-c)\\
\psi_2 &= \tfrac{1}{2}(1-a)\tfrac{1}{2}(1+b) = \tfrac{1}{4}(1-a+b-c)\\
\psi_3 &= \tfrac{1}{2}(1-a)\tfrac{1}{2}(1-b) = \tfrac{1}{4}(1-a-b+c)
\end{aligned} \tag{146}$$

Are step 2 primitive idempotents fulfilling the bilateral orthogonality relations

$$\psi_i \psi_k = \delta_{ik} \psi_k \tag{147}$$

That the η are non primitive idempotents follows from the identities

$$\eta_1 = \psi_1 + \psi_2 \qquad \eta_2 = \psi_1 + \psi_3 \qquad \eta_3 = \psi_1 + \psi_4$$
$$\eta_3^c = \psi_2 + \psi_3 \qquad \eta_2^c = \psi_2 + \psi_4 \qquad \eta_1^c = \psi_3 + \psi_4 \tag{148}$$

The primitive idempotents have a "pacwoman" property. This guarantees that ψ_0 appears as Eigenfield of all three extensions, that is, ψ_0 is an Eigenfield of the maximal Cartan extension $C = span_F\{Id, a, b, c\}$ with some number field F.

$$a\psi_0 = \psi_0 \qquad b\psi_0 = \psi_0 \qquad c\psi_0 = \psi_0 \tag{149}$$

Eigenfields ψ_1, ψ_2, ψ_3 simulate an isospin property

$$\begin{array}{lll} a\psi_1 = +\psi_1 & & \\ & b\psi_1 = -\psi_1 & c\psi_1 = -\psi_1 \\ a\psi_2 = -\psi_2 & & \\ & b\psi_2 = +\psi_2 & c\psi_2 = -\psi_2 \\ a\psi_3 = -\psi_3 & & \\ & b\psi_3 = -\psi_3 & c\psi_3 = +\psi_3 \end{array} \tag{150}$$

We know the s-quark having strangeness $s = -1$ is an isospin singlet whereas the u- and d-quark form an isospin doublet. With a definition $I = \eta_2 - \eta_3$ we obtain eigenvalue equations

$$I\psi_1 = 0\psi_1; \qquad I\psi_2 = +1\psi_2; \qquad I\psi_3 = -1\psi_3 \tag{151}$$

We can even define the hypercharge property of extension eigenfields by

$$Y = \frac{-2a + b + c}{6} \tag{152}$$

Consider the field ψ_1 having isospin 0. It has hypercharge

$$Y\psi_1 = \frac{1}{6}(-2a\psi_1 + b\psi_1 + c\psi_1) = -\frac{2}{3}\psi_1 \tag{153}$$

equal $-2/3$ because of equations (150). That is, the ψ_1 can represent an s-quark with strangeness zero and hypercharge $-2/3$. The isospin doublet, the u- and d-quark both would give us correct hypercharges of $+1/3$. All this can be realized by the elements of the maximal Cartan subalgebra

$$\{a, b, c\} \overset{def}{=} \{e_1, e_{24}, e_{124}\} \tag{154}$$

However, we have to be aware, the functions ψ_1, ψ_2, ψ_3 themselves are not spinors. But they can serve as generators of the ideals for spinor spaces.

As a matter of fact, the constitutive graded Lie group that unfolds the whole idempotent manifolds and additionally performs all the necessary $SU(3)$ rotations of pure states, such as flavour and colour, originally follows from the assumption that the primitive idempotents actually represent the pure states of fermions (Schmeikal 2008). That is, we obtain the correct spinor spaces of higher spin, their associated direction fields in the sense of Cartan and all the typical groups and subgroups of the standard model, provided we assume that the Pauli principle involves some motion which maps the dynamics from a single direction, say, e_1, onto a directed area, say, e_{24} and from there to a directed space-time volume e_{124}. That is, the trajectory is convoluted and deconvoluted in some conformal mapping of a point set space. Space, in this sense, seems discontinuous rather than smooth. There are several ways to cope with this problem.

Once we have taken as a signal that we could derive the standard model as a space-time group beginning with the simple assumption of pure states and transposition involutions (Coxeter reflections) based thereon, we sure will not give up the concept of the constitutive Lie algebra $\mathcal{L}^{(2)}$ in $\mathcal{L}_{3,1}$. But we shall also adhere to the idea of primitive idempotents as physical states. However, the latter requires that we also adhere to the fact that those states are neither spinors nor direction fields. What can we do?

We can take the primitive idempotents as cursors and construct ideals which accommodate the spinors and the direction fields. This problem is solvable as we have shown in chapter »*Isotropic Fields in* $Cl_{3,1}$« where we factored out the primitive idempotents to derivate the isotropic direction fields. All the relations that concern the relative location of a pure state in relation to some other, orthogonality, colour, flavour etc., are now transposed from the primitive idempotent onto the isotropic spinor. That is, if, say, state 1 could be flavour rotated into state 2 this feature is transposed from the primitive idempotents onto their isotropic direction fields. This is absolutely compatible with the traditional approach.

There is a second aspect that may contribute to the solution of this problem. Namely, until today our whole attention has concentrated in the observed object. Though we have understood that the object cannot be a pure object as it depends on the way we observe it, we always say much less about the observer field which is used to observe the object, than about the observed. May be we should always consider two fields, an observer field and an observed field. Both are in some way separate and in some other connected. This duality of separation and connection should be described by our models. That is we describe the interaction, may be, by some geometric product as in the model where we measure quantum numbers by 1-forms of idempotents. This point is suggesting that we develop a new theory of observation.

The only theory that cared about the observer field and still was developed far enough to cope with all the questions posed in this book is the one created by my friend Zbigniew Oziewicz. Unfortunately or fortunately, had to clean up physics and discharge a lot of dogmatic viewpoints. So he also gave a new formulation to special relativity, something, the established science community does not like very much. But as far as I can see, he is right.

Toolbox 7

HIGHER SPIN ISOTROPIC DIRECTION FIELDS

The constitutive algebra $\mathcal{L}^{(2)}$ is the most important tool in my approach to physics. It is given by eight graded elements of the Clifford algebra of the Minkowski space-time with Lorentz metric.

$$\lambda_1 = \tfrac{1}{4}(e_{34} - e_{134}) \qquad \lambda_2 = \tfrac{1}{4}(-e_{23} + e_{123}) \qquad \lambda_3 = \tfrac{1}{4}(e_{24} - e_{124})$$
$$\lambda_4 = \tfrac{1}{4}(e_3 + e_{234}) \qquad \lambda_5 = -\tfrac{1}{4}(e_{13} + j) \qquad \lambda_6 = \tfrac{1}{4}(e_2 + e_{14})$$
$$\lambda_7 = -\tfrac{1}{4}(e_4 + e_{12}) \qquad \lambda_8 = \tfrac{1}{2\sqrt{3}}(-2e_1 + e_{24} + e_{124}) \tag{155}$$

The Gell-Mann matrices in $\mathbb{C}\otimes Cl_{3,1}$ are identical with

$$\{-2\lambda_1, 2i\lambda_2, -2\lambda_3, -2\lambda_4, 2i\lambda_5, 2\lambda_6, -2i\lambda_7, 2\lambda_8\},\ i=\sqrt{-1} \tag{156}$$

Verify, the constitutive algebra $\mathcal{L}^{(2)}$ can be written as a Lie commutator product

$$\mathcal{L}^{(2)} = sl_{Cl}(2) \times so_{Cl}(3) \tag{157}$$

The factor $so_{Cl}(3)$ has generating elements $\{\lambda_2, \lambda_5, \lambda_7\} \in Cl_{3,1}$. These satisfy the characteristic $so(3)$-commutation relations

$$[\lambda_i, \lambda_j] = \tfrac{1}{2}\varepsilon_{ijk}\lambda_k,\ \varepsilon_{ijk} \text{ antisymmetric in } i, j, k = 2, 5, 7 \tag{158}$$

$$[\lambda_2, \lambda_5] = \tfrac{1}{2}\lambda_7, \qquad [\lambda_2, \lambda_7] = -\tfrac{1}{2}\lambda_5, \qquad [\lambda_5, \lambda_7] = \tfrac{1}{2}\lambda_2$$

These relations correspond one-to-one with the well known bivector relations

$$[L_1, L_2] = \tfrac{1}{2}L_3; \qquad [L_1, L_3] = -\tfrac{1}{2}L_2; \qquad [L_2, L_3] = \tfrac{1}{2}L_1 \tag{159}$$

with bivectors $L_1 = e_{23}$, $L_2 = e_{13}$, $L_3 = e_{12}$. Both $so_{Cl}(3)$ algebras satisfy the inter-commutation relations

$$[\lambda_2, e_{23}] = 0 \qquad [\lambda_5, e_{12}] = 0 \qquad [\lambda_7, e_{13}] = 0 \tag{160}$$

We know that the bivector $L_1 = e_{23}$ causes rotations about the x_1-axis, L_2 about the x_2 and L_3 about x_3 in the Minkowski space. But which ones are the axis left invariant by the λ_i ? The answer is a little surprise. Consider the Clifford manifold

$$\begin{aligned} X = \{&x_1 Id + x_2 e_1 + x_3 e_2 + x_4 e_3 + x_5 e_4 + x_6 e_{12} + x_7 e_{13} + x_8 e_{14} + \\ &+ x_9 e_{23} + x_{10} e_{24} + x_{11} e_{34} + x_{12} e_{123} + x_{13} e_{124} + x_{14} e_{134} + x_{15} e_{234} + x_{16} j\} \end{aligned} \tag{161}$$

with directed space-time-volume $j = e_{1234}$.

The element λ_2 leaves the x_2, the x_9 and x_{12} unaltered. That is, it simultaneously rotates the space about base units e_1, e_{23} and directed volume e_{123}. Note those three commute. Thereby it rotates the pairs $\{x_3, x_4\}$ of a plane coordinating base units $\{e_2, e_3\}$ in $\mathbb{R}^{3,1}$. But that is not yet the most general action of λ_2. It has been shown that there are six colour spaces corresponding with isomorphic maximal Cartan subalgebras of the $Cl_{3,1}$, namely

$$\begin{aligned} ch_1 &= \{1, e_1, e_{24}, e_{124}\} & ch_4 &= \{1, e_2, e_{14}, e_{124}\} \\ ch_2 &= \{1, e_1, e_{34}, e_{134}\} & ch_5 &= \{1, e_3, e_{14}, e_{134}\} \\ ch_3 &= \{1, e_2, e_{34}, e_{234}\} & ch_6 &= \{1, e_3, e_{24}, e_{234}\} \end{aligned} \tag{162}$$

Each of those spaces can be generated by two units of grades 1 and 2. Taking out the scalar these are 6 isomorphic Cartan algebras of the rank 3 Lie algebra $\mathcal{L}_{3,1}$ derived from $Cl_{3,1}$. The pure states within those colour-spaces are the primitive idempotents which can represent fermion states.

The second important factor of $\mathcal{L}_{3,1}$ is $sl_{Cl}(2)$. This is given by the elements $\{\lambda_1, \lambda_2, \lambda_3\}$ which fulfil the commutation relations

$$[\lambda_1, \lambda_2] = \lambda_3; \qquad [\lambda_1, \lambda_3] = \lambda_2; \qquad [\lambda_2, \lambda_3] = \lambda_1 \tag{163}$$

Those play back the structure constants of the algebra for the Lie group $Sl(2, \mathbb{R})$ which is isomorphic with the positive components of the spin groups $Spin^+_{2,1}$ and $Spin^+_{1,2}$. So the first isospin module brings in the Lorentz transformations restricted to dimensions p+q = 3 if we forget about grades higher than 1. The constitutive group and its special linear component can of course administrate any desired grade.

To discern the immediate meaning of the isotropic isospin direction fields consider the isospin, u-spin and v-spin, that is, the elements λ_2, λ_5, λ_7 . We consider their antisymmetric matrices in the following standard representation of the basis vectors of the Minkowski space-time $\mathbb{R}^{3,1}$.

$$e_1 = \begin{bmatrix} 1 & 0 & 0 & 0 \\ 0 & -1 & 0 & 0 \\ 0 & 0 & -1 & 0 \\ 0 & 0 & 0 & 1 \end{bmatrix} \qquad e_2 = \begin{bmatrix} 0 & 1 & 0 & 0 \\ 1 & 0 & 0 & 0 \\ 0 & 0 & 0 & 1 \\ 0 & 0 & 1 & 0 \end{bmatrix}$$

$$e_3 = \begin{bmatrix} 0 & 0 & 1 & 0 \\ 0 & 0 & 0 & -1 \\ 1 & 0 & 0 & 0 \\ 0 & 0 & -1 & 0 \end{bmatrix} \qquad e_4 = \begin{bmatrix} 0 & -1 & 0 & 0 \\ 1 & 0 & 0 & 0 \\ 0 & 0 & 0 & -1 \\ 0 & 0 & 1 & 0 \end{bmatrix} \tag{164}$$

$$\lambda_2 = \begin{bmatrix} 0 & 0 & 0 & 0 \\ 0 & 0 & -\frac{1}{2} & 0 \\ 0 & \frac{1}{2} & 0 & 0 \\ 0 & 0 & 0 & 0 \end{bmatrix} \qquad \lambda_5 = \begin{bmatrix} 0 & 0 & 0 & 0 \\ 0 & 0 & 0 & 0 \\ 0 & 0 & 0 & \frac{1}{2} \\ 0 & 0 & -\frac{1}{2} & \end{bmatrix}$$

$$\lambda_7 = \begin{bmatrix} 0 & 0 & 0 & 0 \\ 0 & 0 & 0 & -\frac{1}{2} \\ 0 & 0 & 0 & 0 \\ 0 & \frac{1}{2} & 0 & 0 \end{bmatrix} \tag{165}$$

Among the eigenvectors of these matrices are two real eigenvectors with zero eigenvalues and two isotropic spinors .

$$\begin{aligned} &\lambda_2 : [0,1,-i,0];\ [0,1,i,0];\ [1,0,0,0];\ [0,0,0,1] \\ &\lambda_5 : [0,0,1,i];\ [0,0,1,-i];\ [1,0,0,0];\ [0,1,0,0] \\ &\lambda_7 : [0,1,0,-i];\ [0,1,0,i];\ [1,0,0,0];\ [0,0,1,0] \end{aligned} \tag{166}$$

The eigenvalues of eigenvectors with zero norm are +½, –½ . The real eigenvectors have eigenvalues zero.

It is surprising that the $Cl_{3,1}$ contains a subspace which provides isotropic spinors just like the isospin subalgebra, namely the timespace. The timespace is spanned by

$$\mathcal{U} = span_{\mathcal{F}} \{e_4, j, e_{123}\} \text{ with } \mathcal{F} \text{ equal to } \mathbb{R} \text{ or } \mathbb{C} \tag{167}$$

Base units e4, j, e123 are in 1-1 correspondence with base quaternions *i*, *j*, *k*. The Dirac like operator is given by

$$D \overset{def}{=} \partial_\tau + e_4 \partial_t + j \partial_u + e_{123} \partial_v \tag{168}$$

Consider the matrix representations for $\mathcal{U}$:

$$e_4 = \begin{bmatrix} 0 & -1 & 0 & 0 \\ 1 & 0 & 0 & 0 \\ 0 & 0 & 0 & -1 \\ 0 & 0 & 1 & 0 \end{bmatrix} \qquad e_{123} = \begin{bmatrix} 0 & 0 & 0 & -1 \\ 0 & 0 & -1 & 0 \\ 0 & 1 & 0 & 0 \\ 1 & 0 & 0 & 0 \end{bmatrix}$$

$$j = \begin{bmatrix} 0 & 0 & -1 & 0 \\ 0 & 0 & 0 & 1 \\ 1 & 0 & 0 & 0 \\ 0 & -1 & 0 & 0 \end{bmatrix} \tag{169}$$

Those four base units have the eigenvectors

e_4: [1, -i, 0, 0]; [0, 0, 1, -i]; [1, i, 0, 0]; [0, 0, 1, i]

e_{123}: [1, 0 , 0, -i]; [0, 1, -i, 0]; [0, 1, i, 0]; [1, 0, 0, i]

j: [0, 1, 0, i]; [0, 0, 1, -i]; [1, i, 0, 0]; [0, 1, 0, -i]

with twofold degenerate eigenvalues i and $-i$. In this desing there is the reason why the timespace essentially provides a peculiar angular momentum algebra

A timespace spinor can be written as Cartan spinor

$$\mathrm{Y} = (\xi_0^2 - \xi_1^2)e_4 + i(\xi_0^2 + \xi_1^2)e_{123} - 2\xi_0\xi_1\, j \tag{170}$$

giving square

$$\mathrm{Y}\mathrm{Y} \equiv 0 \tag{171}$$

Therefore the timespace spinor is associated with an isotropic timespace 3-multivector constituted by directed time, volume and space-time-volume. It is connected with a measure of time-volume which is so to say a time-space-time-volume without volume. This isotropic time-like 3-vector covers the time-convolution and de-convolution from time to volume and space-time-volume by the strong force.

Toolbox 8

MATRIX REPRESENTATION OF $CL_{4,1}$

Wallace (2008) investigates a symmetry breakage imposed on the orthogonal directed line elements for the algebra $sl(4, \mathbb{C})$. For this purpose he represents the algebra

$$Cl_{4,1} \cong Cl_{3,1} \oplus i\,Cl_{3,1} \tag{172}$$

by 16+16 matrices $\in$ Mat(4, $\mathbb{R}$) and $\in i$ Mat(4, $\mathbb{R}$), denoted by „ungraded" capital letters S, Z, P, E, L, F, T, Y, M, V, R, X, N, D, Q, U and iS, iZ, iP, iE, iL, iF, iT, iY, iM, iV, iR, iX, iN, iD, iQ, iU for calculations on a spreadsheet which does not accept subscripts.

$$Id = S = \begin{bmatrix} 1 & 0 & 0 & 0 \\ 0 & 1 & 0 & 0 \\ 0 & 0 & 1 & 0 \\ 0 & 0 & 0 & 1 \end{bmatrix} \qquad e_1 = X = \begin{bmatrix} 0 & 0 & -1 & 0 \\ 0 & 0 & 0 & 1 \\ -1 & 0 & 0 & 0 \\ 0 & 1 & 0 & 0 \end{bmatrix}$$

$$e_2 = Y = \begin{bmatrix} 0 & -1 & 0 & 0 \\ -1 & 0 & 0 & 0 \\ 0 & 0 & 0 & -1 \\ 0 & 0 & -1 & 0 \end{bmatrix} \qquad e_3 = Z = \begin{bmatrix} -1 & 0 & 0 & 0 \\ 0 & 1 & 0 & 0 \\ 0 & 0 & 1 & 0 \\ 0 & 0 & 0 & -1 \end{bmatrix}$$

$$e_4 = iV = \begin{bmatrix} 0 & 0 & i & 0 \\ 0 & 0 & 0 & -i \\ -i & 0 & 0 & 0 \\ 0 & i & 0 & 0 \end{bmatrix} \qquad e_5 = T = \begin{bmatrix} 0 & 1 & 0 & 0 \\ -1 & 0 & 0 & 0 \\ 0 & 0 & 0 & 1 \\ 0 & 0 & -1 & 0 \end{bmatrix}$$

$$e_{12} = N = \begin{bmatrix} 0 & 0 & 0 & 1 \\ 0 & 0 & -1 & 0 \\ 0 & 1 & 0 & 0 \\ -1 & 0 & 0 & 0 \end{bmatrix} \qquad e_{13} = -M = \begin{bmatrix} 0 & 0 & -1 & 0 \\ 0 & 0 & 0 & -1 \\ 1 & 0 & 0 & 0 \\ 0 & 1 & 0 & 0 \end{bmatrix}$$

$$e_{14} = iP = \begin{bmatrix} i & 0 & 0 & 0 \\ 0 & i & 0 & 0 \\ 0 & 0 & -i & 0 \\ 0 & 0 & 0 & -i \end{bmatrix} \quad e_{15} = -D = \begin{bmatrix} 0 & 0 & 0 & -1 \\ 0 & 0 & -1 & 0 \\ 0 & -1 & 0 & 0 \\ -1 & 0 & 0 & 0 \end{bmatrix}$$

$$e_{23} = L = \begin{bmatrix} 0 & -1 & 0 & 0 \\ 1 & 0 & 0 & 0 \\ 0 & 0 & 0 & 1 \\ 0 & 0 & -1 & 0 \end{bmatrix} \quad e_{24} = -iQ = \begin{bmatrix} 0 & 0 & 0 & i \\ 0 & 0 & -i & 0 \\ 0 & -i & 0 & 0 \\ i & 0 & 0 & 0 \end{bmatrix}$$

$$e_{25} = E = \begin{bmatrix} 1 & 0 & 0 & 0 \\ 0 & -1 & 0 & 0 \\ 0 & 0 & 1 & 0 \\ 0 & 0 & 0 & -1 \end{bmatrix} \quad e_{34} = iR = \begin{bmatrix} 0 & 0 & -i & 0 \\ 0 & 0 & 0 & -i \\ -i & 0 & 0 & 0 \\ 0 & -i & 0 & 0 \end{bmatrix}$$

$$e_{35} = -F = \begin{bmatrix} 0 & -1 & 0 & 0 \\ -1 & 0 & 0 & 0 \\ 0 & 0 & 0 & 1 \\ 0 & 0 & 1 & 0 \end{bmatrix} \quad e_{45} = iU = \begin{bmatrix} 0 & 0 & 0 & i \\ 0 & 0 & i & 0 \\ 0 & -i & 0 & 0 \\ -i & 0 & 0 & 0 \end{bmatrix}$$

$$e_{123} = -U = \begin{bmatrix} 0 & 0 & 0 & -1 \\ 0 & 0 & -1 & 0 \\ 0 & 1 & 0 & 0 \\ 1 & 0 & 0 & 0 \end{bmatrix} \quad e_{124} = iF = \begin{bmatrix} 0 & i & 0 & 0 \\ i & 0 & 0 & 0 \\ 0 & 0 & 0 & -i \\ 0 & 0 & -i & 0 \end{bmatrix}$$

$$e_{125} = R = \begin{bmatrix} 0 & 0 & -1 & 0 \\ 0 & 0 & 0 & -1 \\ -1 & 0 & 0 & 0 \\ 0 & -1 & 0 & 0 \end{bmatrix} \quad e_{134} = iE = \begin{bmatrix} i & 0 & 0 & 0 \\ 0 & -i & 0 & 0 \\ 0 & 0 & i & 0 \\ 0 & 0 & 0 & -i \end{bmatrix}$$

$$e_{135} = Q = \begin{bmatrix} 0 & 0 & 0 & -1 \\ 0 & 0 & 1 & 0 \\ 0 & 1 & 0 & 0 \\ -1 & 0 & 0 & 0 \end{bmatrix} \quad e_{145} = -iL = \begin{bmatrix} 0 & i & 0 & 0 \\ -i & 0 & 0 & 0 \\ 0 & 0 & 0 & -i \\ 0 & 0 & i & 0 \end{bmatrix}$$

$$e_{234} = iD = \begin{bmatrix} 0 & 0 & 0 & i \\ 0 & 0 & i & 0 \\ 0 & i & 0 & 0 \\ i & 0 & 0 & 0 \end{bmatrix} \quad e_{235} = P = \begin{bmatrix} 1 & 0 & 0 & 0 \\ 0 & 1 & 0 & 0 \\ 0 & 0 & -1 & 0 \\ 0 & 0 & & -1 \end{bmatrix}$$

$$e_{245} = -iM = \begin{bmatrix} 0 & 0 & -i & 0 \\ 0 & 0 & 0 & -i \\ i & 0 & 0 & 0 \\ 0 & i & 0 & 0 \end{bmatrix} \quad e_{345} = -iN = \begin{bmatrix} 0 & 0 & 0 & -i \\ 0 & 0 & i & 0 \\ 0 & -i & 0 & 0 \\ i & 0 & 0 & 0 \end{bmatrix}$$

$$e_{1234} = -V = \begin{bmatrix} 0 & 0 & -1 & 0 \\ 0 & 0 & 0 & 1 \\ 1 & 0 & 0 & 0 \\ 0 & -1 & 0 & 0 \end{bmatrix} \quad e_{1235} = -iZ = \begin{bmatrix} i & 0 & 0 & 0 \\ 0 & -i & 0 & 0 \\ 0 & 0 & -i & 0 \\ 0 & 0 & 0 & i \end{bmatrix}$$

$$e_{1345} = -iY = \begin{bmatrix} 0 & i & 0 & 0 \\ i & 0 & 0 & 0 \\ 0 & 0 & 0 & i \\ 0 & 0 & i & 0 \end{bmatrix} \quad e_{2345} = -iX = \begin{bmatrix} 0 & 0 & i & 0 \\ 0 & 0 & 0 & -i \\ i & 0 & 0 & 0 \\ 0 & -i & 0 & 0 \end{bmatrix}$$

$$e_{12345} = iId = \begin{bmatrix} i & 0 & 0 & 0 \\ 0 & i & 0 & 0 \\ 0 & 0 & i & 0 \\ 0 & 0 & 0 & i \end{bmatrix} \qquad (173)$$

We are using those matrices in our final chapters on topological evolution of subnuclear particles.

Toolbox 9

TOPOLOGICAL EVOLUTION

THE PFAFF SEQUENCE

The reader is strongly advised to read volume 1 on *Non-Equilibrium Thermodynamics* in Kiehns »*Non-Equilibrium Systems and Irreversible Processes*«. Probably this provides, at present, the only significant foundation for someone who wants to derive geometry from topology rather than construct topology within the set frameworks of geometry. As for myself, I have based my thoughts on the experience that the interval between geometry and topology is not as big as is often assumed. Especially in the context of geometric algebra the two are absolutely compatible. However, from a physics viewpoint, that is, from a historic and material perspective, it becomes apparent, the basis of geometry is a vivid topological affair. We can say that geometry with its orthogonalities and idempotents, minimal ideals, orthogonal groups and so forth is originally brought forth in a chaotic topological process which involves both matter and cognitive evolution. The evolution of thought with its logic and geometry is so to say the consolidated outcome of a topological affair that is initially free of all those orderly properties which metric spaces and their algebras do have. This is the reason why, by the time, I began to revert the sequence that led from metric spaces to topology. In this toolbox section I will explain only the main instruments we have been using especially in the last eight chapters, namely, the Pfaff sequence, Cartans topology, frames and 'magic formula'.

If the number of independent variables in the base variety of some geometric algebra is 2^n, as is the case in the Clifford algebra, then the maximum Pfaff dimension would also be 2^n. However, in practise it is smaller. In the ground space $\mathbb{R}^4$ – which, even in the Clifford algebra applications concerns a great many cases – the maximum Pfaff topological dimension is 4. The definition is constructed as follows. Consider a base variety {x, y, z, t} and

$$A = A_k(x,y,z,t)\,dx^k \quad \text{a 1-form of action defined on } \{x,y,z,t\} \qquad (174)$$

We are asking what is the irreducible minimum number of independent functions $\theta(x, y, z, t)$ required to describe the topological properties on a contractible domain that can be generated by the 1-form of action A. This number is defined to be the "*Pfaff Topological Dimension*" (PTD) of the 1-form A. If for example

$$A = A_k dx^k \Rightarrow d\theta(x, y, z, t)_{irreducible} \quad \text{such that} \qquad (175)$$
$$A_k = \partial\theta(x, y, z, t)/\partial x^k$$

then only one function $\theta(x, y, z, t)$ is required to describe the action, not four. Although the geometric dimension of the variety is four, the irreducible Pfaff topological dimension in this case is one. In this sense, the Pfaff topological dimension defines the existence of a smaller domain of topological base variables that can be seen as submersions from the original coordinates. Differential forms are structurally well defined and can be constructed by the submersive mapping. Relative to the Cartan topology, the PTD can be generated by each of the Pfaffian forms of the specific application. For a given 1-form A of Action the Pfaff sequence has the form

$$\text{Pfaff sequence} \stackrel{def}{=} \{A, dA, A\wedge dA, dA\wedge dA\} \qquad (176)$$

The PTD is equal to the number of non-zero terms in the sequence. For example, if the Pfaff sequence in a region $U \subset \{x, y, z, t\}$ for a given 1- form A is $\{A, dA, 0\ 0\}$, then the PTD in that region U is 2. The 1-form A admits in region U a description in terms of only 2 but not less than 2 independent variables, say $\{u^1, u^2\}$. For a differentiable map $\varphi : \{x, y, z, t\} \rightarrow \{u^1, u^2\}$, the exterior differential 1-form defined on the target variety U of two dimensions is

$$A(u^1, u^2) = A_1(u^1, u^2)du^1 + A_2(u^1, u^2)du^2 \qquad (177)$$

It has a functionally well-defined pre-image $A(x, y, z, t)$ on the original base variety of four dimensions. Kiehn (2008a, p. 35) denotes the two dimensions $\{u^1, u^2\}$ as *pre-geometry dimensions*. This pre-image is obtained by functional substitution of u^1, u^2, $d\,u^1$, du^2 in terms of $\{x, y, z, t\}$ as defined by the mapping φ. The process of functional substitution is called the pullback. Kiehn put it in the form

$$A(x, y, z, t) = A_k(x)dx^k \Leftarrow \varphi^*(A(u^1, u^2)) \Leftarrow \varphi^*(A_\sigma du^\sigma) \qquad (178)$$

indicating that the index σ may vary between 1 and 4 depending on discipline, application and region U. It may be that the PTD is globally equal 2 in almost the whole domain U with a few exceptions where it may take other values. These subregions where PDT is equal to 3 or 4 are said to represent topological defects. Notice, a process of topological evolution may begin with Pfaff Topological Dimension σ and end with a different PTD, say σ'. The concept of the Pfaff Topological Dimension can be used to define equivalence classes of processes. The denotation of PTD replaces the old term “class”. In cases where the 3-form $A\wedge dA$ is not zero, the PTD is at least 3. We can define it as a Topo-‘logic prädicate’. Let $U \subset {}^4\mathbb{R} = \{x, y, z, t\}$. A topological system is defined by the existence of a 1-form of action

$$top\ sys\ (U) \overset{def}{\leftrightarrow} \exists_A A = A_k(x,y,z,t)dx^k \ \&\ (x,y,z,t) \in U \tag{179}$$

$$top\ torsion\ (U) \overset{def}{\leftrightarrow} A\wedge dA \neq 0 \tag{180}$$

$$top\ parity\ (U) \overset{def}{\leftrightarrow} dA\wedge dA \neq 0 \tag{181}$$

We derive by observation of systems that

$$top\ torsion\ (U) \vee top\ parity\ (U) \xrightarrow{logic} \neg equilib(S)$$

systems with topological torsion or parity are not in equilibrium; or equivalently

$$equilib(S) \xrightarrow{logic} \neg top\ parity(U)\ \&\ \neg top\ torsion(U) \tag{182}$$

This signifies what I meant by "logic and topologic" of thermodynamic processes. This *predicate logic* of thermodynamics is actually more extensive than I can show here. For instance non zero values for topological torsion imply that the Frobenius unique integrability theorem for the Pfaffian equation $A = 0$ fails. The concept of topological parity $F\wedge F = dA\wedge dA$ has its roots in historic Pfaff's problem. A recognizable 4-dimensional formulation has appeared in Forsyth (1890 – 1959, vol 1, p. 100).

The eigendirection-fields of the antisymmetric matrix of 2-form coefficients, dA, are complex isotropic direction fields of zero length, but with non-zero complex eigenvalues, if the Pfaff Topological Dimension is even. If the PTD is odd then all but one are isotropic eigendirections. These isotropic eigendirection fields are Cartan spinors. Their exterior product gives a finite area, but the inner product with itself gives zero. In this sense Cartan spinors have zero length, but finite area. Kiehn remarks (p. 37, remark 9) Spinors are related to the Pfaff Topological Dimension, and are natural (and to be expected) occurrences when the 1-form of work induced by a thermodynamic process is not zero.

In the fifth volume »Topological Torsion and Macroscopic Spinors« Kiehn gives the essential names and predicates to the various elements of the Pfaff sequence (page 98):

Topological p-form name	*PS element*	*Nulls*	*PTD*	*Thermodynamic system*
Action	*A*	*dA = 0*	*1*	*Equilibrium*
Vorticity	*dA*	*A^dA = 0*	*2*	*Isolated*
Torsion	*A^dA*	*dA^dA = 0*	*3*	*Closed*
Parity	*dA^dA*	-	*4*	*Open*

Ślebodzinski – Cartan – Lie Differential

The zero grade directional Lie differential L_T along a 1-vector field V was invented 1931 by Ślebodziński. The name 'Lie derivation' was introduced later by van Danzig, a collaborator of Schouten. The Ślebodziński formula, also termed Cartan identity, has been called Cartan's 'magic formula' by Marsden and Ratiu (1994, p. 122). The Ślebodziński-derivation was implicit in Cartan (1922). The impact of a Lie differential on differential forms Σ is a decomposition into a transversal and an exact part

$$\mathsf{L}_T \Sigma = i(V) d\Sigma + d\, i(V)\Sigma \tag{183}$$

Kiehn recently demanded exlicitely to replace the word 'derivation' by 'differential'. To calculate with Cartan's formula we must understand both exterior and inner product with respect to a vector space. The exterior algebra of Cartan is based on an associative, non-commutative exterior product, the so called wedge product ^ of exterior differential forms. We begin with 1-forms on some n-dimensional vector space $V^1(\sigma^j)$ with a basis of 1-forms $\{\sigma^j\}$. A general 1-form is given by

$$\omega^1 = A_k(y)\sigma^k \qquad \text{differential 1-form} \tag{184}$$

with a vector space addition rule

$$\omega_{(1)} = A_k(y)\sigma^k \;\&\; \omega_{(2)} = B_k(y)\sigma^k \rightarrow \omega_{(1)} + \omega_{(2)} = (A_k(y) + B_k(y))\sigma^k \tag{185}$$

Example: $\{\sigma^j\} = \{dx, dy, dz\}$

$\omega_{(1)} = 3x\,dx + 4xz\,dy \qquad \omega_{(2)} = 2y\,dy + 17y\,dz$ from which we calculate

$$\omega_{(1)} + \omega_{(2)} = 3x\,dx + (4xz + 2y)\,dy + 17y\,dz$$

The multiplication rule is the same as for the wedge product in Clifford algebra

$$\begin{aligned}
\sigma^{j\wedge}\sigma^k &= -\sigma^{k\wedge}\sigma^j \\
\sigma^{j\wedge}\sigma^j &= 0 \\
dy^{j\wedge}dy^k &= -dy^{k\wedge}dy^j \\
dy^{j\wedge}dy^j &= 0
\end{aligned} \tag{186}$$

By the exterior product we construct vector subspaces of 0-forms $Id = \omega^0$, 1-forms ω^1, 2-forms ω^2, …, n-forms ω^n. We exterior multiply monomials just like in geometric algebra. For example a 1-form, A, multiplied by a 2-form, B, gives a 3-form, C.

$$A^\wedge B=(A_x dx+A_y dy+A_z dz)^\wedge(B_x dy^\wedge dz+B_y dz^\wedge dx+B_z dx^\wedge dy)=$$
$$(A_x B_x+A_y B_y+A_z B_z)dx^\wedge dy^\wedge dz$$

Making use of rules [186) we calculated the exterior product of two 1-forms

$$A^\wedge B=(A_x dx+A_y dy+A_z dz)^\wedge(B_x dx+B_y dy+B_z dz)=$$
$$(A_x B_y-A_y B_x)dx^\wedge dy+(A_y B_z-A_z B_y)dy^\wedge dz+(A_z B_x-A_x B_z)dz^\wedge dx$$

This expansion has coefficients equivalent to the Gibbs cross product. Kiehn pointed out that Sommerfeld used it more than 60 years ago when he investigated electromagnetic systems similar as we have shown in the chapter on *topological evolution of particles*.

The exterior differential shifts the p in the differential p-form:

$$d(\omega^p)\to\omega^{p+1} \tag{187}$$

with the modified Leibnitz rule

$$d(A^\wedge B)=dA^\wedge B-A^\wedge dB$$
$$d(d(\omega^p))\to 0 \tag{188}$$

The product between a p- and a q-form gives

$$d(\omega^{p\wedge}\omega^q)=d\omega^{p\wedge}\omega^q --^p\omega^{p\wedge}d\omega^q \tag{189}$$

The exterior differential has the property of the total differential which takes a scalar function or 0-form $\omega^0=\theta(y^k)$ to the 1-form, $\omega^1=A_j\,dy^j$,

$$d(\omega^0)=d(\theta(y^k))\to\partial\theta(y^k)/\partial y^j)dy^j=A_j dy^j=\omega^1 \tag{190}$$

The function can be constrained such that the set $\theta(y^j)$ = const defines an implicit surface. In that case the total differential also turns out zero $d\{\theta(y^j)\}=0$. An object whose exterior differential is null is said to be closed – not in an algebraic sense but with respect to the exterior differential. For reasons of logic consistency the differential base elements symbolized by dy^j are closed by definition. Thus we must have $d(dy^j)=0$. The exterior differentials of an arbitrary basis, σ^k are not necessarily zero, but in any case $d(\sigma^k)$ is closed as we must have $dd(\sigma^k)=0$.

The exterior differential of a 1-form is defined by

$$d\omega^1=d(A_j\,dy^j)=(dA_j)^\wedge dy^j+A_j d(dy^j)=$$

$$= \left(\frac{\partial A_j}{\partial y^k} dy^k \right) \wedge dy^j + 0 = (\partial A_j / \partial y^k - \partial A_k / \partial y^j) dy^k \wedge dy^j =$$
$$= F_{[k\,j]} dy^{[k\,j]} = F_H dy^H \tag{191}$$

We used the plain collective index notation and generalize the exterior differentiation

$$d\omega^p = d(A_H\, dy^H) = (dA_H) \wedge dy^H \tag{192}$$

Note that the special 1-form with gradient coefficients $\omega^1 = d\{\theta(y^j)\}$, possesses the exterior differential

$$d\omega = d(d\{\theta(y^j)\}) = \ldots \{\partial^2\theta(y^j)/\partial y^k \partial y^l\} dy^l \wedge dy^k + \ldots + \{\partial^2\theta(y^j)/\partial y^l\, \partial y^k\} dy^k \wedge dy^l =$$
$$= \ldots \{\partial^2\theta(y^j)/\partial y^k \partial y^l - \partial^2\theta(y^j/\partial y^l\, \partial y^k\} dy^l dy^k = 0 \tag{193}$$

for C2-functions θ. Given the assumption that the coefficient functions are twice differentiable, this peculiar 1-form is a closed 1-form. However, as the $\omega^1 = d\{\theta(y^j)\}$ is constructed by a unique primitive function, $\theta(y^j)$, whose exterior differential brings forth $\omega^1 = d\theta$, the 1-form ω is not only said to be closed. But it is also exact. We have used this in our chapter on conformal time-space and time-space expansion shears.

The same predicate holds for all p-forms. A p-form is closed if the exterior differential vanishes. Further, the p-form is exact if it can be constructed by the exterior differential operation applied to a (p-1)-form. There are forms that are closed but not exact, and there are such which are neither closed nor exact. The importance of these predicates is that they imply topological statements about the domain of definition. In a 2-dimensional surface every hole is associated with a unique 1-form that is closed and not exact. The number of closed but not exact 1-forms on a domain counts the number of holes. Kiehn (2008a, p. 425) points out that *this fact is the basis of the Bohm-Aharonov idea in EM theory, and is at the foundation of the theory of flight in terms of the Joukowski transformation*. Suppose the given exterior differential p-form is expressed in terms of a non-integrable basis set σ^H. Then the exterior differential formula becomes

$$d\omega^p = d(A_H \sigma^H) = (dA_H) \wedge \sigma^H + A_H (d\sigma^H). \tag{194}$$

Notice the term $(d\sigma^H)$ is not necessarily zero. This complication arises when the frame-matrix generates 1-forms (σ^k) which are not closed. Then the basis frame is not uniquely integrable. As an example consider a function whose zero set is a 2-sphere

$$\theta(y^j) = (y^1)^2 + (y^2)^2 + (y^3)^2 - 1$$

The exterior differential of this function is

$\omega^1 = d\theta(y^j) = 2(y^1 dy^1 + y^2 dy^2 + y^3 dy^3)$ which is closed as we have

$$d\omega = 2(dy^{1\wedge} dy^1 + dy^{2\wedge} dy^2 + dy^3 dy^3) = 0$$

Kiehn gives the following example to explain the necessity for the modification of the Leibnitz rule that alternates sign for every additional odd factor. Compute the exterior differential of

$(A_x dx + A_y dy + A_z dz)^\wedge (B_x dx + B_y dy + B_z dz)$ and calculate the exterior differential $d(A^\wedge B) = dA^\wedge B - A^\wedge dB = d(C_{[mn]} dy^{[mn]}) = d(C_{[mn]})^\wedge dy^{[mn]}$

which is a 3-form with completely antisymmetric indices. The modification of the Leibnitz rule is required to make the two different ways of computing the resultant compatible.

The exterior differential of an arbitrary 1-form can be defined as

$$d(A_k(y^\alpha)\sigma^k = \{d(A_k(y^\alpha)\}^\wedge \sigma^k + A_k(y^\alpha)^\wedge \{d\sigma^k\} = \\ = \left(\frac{\partial A_k(y^\alpha)}{\partial y^\beta} dy^\beta\right)^\wedge \sigma^k + A_k(y^\alpha)^\wedge \left(d\sigma^k\right) \tag{195}$$

Consider a case where the arbitrary 1-forms σ^k are known to be linearly related to the differentials by means of the linear frame formulas

$\left|\sigma^k\right\rangle = \left[F^k_\alpha(y)\right]\left|dy^\alpha\right\rangle$ then we obtain the exterior differential of the 1-form (196)

$$d(A_k(y^\alpha)\sigma^k = \left(\frac{\partial\left(A_k F^k_\alpha\right)}{\partial y^\beta}\right) dy^{\beta\wedge} dy^k = \left(\frac{\partial \hat{A}_\alpha}{\partial y^\beta} - \frac{\partial \hat{A}_\beta}{\partial y^\alpha}\right) dy^{\beta\wedge} dy^k = \\ = \hat{F}_{[\alpha\beta]}(y) dy^{\beta\wedge} dy^k \quad \text{with} \quad \hat{A}_\alpha(y) = A_k(y) F^k_\alpha(y) \tag{197}$$

This formula is valid on the initial variety $\{y^\alpha, dy^\alpha\}$, whether the frame matrix has an inverse or not. The coefficients of the 2-form correspond to the antisymmetric components of the '*curl*' when n = 3.

The exterior differential of a 1-form where the basis frame σ^k is not integrable is determined by a summand that can be expanded in terms of the paired base elements $\sigma^{[mn]}$

$$d(A_k \sigma^k) = dA_k{}^\wedge \sigma^k + A_k \Lambda^k_{[mn]} \sigma^{[mn]} \tag{198}$$

When the basis forms are closed with respect to differential, the coefficients $\Lambda^k_{[mn]}$ vanish.

The vector space $\Lambda^1\{\sigma^k\}$ has dimension n. So the n basis 1-forms must be linearly independent. They are constructed from a frame matrix according to the equation

$$\left|\sigma^k\right\rangle = \left[F^k_\alpha(y)\right]\left|dy^\alpha\right\rangle \tag{199}$$

The n-fold product is a director based on the non-zero determinant of the frame matrix

$$\omega^n = \sigma^{1\wedge}\sigma^{2\wedge}...^\wedge\sigma^n = \det[F]dy^{1\wedge}dy^{2\wedge}...^\wedge dy^n \neq 0 \tag{200}$$

which is either positive or negative. To obtain the Ślebodziński –Cartan-Lie derivation we need the interior product which lowers the degree of a p-form by unity. The interior product requires a direction vector field *V*. The interior multiplication is denoted by the somewhat lengthy symbol $i(V)$. The interior product of a vector field and an exact base element equal to the differential of a coordinate dy^α is defined to be equal to the αth component of *V*. Definitions are written

$$i(V)\theta(y^\alpha) = 0 \qquad \text{and} \qquad i(V)dy^\alpha = V^\alpha \tag{201}$$

The inner product acting on a 1-form in 3-space gives

$$i(V)A = i(V)(A_x dx + A_y dy + A_z dz) = A_x V^x + A_y V^y + A_z V^z \tag{202}$$

An additional rule for an inner product with respect to V of an outer product of two 1-forms is demanded to care for the antisymmetries of differential forms.

$$i(V)(A^\wedge B) = (i(V)A)^\wedge B) - A^\wedge(i(V)B) \text{ and}$$
$$i(V)i(V)A = 0 \tag{203}$$

This is analogous to the modified Leibniz rule for outer differentials. Use those rules to elaborate higher p-forms

$$i(V)i(V)\omega^p = 0, \qquad i(V)i(W)\omega^p \neq i(W)i(V)\omega^p \tag{204}$$

The Lie Differential

With respect to a vector field generates a new p-form ϑ^p from a p-form ω^p. It is constructed from the raising- and lowering operators *d* and $i(V)$. The formula is

$$\omega^p \mapsto \vartheta^p: \qquad \mathrm{L}_{(V)}\omega^p = i(V)d\omega^p + d(i(V)\omega^p) \qquad (205)$$

Marsden and Kiehn use to call this Cartan's magic formula. If we define the 0-form of internal energy as $U = i(V)\,A$, the 1-form of work as $W = i(V)\,dA$ and the resulting 1-form ϑ as Q, Cartan's magic formula becomes

$$\mathrm{L}_{(V)}A = i(V)dA + d(i(V)A) = W + dU = Q. \qquad (206)$$

This beautiful equation discovered by Kiehn can be recognized as a non-stochastic, dynamic form of the first law of thermodynamics applied to a physical system with action 1-form A.

LITERATURE

Ablamowicz, R.; Fauser, B. Hecke algebra representations by q-Young idempotents; In *Clifford algebras and their applications in Mathematical Physics*; Ablamowicz, R.; Fauser, B.; Eds.; Birkhäuser: Boston, 2000.

Ablamowicz, R.; Ed. *Clifford Algebras – Applications to Mathematics, Physics, and Engineering*; Birkhäuser: Boston, 2004.

Acevedo, M.; López-Bonilla, J.; Sánchez-Meraz, M. *Apeiron*. 2005, *12*, No. 4.

Alpher, R. A.; Bethe, H.; Gamow. G. *Phys Rev*. 1948, *73*, 803.

Ashby, W.R. *An Introduction to Cybernetics*; New York, London, 1956.

Assagioli, R. *Psychosynthesis. A Manual of Principles and Techniques*; Hobbs, Dormann and Company: New York, 1965.

Baker, A. *Matrix Groups: An Introduction to Lie Group Theory*; New York, 2002.

Barendregt, H.P. *The Lambda Calculus – Its Syntax and Semantics*; Amsterdam, 1985.

Bargmann, V.; Wigner, E. P. *Proceedings of the National Academy of Sciences*. 1948 (USA) *34*, 211.

Betz, O. *Licht vom unerschaffnen Lichte. Die kabbalistische Lehrtafel der Prinzessin Antonia in Bad Teinach;* 2. edition; Sternberg Verlag: Metzingen/Württ., 2000.

Bohm, D.; Hiley B. *The undivided Universe*; Routledge: New York, 1993.

Bohm, D. P*hys Rev*. 1952, *85*, 166-179 (part 1); 180-193 (part 2).

Bondi, H.; Gold, T. *Monthly Notices of the Roy Astronom Soc*. 1949, *108*, 3.

deBroglie, L.-V. *Compt Rend Acad Sci Paris*. 1932, *195*, 536, 862.

deBroglie, L.-V. *Une Nouvelle Conception de la Lumiére*; Paris, 1934, 1940.

deBroglie, L.-V. La structure atomique de la matiere et du rayonnement et la mechanique ondulatoire (1927); reprinted in *La physique quantique restera-t-elle indéterministe*; Gauthier Villars: Paris, 1953.

Browand, H. K. *Physica D* 1986, *118*, 173.

Bruns, E. H. *Das Eikonal*; Leipzig, 1895.

Evagrios Pontikos. *Briefe aus der Wüste*; Bunge, B., Ed.; Paulinus: Trier, 1986.

Cartan, E. *Les systèmes différentielles extérieurs et leurs applications géometriques*; Hermann: Paris, 1922.

Cartan, E. *La théorie des groupes finis et continus et l'analysis situs*, Memorial des Sciences Mathématiques 42 ; Paris, 1930 (Euvres Complètes, vol. I) pp 1165 – 1225.

Carter, B. *Phil Trans R Soc Lond*. 1983, *A 310*, 347–363.

Casimir, H.G. B. *Proc Con Ned Akad van Wetensch.* 1948, *B51* (7), 793-796.

J. Chen, J. Li. *Differential Geom* 2000, *55*, 355-384.

Chevalley, C. The Algebraic Theory of Spinors; New York, 1954.

Chisholm, J. S. R. Unified Spin Gauge Theories for the Four Fundamental Forces; In *Clifford Algebras and their Applications in Mathematical Physics*; Micali, A. et al; Eds.; Kluwer: Dortrecht, 1992; pp 363-370.

Church, A. The Calculi of Lambda Conversion; *Annals of Math Studies 6*; Princeton, 1941; 77 p.

Church, A. *Am J Math.* 1936, *58*, 345-363.

Crasmareanu, M. C.; Oziewicz, Z. *Divergence operator is a derivation of the Schouten Nijenhuis algebra*; 2002; private paper.

Cruz Guzmán, J. de J., Oziewicz, Z. *Föhlicher-Nijenhuis Algebra and four Maxwell's Equations for Non-Inertial Observer.* Bulletin de la Société de Sciences et des Lettres de Lódź, *LIII;* Série : Recherches sur les Déformations, *XXXIX*, 2003, pp 107-140.

Curry, H.B. *Combinatory logic*; North-Holland: Amsterdam, 1958, vol 1.

de Solla Price, D. J. *Little Science, Big Science;* Von der Studierstube zur Großforschung; Frankfurt, 1974.

Derwas, J. C. *The desert a city.* An introduction to the study of Egyptian and Palestinian monasticism under the Christian Empire; London, 1977.

Dodel, F. *Das Sitzen der Wüstenväter.* Eine Untersuchung anhand der Apophthegmata Patrum; Fribourg, CH, 1997.

Dvoeglazov, V. V. *Fizika.* 1997, *B 6* (2), 75-80.

Dvoeglazov, V. V. Majorana-like models in the physics of neutral particles, In: *The Theory of the Electron*; Keller, J.; Oziewicz, Z.; Eds.; Mexico, 1997; pp 303-319.

Dvoeglazov, V.V. *Speculations in Science and Technology.* 1998, *21* (2), 111-115; hep-th/9712037

Fauser, B.; Stumpf H. Positronium as an example of algebraic composite calculations. In *The Theory of the Electron*; Keller, J.; Oziewicz, Z.; Eds.; Advances in Applied Cifford Algebra 7 (suppl); UNAM: Mexico, 1997; pp 399-418; hep-th/9510193

Fauser, B. *Journal of Physics A: Mathematical and General.* 1997, *32*, 1919-1936, q-alg/9710020.

Fauser, B.; Oziewicz, Z. *Miscellanea Algebraicae.* In series: Studia i Materialy. 2001, Rok **5** (2), 31-42. Akademia Świętokrzyska, Wydział Zarządzania i Administracji, Kielce, Poland.

Fauser, B. Grade Free Product Formulae from Grassmann-Hopf Gebras. In *Clifford Algebras – Applications to Mathematics, Physics and Engineering*; Ablamowicz, R.; Ed.; Birkhäuser: Boston, 2004; pp 351-372.

Faustmann, C.; Neufeld H.; Thirring, W. *When time emerges.* Institut für Theoretische Physik, Universität Wien: Vienna, Austria, draft dated 1st of December 2006, personal copy.

Feoli, A.; Rampone, S. Is the Strong Anthropic Principle too Weak. *Nuovo Cim.* 1999, *B114*, 281-289.

Fermi, E.; J. Pasta; S. Ulam. *Studies of Nonlinear Problems.* Document LA-1940; May 1955.

Finkelstein, D. *Elementary operation in physics.* Presented at 16th Int Conf on Clifford Algebras in Cookville, May 2002.

Foerster, H. von. Cybernetics of Cybernetics; In *Communication and Control*; Krippendorf, K.; Ed.; New York, 1979.

Foerster, H. von. *Understanding Understanding*: Essays on Cybernetics and Cognition. New York, 2003.

Forsyth, A. R. *Theory of differential equations*, vol 1 and vol 2; Dover: New York, 1959.

Frey, G. *Gesetz und Entwicklung in der Natur*; Meiner: Hamburg, 1958.

Frey, G. *Anthropologie der Künste*; Alber: Freiburg-München, 1994.

Frey, G. *Erkenntnis der Wirklichkeit*. Philosophische Folgerungen der modernen Naturwissenschaften; Kohlhammer: Stuttgart, 1965.

Fueter, R. *Comment Math Helv*. 1935, *7*, 307-330.

Gendün, Lama Rinpoche. *Der Grosse Pfau*. Die Umwandlung der Emotionen im Tibetischen Buddhismus; Zürich-München-Berlin, 1995.

Glanville, R.; Riegler A. (eds.); *Gordon Pask – Philosopher Mechanic*, An Introduction to the Cybernetician's Cybernetician; edition echoraum: Wien, 2007.

Glanville, R.; Riegler A.; Eds.; *The Importance of Being Ernst*. Festschrift for Ernst von Glasersfeld; edtion echoraum: Wien, 2007.

Govinda, Lama Anagarika; Evans-Wentz, W. *Das tibetanische Totenbuch*; Bern-München-Wien, 1980.

Greider, K.; Weiderman, T. Generalized Clifford algebras as special cases of standard Clifford algebras. *IUCD Preprint* 16, 1988.

Greider, K. *Found Phys*. 1984, *14*, 467-506.

Hahn, A. The Clifford Algebra in the Theory of Algebras, Quadratic Forms, and Classical Groups. In: *Clifford Algebras – Applications to Mathematics, Physics and Engineering*; Ablamowicz, R.; Ed.; Birkhäuser: Boston, 2004, pp 305–322.

Hanakam, von H.; Ed.; Hilarion von Gaza, Arsenius der Große, Makarios von Ägypten, Anophrios der Große, Antonios der Grosse. Stern der Wüste; Freiburg im Breisgau, 1989.

Heidegger, M. *Gesamtausgabe*. GA 9: Wegmarken; Herrmann, F.-W. von; Ed.; 2. ed., Frankfurt am Main, 1996.

Hell, D. *Die Sprache der Seele verstehen*. Die Wüstenväter als Therapeuten; Freiburg im Breisgau, 2005.

Helmstetter, J; Lipschitz's Methods of 1886 Applied to Symplectic Clifford Algebras; In *Clifford Algebras – Applications to Mathematics, Physics and Engineering*; Ablamowicz, R.; Ed.; Birkhäuser: Boston, 2004, pp 323-333.

Hestenes, D. *Acta Appl. Math*. 1991, *23*, 65-93.

Hestenes, D. *Found. Phys*. 1983, 15, No. 1, 63-87.

Hestenes, D. Zitterbewegung in Radiative Processes; In *The Electron*; Hestenes, D.; Weingartshofer, A.; Eds.; KLuwer: Dordrecht, 1991; pp21-36.

Hestenes, D. *Quarterly J. Pure and Appl. Math.*, 1988, Vol 62, No 3-4.

Heussi, K. *Der Ursprung des Mönchtums*; Tübingen, 1936.

Hildegard von Bingen. (1200). *Liber Scivias*; Cod. Sal. X 16; http://gutenberg.spiegel.de /autoren/hildegar.htm

Holze, H. *Erfahrung und Theologie im frühen Mönchtum*. Untersuchungen zu einer Theologie des monastischen Lebens bei den ägyptischen Mönchsvätern, Johannes Cassian und Benedikt von Nursia; Forschungen zur Kirchen- und Dogmengeschichte 48; Göttingen, 1992.

Husserl, E. *Cartesianische Meditationen*; Husserliana 1; Springer: Dordrecht, 1950.

Husserl, E. *Vorlesungen zur Phänomenologie des inneren Zeitbewußtseins*. Jahrbuch für Philosophie und phänomenologische Forschung 9; Halle/Saale, 1928.

Jantsch, E. *Technological Planning and Social Futures*; London, 1972.

Jendrzejzyk, J. *Mystik und Meditation am Beispiel der Mechthild von Magdeburg*; Stockach, 1992.

Jordan, P. *Die Herkunft der Sterne*; Stuttgart, 1947.

Jordan, P. *Z Phys*. 1926, *45*, 765.

Jordan, P. et al. *Z Phys*. 1928, *47*, 151.

Kanitscheider, B. *J Gen Philos Scie*. 2004, 35, 1-12.

Kauffman, L.H. *Kybernetes*. 2005, *34,* No. 1/2, 129-150.

Kähler, E. *Innerer und äußerer Differentialkalkül*. Abh Dtsch Akad Wiss Berlin, Kl für Math Phys Tech. 1960, *4,* pp 1-32.

Kähler, E. *Die Dirac-Gleichung*. Abh Dtsch Akad Wiss Berlin, Kl für Math Phys Tech. 1961, *1*, pp 1-38.

Kern, U. Eckhart, Meister. In: *Theologische Realenzyklopädie*; Bd 9, de Gruyter: Berlin, 1982, pp 258–264.

Kiehn, R. M. Instability Patterns, Wakes and Topological Limit Sets; In *Eddy Structure Identification in Free Turbulent Shear Flows*, J. P. Bonnet, J. P.; Glauser, M. N.; Eds.; Kluwer: Dordrecht, 1993, 363 p.

Kiehn, R. M. *Plasmas and Non-equilibrium Electrodynamics, Non-equilibrium Systems and Irreversible Processes*; Lulu Enterprises: Morrisville, NC, 2004; vol. 4, http://www.lulu.com/kiehn

Kiehn, R. M. *Non-equilibrium Systems and Irreversible Processes - Adventures in Applied Topology, Non-Equilibrium Thermodynamics*; Lulu Enterprises: Morrisville, NC, 2004; vol 1. [private copy from author, Dec. 2008a]

Kiehn, R. M. *Non-Equilibrium Systems and Irreversible Processes – Adventures in Applied Topology, Falaco Solitons, Cosmology, and the Arrow of Time*; Lulu Enterprises: Morrisville, NC, 2004; vol 2. [private copy from author,Oct. 2008b]

Kiehn, R. M. *Topological Torsion and Macroscopic Spinors, Non-equilibrium Systems and Irreversible Processes - Adventures in Applied Topology*; Lulu Enterprises: Morrisville, NC, 2010; vol 5. [private copy from author, Dec. 2008c]

Knapp, A. W. *Lie Groups, Lie Algebras, and Cohomology*; Princeton, 1988.

Knapp, A. W. *Lie Groups beyond an Introduction*; Boston, 2002.

Köhler, W. Closed and Open Systems; In *Systems Thinking*; Emery, F. E.; Ed.; Penguin; Harmondsworth, 1969, pp 65–69.

Kopczyński, W. *J Math Phys*. 1989, *30* (2), 243-248.

Kopczyński, W. *Zb MATH* 2006, Review 0673.15014.

Kraft, V. Das Universalienproblem; In *Sprache und Erkenntnis. Festschrift für Gerhard Frey zum 60. Geburtstag*; Innsbrucker Beiträge zur Kulturwissenschaft 19; Kanitscheider, B.; Ed.; Innsbruck,1976, pp 33-36.

Krishnamurti, J. *Das Notizbuch*; Frankfurt am Main, 1976.

Krüger, R. *Ein Versuch über die Archäologie der Globalisierung: die Kugelgestalt der Erde und die globale Konzeption des Erdraums im Mittelalter*; Stuttgart, 2007.

Laing, R.D. *Die Stimme der Erfahrung, Wissenschaft und Psychiatrie*; Köln, 1983.

Lamoreaux, S.K. *Phys Rev Letters*. 1997, *78* (1), 5-8.

Lanczos, C. *Z Phys*. 1929, *57*, 447–473.

Liber vitae meritorum (1148-1163)http://www.medieval.org/emfaq/composers/hildegard.html (englische Diskographie)

Lichnerowicz, A. *Symp. Math. III.* 1970, Bologna, 341-402.

Lipschitz, R. *Untersuchungen über die Summe von Quadraten*; Bonn, 1886.

Lounesto, P. *Clifford algebras and Spinors*; Cambridge, 2001.

Mac Lane, S. *Categories for the Working Mathematician*; New York, 1997.

Maclay, J.; Hammer, J.; George, M.A.; Ilic, R.; Leonard, Qu.; Clark, R. *Measurement of Repulsive Quantum Vacuum Forces*, AIAA/ASME/SAE/ASEE, 37th Joint Propulsion Conference, Salt Lake City, July 8, 2001.

Magnea, U. (2002) *An Introduction to symmetric spaces*. Department of Mathematics, University of Torino, Italy, arXiv:cond-mat/0205288 v1.

Majorana, E. *Nuovo Cimento*. 1937, 14, 171.

Marsden, J. E.; Riatu, D. S. *Introduction to Mechanics and Symmetry*; Springer: Berlin, 1994.

Mechthild von Magdeburg. *Ein vliessende lieht miner gotheit. Offenbarungen der Schwester Mechthild von Magdeburg oder Das fließende Licht der Gottheit*; Unveränd. reprografischer Nachdruck der Ausgabe, aus der einzigen Handschrift des Stiftes Einsiedeln. Wiss Buchgesellschaft: Darmstadt 1989.

Mechthild von Magdeburg. *Das fließende Licht der Gottheit*; Gisela Vollmann-Profe; Ed.; Deutscher Klassiker Verlag: Frankfurt, 2003.

Měska, J. *Czech Math J.* 1984, *34*, 109.

Mertel, von H.; Ed.; *Leben des Heiligen Antonius*; Bibliothek der Kirchenväter 13. München 1917.

Mittelstaedt, P. *Philosophische Probleme der Modernen Physik*. BI-Hochschultaschenbücher. Bibliographisches Institut: Mannheim, 1963.

Morales-Luna, G. (2007) Basic Calculations on Clifford Algebras. arXiv:math/0702896v1 [math.AG]

Müller, A.; Müller K.H. *An Unfinished Revolution*? Heinz von Foerster and the Biological Computer Laboratory | BCL 1958-1976; edition echoraum: Wien, 2007.

Müller, K. H. *A Period of High Trans-Disciplinarity*, 1948-1958, In: An Unfinished Revolution? Müller, A.; Müller, K.H.; Eds.; edition echoraum: Wien, 2007; p. 228ff.

Narnhofer, H.; Thirring, W. *Why Schrödinger's Cat is Most likely to be Either Alive or Dead*; Preprint ESI 343, Vienna, 1996.

Neumann, J.v. *Mathematische Grundlagen der Quantenmechanik*; Berlin, 1932.

Nōno, K. *Bull. Fukuoka Univ. Ed. III.* 1985, *35*, 11-17.

Norretranders, T. *The User Illusion*. Penguin Books: Harmondsworth, 1998.

Nottale, L. Scale relativity, fractal space-time and quantum mechanics; In *Quantum Mechanics, Diffusion and Chaotic Fractals*; El Naschi, M.S.; Rossler, O.E.; Prigogine, I. et al.; Eds.; Oxford, 1995; pp 51-78.

Nyanatiloka Mahathera; *The noble truth of the extinction of suffering*; compiled, translated and explained from the Pali Canon; Colombo, 1952.

Nyanatiloka Mahathera; *Das Wort des Buddha*. Konstanz, 1989.

Pezzaglia Jr., W. M. Dimensionally democratic calculus and principles of polydimensional physics; In *Clifford Algebras and their Application in Mathematical Physics*; Eds.; Ablamowicz, R.; Fauser, B.; Birkhäuser: Boston, 2000, pp 101-124.

Perotti, A. *Holomorphic Functions and Regular Quaternionic Functions on the Hyperkähler Space* $\mathbb{H}$; Reg Isaac Procs, October 4, 2005.
Planck, M. *Zur Theorie des Gesetzes der Energieverteilung im Normalspektrum*, Verhandlungen der deutschen physikalischen Gesellschaft 2; Berlin, 1900; Nr. 17, pp 237 –245. (Vortrag am 14. 12. 1900)
Porete, M. *Der Spiegel der einfachen Seelen: Wege der Frauenmystik*. Aus dem Altfranzösischen übertragen und mit einem Nachwort und Anmerkungen versehen von Louise Gnädinger; Zürich, 1987.
Osserman, R. *A Survey of Minimal Surfaces*; Cambridge, 1986.
Oziewicz, Z. Clifford algebras of multivectors: The minimal polynomials of the tensor product Dirac matrices and the opposite Clifford algebra. In: *The theory of the electron*; Keller, J.; Ed.; Universidad Nacional Autónoma de México: México, 1997; pp 467-486.
Oziewicz, Z. *Clifford coalgebra, and linear energy-momenta addition*, private draft: October 10, 2001, After the Conference in Toulouse 2005, few references added.
Oziewicz, Z. Guest Editor's Note: Clifford algebras and their applications, *Int J Theor Phys.* 2001, *40* (1), 1-13.
Oziewicz, Z. *Contemp Math.* 2003, *318*, 175-197.
Oziewicz, Z., *Letter 05 to Bernd Schmeikal*, November 13, 2005.
Oziewicz, Z. *Int J Geometric Methods Mod Phys.* 2007, *4*, (1) math. CT / 0608770
Pierce, R. *Associative Algebras*; Springer: New York, 1982.
Reichenbach, H. Axiomatik der relativistischen Raum-Zeit-Lehre, Nachdruck; Vieweg: Braunschweig, 1965.
Reichenbach, H. *Die philosophische Bedeutung der Relativitätstheorie*; In Werke Bd. 3, Vieweg: Braunschweig, 1979.
Riesz, M. *Clifford Numbers and Spinors*; University of Maryland: Maryland, 1958.
Riesz, M. *Lecture series given by Professor Marcel Riesz October 1957 – January 1958*; Kluwer: Dordrecht, 1957; Facsimile reprint, 1993.
H. P. Robertson, Postulate versus Observation in the Special Theory of Relativity, *Rev Mod Phys*, 1949, *21*, 378-382.
Rönn, St. (2001). Bicomplex algebra and function theory. arXiv:math/0101200v1 [math.CV]
Sauter, F. *Z Phys.*1930, *63*, 803-814.
Sattinger, D.H.; Weaver, O.L. *Lie Groups and Algebras with Applications to Physics, Geometry and Mechanics*; New York, 1986.
Sciascia, L. *La scomparsa di Majorana*; Einaudi: Turin, 1975.
Sciascia, L. *Das Verschwinden des Ettore Majorana*; Wagenbach: Berlin, 2003.
Schmeikal B. The generative process of space-time and strong interaction – quantum numbers of orientation. In *Clifford Algebras with Numeric and Symbolic Computations*; Ablamowicz, R.; Lounesto, P; Parra, J. M.; Eds.; Birkhäuser: Boston, 1996; pp 83-100.
Schmeikal, B. *Zeitumkehr. Konstruktion und Dekonstruktion des Zeitkonzeptes*; Sonderzahl: Wien, 2000.
Schmeikal, B. Clifford Algebra of Quantum Logic; In *Clifford Algebras and their Application in Mathematical Physics*; Ablamowicz, R.; Fauser, B.; Eds.; Birkhäuser: Boston, 2000, pp 219-241.

Schmeikal, B. Transposition in Clifford Algebra; In *Clifford Algebras – Applications to Mathematics, Physics and Engineering;* Ablamowicz, R.; Ed.; Birkhäuser: Boston, 2004, pp 351-372.

Schmeikal, B. *Adv Appl Cliff Alg.* 2005, *15*, 271-290.

Schmeikal, B. *Adv Appl Cliff Alg.* 2006, *17*, 107–135.

Schmeikal, B. *Space-time Matter – Standard Model of Space-time*; Garsten, 2007.

Schmeikal, B. Clifford Algebra of Quantum Logic; In *Clifford Algebras and their Application in Mathematical Physics*; Ablamowicz, R.; Fauser, B.; Eds.; Birkhäuser: Boston, 2000, pp 219-241.

Schmeikal, B. *Al-muqabala des Nilpotentes - Adjustment of nilpotent algebra and manifolds*; unpublished 2005.

Schmeikal, B. *Adv Appl Cliff Alg.* 2006, *16*, 69-83.

Schmeikal, B. Lie Group Guide to the Universe. In: Lie Groups - New Research; Canterra, A.B.; Ed.; Nova Science Publishers: New York, 2009, pp 1-59.

Schmidt, S. J. *Heinz von Foerster: Heritage and Commitment*; In Müller and Müller. 2007, pp 323-335.

Schrader, M.; Führkötter, A. *Die Herkunft der Heiligen Hildegard*; Quellen und Abhandlungen zur mittelrheinischen Kirchengeschichte 43; 2. edition; Mainz, 1981.

Scott, B. The Cybernetics of Gordon Pask; In *Gordon Pask – Philosopher Mechanic, An Introduction to the Cybernetician's Cybernetician*; Glanville, R.; Riegler, A.; Eds.; edition echoraum: Wien, 2007, pp 31-51.

Segre, C. *Mat Amm.* 1892, *40*, 413-467.

Shapiro, M.V.; Vasilevski, N. L. *Complex Variables Theory Appl.* 1995, *27*, no. 1, 17-46.

Shaw, R. Finite geometry, Dirac groups and the table of real Clifford Algebras; In *Clifford Algebras and Spinor Structures*; Ablamowicz, R.; Lounesto, P.; Eds.; Kluwer: Dordrecht, 1995, pp 59-99.

Ślebodziński, W. *Bulletin de l'Académie Belgique*. 1931, *17*, 5, 864-870.

Ślebodziński, W. *Exterior Forms and their Applications*, PWN, Warsaw, 1970.

Spaarnay, M.J. *Physica*. 1958, *24*, 751.

Sudbery, A.. *Mat Proc Camb Phil Soc.* 1979, *85*, 199-225.

Sullivan, D. *The spinor representation of minimal surfaces in space*; unpublished notes, 1989; [mentioned in Kiehn 2008, Vol. 5, p. 21: "Kusner and Schmitt report that the topologist D. Sullivan, in 1989 constructed some lecture notes"].

Thanissaro Bhikkhu (Geoffrey DeGraff). *The Wings to Awakening*; Dhamma Dana Publications: Taipei, 1996.

Thirring, W. *Lust am Forschen – Lebensweg und Begegnungen*; Seifert: Wien, 2008.

Tritton, D. J. *Physical Fluid Dynamics*; Berkshire, 1977.

Trungpa, Chögyam. *Der Mythos Freiheit und der Weg der Meditation*; Boston, 1994.

Vargas, J.G.; Torr, D. *New Perspectives on the Kähler Calculus and Wave Functions*; 2004, 15 p.

Vargas, J.G. *Found Phys*. 2008, *38*, 610-647.

Vargas, J.G. *Differential geometry as calculus of vector-valued differential forms*; 2010, forthcoming.

Wallace, R. G. *Adv Appl Cliff Alg.* 2008, *18*, 115-133.

Whitehead, A. N. *The Concept of Nature*; Cambridge, 1920.

Weber-Brosamer, B.; Back, D. M.; Eds. *Die Philosophie der Leere*. Nāgārjunas Mulamadhyamaka-Karikas; Übersetzung des buddhistischen Basistexts mit kommentierenden Einführungen; 2. Aufl.; Beiträge zur Indologie 28; Harassowitz: Wiesbaden, 2005, 130 p.

Weizsäcker, C. F. von; Gopi Krishna. *Biologische Basis religiöser Erfahrung*; Weilheim, 1971.

Weizsäcker, C. F von. Die Quantentheorie der einfachen Alternative (Komplementarität und Logik II); Aus dem Philosophischen Seminar der Universität Hamburg; *Z f Naturforschg*. 1958, *13a*, 245-253.

Weizsäcker, C. F. von; Scheibe, E.; Süssmann, G. *Z f Naturforschg*. 1958, *13a*, 705-721.

Weyl, H. *Gruppentheorie und Quantenmechanik*; Nachdruck der 2. Auflage; Darmstadt, 1967.

Wiener, N. *Cybernetics, or Communication and Control in the Animal and the Machine*; second edition; MIT Press: Cambridge, 1985.

Wilczek, F.; Zee, A. *Phys Rev*. 1982, *D 25*, 553.

Woronowicz, S. L. *Commun Math Phys*. 1989, *122*, 125-170.

INDEX

A

B

C

D

E

F

G

H

I

J

K

L

M

N

O

P

Q

R

S

T

U

V

W

Y